Contemporary Computing

for Engineers and Scientists Using Fortran 90

Chester Forsythe
Michigan State University

PWS Publishing Company

I(T)P

An International Thomson Publishing Company

Boston • Albany • Bonn • Cincinnati • Detroit • London • Madrid • Melbourne • Mexico City
New York • Pacific Grove • Paris • San Francisco • Singapore • Tokyo • Toronto • Washington

PWS Publishing Company
20 Park Plaza, Boston, MA 02116-4324

Copyright © 1997 by PWS Publishing Company, a division of International Thomson Publishing Inc.

All rights reserved. No part of this work may be reproduced, stored in a retrieval system, or transcribed in any form or by any means—electronic, mechanical, photocopying, recording, or otherwise—without the prior written permission of PWS Publishing Company.

International Thomson Publishing
The trademark ITP is used under license.

For more information, contact:
PWS Publishing Company
20 Park Plaza
Boston, MA 02116-4324

International Thomson Publishing Europe
Berkshire House 168-173
High Holborn
London WC1V 7AA
England

Thomas Nelson Australia
102 Dodds Street
South Melbourne, 3205
Victoria, Australia

Nelson Canada
1120 Birchmont Road
Scarborough, Ontario
Canada M1K 5G4

Printed and bound in the United States of America.
97 98 99 00—10 9 8 7 6 5 4 3 2 1

Sponsoring Editor: *Jonathan Plant*
Editorial Assistant: *Monica Block*
Marketing Manager: *Nathan Wilbur*
Production Editor/Cover Designer: *Pamela Rockwell*
Interior Designer: *Julia Gecha*
Manufacturing Buyer: *Andrew Christensen*
Compositor: *Pre-Press Company, Inc.*
Text Printer: *R.R. Donnelley & Sons—Crawfordsville*
Cover Printer: *Mid-City Lithographers*
Cover Image: © *Ivan Chermayeff, 1995/Nonstock.*
All rights reserved

International Thomson Editores
Campos Eliseos 385, Piso 7
Col. Polanco
11560 Mexico D.F., Mexico

International Thomson Publishing GmbH
Königswinterer Strasse 418
53227 Bonn, Germany

International Thomson Publishing Asia
221 Henderson Road
#05-10 Henderson Building
Singapore, 0315

International Thomson Publishing Japan
Hirakawacho Kyowa Building, 31
2-2-1 Hirakawacho
Chiyoda-ku, Tokyo 102
Japan

Library of Congress Cataloging-in-Publication Data
Forsythe, Chester.
    Contemporary computing for engineers and scientists using Fortran 90 / Chester Forsythe.
        p.   cm.
   Includes index.
   ISBN 0-534-93139-1
    1. FORTRAN 90 (Computer program language)   I. Title.
QA76.73.F28F66   1997
005.13'3—dc21
                                   96-49665
                                      CIP

This book is dedicated to Linda whom I am the Eternal of, and our two sons, Al and Matt, who brought me up.

Contents

Preface xi

1 An Intuitive Walk Through the World of Computers 1

Introduction 1
- **1.1** What Is a Computer Program? 2
- **1.2** What a Computer Can't Do 3
  - *Important Concepts Review* 5
- **1.3** Altering Program Flow 6
- **1.4** Top-Down Design 12
  - *Important Concepts Review* 14
- **1.5** High-Level Languages 14
- **1.6** Applications 16
- **1.7** Networks 19
  - *Important Concepts Review* 21
  - Exercises 21

2 Learn to Design First 23

Introduction 23
- **2.1** Program Design 24
  - *Important Concepts Review* 32
- **2.2** Improve the Program by Running It 32
  - *Important Concepts Review* 41
- **2.3** Program Form—For Each Structure Chart Block 42
- **2.4** Program Format 44
- **2.5** Identifiers 45
  - *Important Concepts Review* 46
  - Exercises 47

3 Subroutines 49

Introduction 49
- **3.1** Writing a Subroutine 50
- **3.2** Subroutine Definition 52
- **3.3** Scope of Subroutines 52
- **3.4** Arguments 54
  Important Concepts Review 58
- **3.5** Flow of Data Through Arguments 59
  Important Concepts Review 63
- **3.6** Local Variables 64
- **3.7** Entering a Modular Program into a Computer 65
- **3.8** Subroutine Libraries 66
  Important Concepts Review 69
  Exercises 70

4 Data, Variables, Constants and Expressions 72

Introduction 72
- **4.1** What Are Data? 73
- **4.2** Variables 74
- **4.3** Constants 75
  Important Concepts Review 78
- **4.4** Defining and Using Variables 79
- **4.5** Variable Declarations 82
  Important Concepts Review 87
- **4.6** Arithmetic Expressions 88
  Important Concepts Review 94
  Exercises 95

5 Derived Data Types, MODULEs and Data Representations 98

Introduction 98
- **5.1** Derived Data Types and MODULEs 100
  Important Concepts Review 109

- **5.2** Special Operations for CHARACTER Type Data   110
  Important Concepts Review   115
- **5.3** How Numeric Values Are Stored in Memory   115
  Important Concepts Review   120
  Exercises   120

Selection Structures   123

Introduction   123
- **6.1** Conditions, Blocks and Block Statements   124
- **6.2** Conditions Are Logical Expressions   127
- **6.3** Logical Data   130
  Important Concepts Review   130
- **6.4** Block IF Statement   131
- **6.5** Program Design Example   132
- **6.6** Multialternative IF   142
- **6.7** SELECT CASE Statement   146
  Important Concepts Review   151
  Exercises   152

Looping Structures   156

Introduction   156
- **7.1** DO WHILE loops   157
- **7.2** Error Checking   160
- **7.3** Reading an Unknown Number of Values   162
  Important Concepts Review   164
- **7.4** Series Summation   165
- **7.5** Program Design Example   168
- **7.6** Libraries Revisited   177
  Important Concepts Review   181
  Exercises   181

Spreadsheets: Special Section 1   185

Introduction   185
Elements of a Spreadsheet   187

Navigating a Spreadsheet   189
Cells and What They Can Hold   191
The Active Cell   192
Making an Entry in a Cell   192
Editing a Cell Entry   193
Cell Attributes   195
Important Concepts Review   196
Working with Groups of Cells   198
Moving Groups of Cells   199
Copying Cells   200
Creating a Series   202
Important Concepts Review   204
Exercises   205

Spreadsheets: Special Section 2   207

Introduction   207
Working with Formulas   208
Functions in Formulas   210
Copying Formulas   211
Copying Relative Cell References   212
Copying Absolute Cell References   215
Important Concepts Review   216
Graphing   216
Importing Data into a Spreadsheet   225
Importing Free Format Data   226
Importing Tabular Data   227
Important Concepts Review   229
Exercises   230

8   Functions   232

Introduction   232
8.1   What Is a Fortran Function?   233
8.2   Styles of Functions   236
Important Concepts Review   241

8.3 Program Design Example 241
Important Concepts Review 258
Exercises 258

One-Dimensional Arrays 262

Introduction 262
9.1 One-Dimensional Arrays 264
Important Concepts Review 272
9.2 Array Sections 272
Important Concepts Review 280
9.3 Array Variables as Arguments 281
9.4 Array Arguments in Intrinsic Functions 286
9.5 Program Design Example 288
Important Concepts Review 298
9.6 DO Loops for Arrays 298
9.7 Array Masking Operations—The WHERE statement 300
Important Concepts Review 303
Exercises 303

Styles of Arrays 306

Introduction 306
10.1 Assumed Shape Arrays 307
10.2 Allocatable Arrays 313
10.3 Automatic Arrays 318
Important Concepts Review 322
10.4 Parallel Arrays 322
10.5 Derived Data Type Arrays 323
10.6 Program Design Example 329
Important Concepts Review 340
Exercises 341

Multidimensional Arrays 347

Introduction 347
11.1 Multidimensional Arrays 349

11.2 Assignment and Expressions   354
   Important Concepts Review   356
11.3 Array Sections   356
11.4 Multidimensional Arrays and Loops   359
11.5 Program Design Example   363
   Important Concepts Review   382
   Exercises   382

12   Data Files   388

Introduction   388
12.1 I/O . . . What Is It?   389
12.2 List-Directed I/O   390
12.3 Controlled I/O   396
12.4 Data Files   396
   Important Concepts Review   398
12.5 Basic File Manipulation Statements   399
   Important Concepts Review   404
12.6 Styles of Data Files   405
   Important Concepts Review   413
   Exercises   414

13   Advanced Functions and Modules   417

Introduction   417
13.1 Creating Custom Dot Operators   418
13.2 Overloading Existing Operators   422
13.3 Overloading Function/Subroutine Interfaces   428
   Important Concepts Review   432
   Exercises   432

Appendix A: ASCI Collating Sequence   434

Appendix B: Generic Intrinsic Functions   443

Appendix C: The Art of Debugging—Error Removal   450

Introduction   450

- **C.1** Debugging 451
- **C.2** Antibugging 453
- **C.3** Syntax Errors 454
- **C.4** Runtime Errors 457
  - *Important Concepts Review* 460
- **C.5** Logic Errors 461
- **C.6** Testing Programs 465
- **C.7** Debugging in Different Computing Environments 467
- **C.8** Debuggers and Tracers 468
  - *Important Concepts Review* 468

Appendix D: Formats 469

Introduction 469
- **D.1** What Are Formats? 470
- **D.2** Format Descriptors 471
- **D.3** Using Formats to Control I/O 473
  - *Important Concepts Review* 475
- **D.4** Repeating Format Descriptors 476
- **D.5** Stopping the Output of Unwanted Strings 477
- **D.6** Output Considerations for Descriptors 478
- **D.7** Carriage Control 481
- **D.8** Input Considerations for Descriptors 482
  - *Important Concepts Review* 484

Appendix E: I/O Controls 485

Quick Reference Syntax Guide 489
Glossary 493
Index 509

Preface

The pedagogical tactics used in *Contemporary Computing for Engineers and Scientists Using Fortran 90* are the result of a successful ten-year classroom-based metamorphosis of traditional teaching methods. Year after year I found myself increasingly frustrated as I wrote reading assignments from traditional FORTRAN 77 texts that seemed as though they had been chosen by a random number generator. My students had to play chapter leapfrog as I assigned a few pages from Chapter *X* on design followed by some sections in Chapter *Y* on subroutines while carefully avoiding all flowcharted pages and those with statements such as GO TO. The path to producing the best problem solvers was not offered by contemporary Fortran texts.

In addition, the standard problem-solving strategies endorsed by existing 100-level I-want-to-be-an-engineer-or-scientist Fortran texts do not lead to "think before you act" problem-solving program designers who use existing subroutine libraries whenever possible. Instructional methodology must move to a new, more effective paradigm to get budding engineers and scientists off to the best start possible. This will lead students toward careers of more efficient data analysis while spending less time recoding and untangling inherited monolithic programs written by poorly trained program designers.

Contemporary Computing for Engineers and Scientists is written for 100-level courses whose goals are to provide a foundation in problem solving through procedural program design in the context of engineering and scientific examples. Additionally, this text implicitly helps form an attitude of "choose the right tool for the job" by presenting two "Special Sections" on spreadsheets. Increasingly, software packages such as spreadsheets can "get answers" faster than the time it takes to design and code a procedural program to accomplish the same task.

The level of difficulty of this text presupposes that its readers don't need to have a keyboard explained to them or be taught how to turn a computer on—students capable of understanding a 100 course in engineering or science are assumed. Although this text *is* rigorous, it does start from square one in Chapter 1 with a general overview of computing.

The relevant limitations and scope of *Contemporary Computing* are listed below:

1. This text is not intended to be a reference manual.

2. Known-to-be-bad constructs including obsolescent Fortran features have been deliberately omitted.

3. Kind_parameters, which are used to improve portability through parameterizing intrinsic data structures, have been omitted because of the confusing nature of the information for 100-level students.

4. Pointers are fully supported in Fortran 90 but are considered too confusing for students in a 100-level computing course, so they have been omitted.

Organization

Contemporary Computing for Engineers and Scientists has several unique organizations that effectively facilitate learning to design and solve problems. One of the many unique schemes puts subroutines as the first "code" chapter (Chapter 3), which immediately follows the chapter that completely lays down the principles of program design. Students see main programs that primarily consist of calls to subroutines. It is a wonderful teaching experience seeing students' first simple programs taking the following well-designed form:

```
      PROGRAM   . . .
                     IMPLICIT NONE
         . . .
            CALL  . . .
            CALL  . . .
            CALL  . . .
      END PROGRAM   . . .
!=========================================
      SUBROUTINE . . .
                     IMPLICIT NONE
            . . .
      END SUBROUTINE   . . .
!=========================================
      SUBROUTINE . . .
                     IMPLICIT NONE
            . . .
      END SUBROUTINE   . . .
!=========================================
            . . .
```

All nontrivial example programs will involve subroutines, and I recommend that procedural programming be taught that way. It is better to have the problem of students using too many subroutines than not enough or none. Many other valuable possibilities become available when subroutines are studied early.

The perpetual problem, "Students don't know enough to do that yet" is solved in a unique way by supplying black boxes (subroutines) to accomplish tasks that require advanced knowledge. This approach reinforces the idea of using library routines and empowers students with the ability to write significant programs before they have sophisticated knowledge.

Functions, on the other hand, are taught much later in the text. Students are traditionally confused when Fortran subroutines and functions are lumped together; pupils always ask the question, "How do I call a function?" Functions are, therefore, taught as being *variables* whose values are dynamically derived from the function's algorithm when the function is invoked. Both intrinsic and program unit functions are completely explained.

I/O concepts such as FORMATs, READ, WRITE and PRINT are introduced in a "trickle-down" fashion with minimal explanation. Students become functional with basic I/O by seeing many specific examples. It is important that students don't become entangled in a sea of details when they first learn to design programs. This trickle-down technique is particularly useful with formats. Formats need to take a back seat to more important ideas such as design, subroutines, control structures, functions, arrays, etc., but it is frustrating to display answers according to the whims of a Fortran compiler. The philosophy of introducing specific instances of formats throughout the text will encourage and begin to empower students to present their answers more professionally. This will implicitly cause students to want to know more about formats to gain control over the appearance of their output. To further support this position, formats are put in an appendix, Appendix D. This appendix is very referential in design so that students can jump to it and learn whatever they choose at any time. This also provides professors with the flexibility to teach formats at the point in their course where they feel it is appropriate.

After the fundamentals of program design, data and control structures are introduced two chapter-sized special sections on spreadsheets are presented. Increasingly, engineers and scientists are using software packages for data analysis rather than their own customized programs. As mentioned previously, this implicitly suggests to students to "pick the right tool for the job." There are several reasons for the central placement of these special sections:

1. A spreadsheet is a two-dimensional grid of cells that can hold values, and it provides a comfortable transition to the basic ideas used in the study of arrays.

2. Functions are used continually in spreadsheets and come very naturally to students in that context. This makes the study of functions in Chapter 8 (which immediately follows the spreadsheets special sections) much easier.

3. At the point that spreadsheets are introduced, students have just completed seven chapters of challenging material and can benefit from a change of focus. Spreadsheets are entertaining to students (in addition to being extremely useful) with all the amazing features such as graphing functions with which graphs can be quickly created.

Pedagogical Features

There are four additional valuable features of *Contemporary Computing* that should be pointed out:

1. *Important Concepts Reviews*—brief lists of definitions and other explanations that help students focus on the salient issues of the material just studied. These reviews are presented after any body of information that approaches a learning saturation level.

2. *Quick Reference Section*—basic syntactical definitions of the Fortran 90 programming language. The Quick Reference Section is not intended to be the complete ANSI Fortran 90 language definition, but rather a concise explanation of the most often used and useful constructs.

3. *Glossary*—essentially the sum total of all of the Important Concepts reviews.

4. *Appendix C, Debugging*—this is a chapter-length explanation of the art of debugging programs including antibugging techniques that help prevent bugs from ever getting into a program.

There are three chapters that should be given careful consideration by instructors using *Contemporary Computing*. Some of the topics in these chapters are included for completeness and may not be appropriate in a particular 100-level course implementation. The chapters are

1. Chapter 5, "Derived Data Types and Data Representations." The author does not feel that this material is too demanding for most 100-level courses, but does realize in certain academic environments that the material may be deemed inappropriate. This chapter does include a section on CHARACTER type data manipulations that *should not* be omitted.

2. Chapter 10, "Styles of Arrays." The discussion on parallel arrays does not involve any of the more abstruse array topics such as allocatable, assumed shape and automatic arrays. Parallel arrays are well worth including particularly if the study of derived data types is omitted.

3. Chapter 13, "Advanced Functions and Modules." This chapter discusses the object-oriented features of Fortran 90, MODULEs. It is a reasonably brief chapter but it may be considered too demanding for some 100-level courses.

Acknowledgments

Long overdue thanks to Mr. Irwin Hoffman whose heroic efforts on my behalf as my high school mathematics teacher showed me the clarity and truth in knowledge.

Colleagues:

Dr. Vhibavasu Vuppala, Michigan State University
Dr. Tyan-Shu Jou, Michigan State University
Dr. Donald J. Weinshank (Professor Don), Michigan State University
Dr. Edwin Reilley, State University of New York at Albany

And thanks to the exceptional graduate assistants that staffed my courses over the years and to Ms. Michelle M. Sidel without whom nothing gets done.

Thank you Susan Aussicker!

Special recognition must be given to the Numerical Algorithms Group for having the audacity to actually write the first compiler that was a full implementation of the ANSI Fortran 90 standard.

I am especially grateful to the many students who not always knowingly participated in my teaching research over the last ten years.

And thanks to the following manuscript reviewers:

Richard Albright
University of Delaware

William Beckwith
Clemson University

Ramzi Bualuan
University of Notre Dame

Thomas Casavant
University of Iowa

D. D. Hearn
University of Illinois

Duayne McCalister
University of California

R. Papannareddy
Purdue University

Joseph Saliba
University of Dayton

Thomas Walker
Virginia Polytechnic Institute

C.F.

Contemporary Computing

An Intuitive Walk Through the World of Computers

Introduction

It is vital to understand the answers to the questions: What is a computer program? What are instructions? What is program flow? to begin learning how to design computer programs. These answers define the fundamental rules of how Fortran approaches programs and will help you understand how Fortran "thinks."

When you engage in any activity that is governed by strict rules, such as playing Bridge or filling out an income tax return, not understanding the regulations can lead to frustration, delays, and penalties. Consider a baseball player who has great raw talent and can hit a baseball out of the park nearly every time he is at bat. If he runs across the pitcher's mound and stands on second base after hitting the ball over the left-field fence because he doesn't know the rule about running to first base before second, HE'S OUT. An understanding of fundamental rules is essential for playing baseball correctly and understanding Fortran's rules is essential for designing Fortran computer programs that work correctly.

Program designers must also learn what a computer can do and how to make it do those things without investing more time

1.1 What is a computer program?

1.2 What a computer can't do

Important Concepts Review

1.3 Altering program flow

1.4 Top-down design

Important Concepts Review

1.5 High-level languages

1.6 Applications

1.7 Networks

Important Concepts Review

Exercises

than necessary. Unfortunately, when one is just beginning to study program design, there is so much knowledge needed to do anything significant that it becomes overwhelming to absorb it all.

The purpose of this chapter is to give you an overview of what is and is not possible in the world of Fortran programming and to briefly explain several powerful computer applications and facilities. You are not expected to grasp every detail presented in this chapter; you will get an in-depth understanding of all the concepts by studying subsequent chapters. For now, just try to get an intuitive feel for what possibilities are offered by computers.

● ● ●

1.1 What Is a Computer Program?

Computer programs manipulate data to produce useful results. A program consists of a collection of instructions that are put together in a specific order to make a computer accomplish some task. Computer programs are called **software**. Computers are governed by **operating systems**, which are sophisticated computer programs that manage all the various functions of the computer.

Learning a computer language such as Fortran so that you can write computer programs is a lot like learning a foreign language. The "words" and symbols used in a computer language are its vocabulary, and the way the words are put together is the language's grammar. Grammar is called **syntax** in computer languages. Look at the simple Fortran program in Figure 1.1 to begin understanding how computer programs work. The program in Figure 1.1 is processed as described below:

The first line: `PROGRAM Sq_Root` lets Fortran know that a program called `Sq_Root` is beginning.

Next line: `REAL :: Square_Root` explains to Fortran that the program is going to need a place to store a decimal number and that place is named `Square_Root`.

Third line: `Square_Root = SQRT ( 2.0 )` tells Fortran to calculate the square root of `2.0` and store the result in `Square_Root`.

Fourth line: `PRINT *, Square_Root` displays the square root of `2.0` (previously stored in `Square_Root`) on the screen.

The final line: `END PROGRAM Sq_Root` tells Fortran that `PROGRAM Sq_Root` is finished.

Figure 1.1

```
PROGRAM Sq_Root
    REAL      ::      Square_Root
    Square_Root = SQRT( 2.0 )
    PRINT *, Square_Root
END PROGRAM  Sq_Root

*************** PROGRAM RESULTS *************

    1.4142135
```

This explanation of how the program in Figure 1.1 works is exactly how Fortran approaches all programs. Fortran analyzes a program by reading it as a person reads a book: from left to right and top to bottom. Computers proceed line by line from top to bottom doing exactly what they are told. This movement from line to line (instruction to instruction) is called **program flow**.

Each line of the program in Figure 1.1 is an **instruction** telling Fortran to do something. In program `Sq_Root`, the following instruction tells Fortran to calculate the square root of `2.0` and store it in `Square_Root`:

$$\texttt{Square\_Root = SQRT(2.0)}$$

Then the following instruction asks the computer to display the value contained in `Square_Root`, which is `1.4142135`, on the computer screen:

$$\texttt{PRINT *, Square\_Root}$$

Notice the importance of program flow and instructions. The outcome of a program would be completely unpredictable if program flow was ill-defined and the computer **executed** (performed) instructions in some random way. As well-defined flow of control is vital to programs, so are instructions. Without instructions, programs wouldn't do anything. Instructions systematically explain to Fortran what actions need to be taken.

1.2 What a Computer Can't Do

Computers have no built-in intelligence. They have amazing potential in that they can do many millions of instructions per second. The speed of a computer is often described in terms of **MIPS** (**M**illions of **I**nstructions **P**er **S**econd; pronounced like: "rips"). Although computers can perform remarkable tasks extremely fast, **they are not smart!** *Computer programs* are what make computers appear to perform intelligently.

Programmers design and **code** (enter correct instructions) programs, and computers follow programs' instructions *blindly*. Computers have no built-in ability to think intuitively and fix a programmer's mistakes. For example, a computer would never say the following to itself:

"Oh, I understand. The programmer really wanted an approximation for the value of π. I'll just substitute 3.14159265358979 for the obviously mistyped value, 3.04159265358979."

Program designers need to understand instructions and program flow to be able to *think like the computer*. The value of thinking like the computer is immeasurable when programs become long and complicated and aren't giving correct answers. Knowing how a computer will step through a program helps the programmer narrow in on subtle errors. Examine Figure 1.2, which contains an incorrect version of the program from Figure 1.1. Thinking like a computer, it is possible to figure out why the computer printed "?????????". Following the think-like-the-computer scenario below isolates the problem.

Line 1   tells the computer that program `Sq_Root` is about to begin.

Line 2   creates variable `Square_Root` and says that it may have decimal places (is of type `REAL`).

Line 3   writes the value contained in `Square_Root` on the screen. However, `Square_Root` hasn't received a value at this point in the program so a series of question marks appears on the screen indicating that the computer has no value for `Square_Root`; it is undefined.

Line 4   will calculate the square root of `2.0` and store it into variable `Square_Root` ... TOO LATE; the undefined value of `Square_Root` was already displayed.

Line 5   stops the program.

By thinking sequentially like computers do, it becomes obvious that the programmer printed the contents of `Square_Root` before giving it the value, `SQRT (2.0)`. Therefore, **lines 3 and 4 are out of order**. The computer flows through every line of program `Sq_Root`, blindly obeying each instruction and having no idea that the programmer should have switched lines 3 and 4.

A final point regarding the limitations of computers is that they are not good at **"dreaming the impossible dream."** If an instruction in a program tries to do something that is mathematically impossible, the program will stop immediately (crash). The example program in Figure 1.3 has two such errors. Line 1 tells the computer that program `No_Can_Do` is starting. Line 2 creates three variables: `X`, `Div_By_Zero` and

Figure 1.2

```
PROGRAM Sq_Root
     REAL        ::        Square_Root
     PRINT * , Square_Root
     Square_Root = SQRT ( 2.0 )
END PROGRAM Sq_Root

*************PROGRAM RESULTS***************

    ?????????
```

Figure 1.3

```
PROGRAM No_Can_Do
    REAL       ::      X, Div_By_Zero, Log_Of_Neg_X
    Div_By_Zero = 5.0/0.0
    X = -7.9
    Log_Of_Neg_X = LOG( X )
    WRITE ( UNIT = * , FMT = * ) X, Log_Of_Neg_X, Div_By_Zero
END PROGRAM No_Can_Do
```

Log_Of_Neg_X that can each hold a number. These variables are of type REAL, which simply means that they can have decimal parts. Line 3 attempts division by zero, which is mathematically undefined. Program No_Can_Do will **crash** at this point; it will end at once without doing any more instructions.

After correcting that problem and running the program again, program flow will ultimately get to the fourth line:

$$X = -7.9$$

This instruction will take the value, -7.9, and store it in X. Line 5 attempts to use Fortran's built-in LOG function to calculate the natural log of X. (There are many such built-in mathematical functions in Fortran. See Appendix B.)

$$Log\_Of\_Neg\_X = LOG(X)$$

This will cause the program to crash because it is mathematically impossible to take the logarithm of a negative number (remember the instruction X = -7.9). If it can't be done mathematically, Fortran can't do it either.

Important Concepts Review

- **Computers are dumb**; programs are what make computers appear smart.
- The **operating system** is a collection of sophisticated computer programs that manage the basic functions of a computer.
- **Instructions** cause the computer to take some action (e.g., print a value, perform a calculation, etc.).
- **Program flow** is the order that a computer executes (performs) a program's instructions.
- **Computers aren't mystical.** They can't do the impossible such as dividing by zero.
- PRINT displays information on the computer screen/monitor.

1.3 Altering Program Flow

In the previous examples, program flow was likened to reading a book. The execution of program instructions went from the first line of a program to the last without skipping any lines. This organization of program instructions is called a **sequential statement structure**. Often, however, problems are impractical or impossible to solve with a computer program if flow of control is limited to this sequential style only. There are several desirable ways of altering program flow, which are discussed below.

Selection Structures

One situation that frequently arises is the necessity to execute groups of statements selectively ("**statements**" is another way of expressing the idea of "instructions"). For instance, if a programmer is confronted with the problem of finding the roots of a general quadratic equation, $y = ax^2 + bx + c$, there is the problem of imaginary roots to consider. The roots of a general quadratic equation are shown below:

$$x = \frac{-b \pm \sqrt{b^2 - 4ac}}{2a}$$

If the discriminant, $b^2 - 4ac$, is less than zero, the programmer must deal with imaginary roots and must not try to take the square root of $b^2 - 4ac$. Calculating the square root of a negative number is mathematically impossible and would cause the program to crash. A **selection structure** can solve the problem.

Selection structures affect program flow by letting a programmer selectively execute groups of program statements. In the next program (Figure 1.4), the discriminant is tested with a Fortran selection structure called an IF statement to find out whether $y = ax^2 + bx + c$ has real or imaginary roots.

The first line in Figure 1.4, PROGRAM Test_Discrim, lets Fortran know that a program called Test_Discrim is beginning. Next, variables A, B, C and Discrm are declared (created) as type REAL. The next two lines in the program,

Figure 1.4

```
PROGRAM Test_Discrim
      REAL     ::      A, B, C, Discrm
      PRINT * , "PLEASE ENTER A, B AND C"
      READ * , A, B, C
      Discrm = B ** 2 - 4 * A * C
      IF ( Discrm < 0.0 ) THEN
           PRINT * , "IMAGINARY ROOTS"
      ELSE
           PRINT * , "REAL ROOT(S)"
      ENDIF
END PROGRAM Test_Discrim
```

PRINT . . . and READ . . ., do something very interesting and useful and are worth a brief digression. The PRINT statement is shown below:

> PRINT * , "PLEASE ENTER A, B AND C"

The PRINT statement displays PLEASE ENTER A, B AND C on the computer screen. This is called a **prompting message** because it prompts the person using the program (the **user**) to enter three numeric values from the keyboard.

> READ * , A, B, C

This READ statement stops the program temporarily to let the user type in three values from the keyboard, which are then stored into A, B and C respectively. A prompting message followed by a READ statement is THE standard way of getting values from a user.

Back to the selection structure issue . . . after the numbers are stored into A, B and C from the READ statement, execution will continue, starting with the fifth line of the program:

> Discrm = B ** 2 - 4 * A * C

This will calculate $b^2 - 4ac$ for the just-read values saved in A, B and C and store the result in variable Discrm. The next instruction asks the question: is Discrm less than zero? (the IF statement is set off by a square bracket in Figure 1.4. *Note: the bracket would not appear in the actual program.*)

> IF (Discrm < 0.0) THEN

If the answer to the question is "yes," the instruction, PRINT * , "IMAGINARY ROOTS", is performed (which makes sense), but PRINT * , "REAL ROOT(S)" is NOT performed. If, on the other hand, Discrm is greater than or equal to zero, the reverse is true: PRINT * , "REAL ROOT(S)" is performed and PRINT * , "IMAGINARY ROOTS" is not. Examine some program results (Figure 1.5, p. 8) from running the program in Figure 1.4 three different times with a variety of values for A, B and C. Using an IF statement to selectively execute groups of instructions is a very useful way to alter program flow. It gives a program the power of choice.

Remember that you cannot expect to understand everything about IF statements from the example in Figure 1.4. Just understand that it is possible to ask questions such as, "Is *A* greater than *B*?" and selectively execute Fortran statements based on the answer to that question.

Looping Structures

Another way to change program flow is with **looping structures.** Looping structures **repeat program instructions over and over** *in a controlled way.* Putting program instructions inside a loop to repeat them is one of the computer's greatest strengths. Loops simplify programs and make it possible to solve problems that would otherwise be cumbersome or even impossible to work out with a computer program. The basic looping structure in Fortran is the DO WHILE loop.

Figure 1.5

```
****************** PROGRAM RESULTS **************

PLEASE ENTER A, B AND C    <-- Prompting message printed by
                               the program.
1.0, 1.0, 1.0              <-- Values for A, B and C typed
                               in at the keyboard by the
                               user for the READ statement.
IMAGINARY ROOTS            <-- The IF statement printing
                               "IMAGINARY ROOTS" because
                               b²-4ac is less than zero.

****************** PROGRAM RESULTS **************

PLEASE ENTER A, B AND C
3.786,229.912,-11.01
REAL ROOT(S)

****************** PROGRAM RESULTS **************

PLEASE ENTER A, B AND C
10.1,-3.7,0.45
IMAGINARY ROOTS

****************** PROGRAM RESULTS **************

PLEASE ENTER A, B AND C
2.0,-8.0,8.0
REAL ROOT(S)
```

The top of the loop in Figure 1.6 is

```
DO WHILE ( BASE <= 4.0 )
```

and the bottom of the loop is

```
END DO
```

Figure 1.6's `DO WHILE` loop is set off by a square bracket. (*Again, this bracket would NOT appear in the actual program.*) There are two statements that are repeated by the `DO WHILE` statement:

```
PRINT *, Base, Base ** 2, SQRT ( Base )
Base = Base + 1
```

Here is how the loop works: just before the loop begins, `Base` is set to an **initial value** of 1.0. This is accomplished with the statement, `Base = 1.0`. Then the top of

Figure 1.6

```
PROGRAM Try_DO
    REAL    ::      Base
    Base = 1.0
    DO WHILE ( Base <= 4.0 )
        PRINT *, Base, Base ** 2, SQRT ( Base )
        Base = Base + 1.0
    END DO
END PROGRAM Try_DO

******************* PROGRAM RESULTS ***************

    1.0000000    1.0000000    1.0000000
    2.0000000    4.0000000    1.4142135
    3.0000000    9.0000000    1.7320508
    4.0000000   16.0000000    2.0000000
```

the loop is encountered and `Base` is examined to make sure that it is less than or equal to `4.0` (`Base <= 4.0`). Since `Base` has a value of `1.0`, it IS less than `4.0` and flow of control enters the loop. The `PRINT` statement is executed, which displays the following on the screen:

$$1.0000000 \quad 1.0000000 \quad 1.0000000$$

These values represent `Base`, `Base`^{2} and $\sqrt{Base}$ (`SQRT( )` calculates square roots) when `Base` is equal to `1.0`. Then the instruction, `Base = Base + 1.0` is performed. This statement simply adds `1.0` to `Base`. The next instruction is `END DO`, which is the bottom of the loop. `END DO` has only one job: it passes flow of control back to the top of the loop.

At the top of the loop, `Base` (which now has a value of `2.0`), is tested again to make sure that it's still less than or equal to `4.0`. Since it is, flow enters the loop a second time where the `PRINT` statement displays the following:

$$2.0000000 \quad 4.0000000 \quad 1.4142135$$

These numbers represent `Base`, `Base`^{2} and $\sqrt{Base}$. `Base` is changed from `2.0` to `3.0` by the statement `Base = Base + 1.0`, and the bottom of the loop is met again. This looping process goes on until `Base` gets bigger than `4.0` at which point program flow will "fall through" the `END DO` instruction ending the loop.

So why use loops? This program could easily have been written without a loop as shown in Figure 1.7 (p. 10).

For a small number of values for `Base`, there is no real problem with the solution in Figure 1.7, but what if it is necessary to print out the first 1,000 squares and square roots? This version of the program would need 1,000 `PRINT` statements, which would require an enormous, cumbersome program prone to typing mistakes. Using the original

Figure 1.7

```
PROGRAM Same_Prog_With_No_Loop
      PRINT *, 1.0, 1.0 ** 2, SQRT ( 1.0 )
      PRINT *, 2.0, 2.0 ** 2, SQRT ( 2.0 )
      PRINT *, 3.0, 3.0 ** 2, SQRT ( 3.0 )
      PRINT *, 4.0, 4.0 ** 2, SQRT ( 4.0 )
END   PROGRAM Same_Prog_With_No_Loop

******************* PROGRAM RESULTS ***************

    1.0000000    1.0000000    1.0000000
    2.0000000    4.0000000    1.4142135
    3.0000000    9.0000000    1.7320508
    4.0000000   16.0000000    2.0000000
```

program from Figure 1.6, however, the looping structure can easily be modified to print 1,000 squares and square roots without increasing the program's size at all (Figure 1.8).

For now it is enough to understand from the examples in Figures 1.6 and 1.8 that there is a method (looping structures) by which a programmer can repeat a group of program statements over and over in a controlled way.

Figure 1.8

```
PROGRAM Try_DO
      REAL      ::      Base
      Base = 1.0
      DO WHILE ( Base <= 1000.0 )
           PRINT *, Base, Base ** 2, SQRT ( Base )
           Base = Base + 1.0
      END DO
END PROGRAM Try_DO

******************* PROGRAM RESULTS ***************

    1.0000000       1.0000000      1.0000000
    2.0000000       4.0000000      1.4142135
    3.0000000       9.0000000      1.7320508
    4.0000000      16.0000000      2.0000000
            .            .
            .            .
            .            .
  9.9800000E+02   9.9600400E+05   31.5911388
  9.9900000E+02   9.9800100E+05   31.6069622
  1.0000000E+03   1.0000000E+06   31.6227760
```

Note: E+05 means 10^5 - this is Fortran's way of representing scientific notation.

Figure 1.9

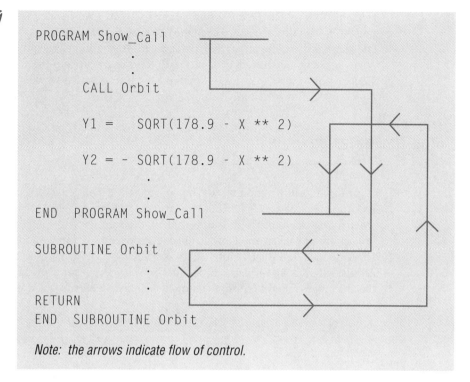

Note: the arrows indicate flow of control.

Calling Subroutines

The final technique for altering program flow is the CALL statement, which transfers control to a Fortran **subroutine**. A subroutine is a mini Fortran program that has one simple well-defined purpose. After a subroutine is finished performing its task, flow of control returns to the program statement immediately following the CALL statement that transferred control to the subroutine. Figure 1.9 shows how calling a subroutine works.

Fortran begins running the program at PROGRAM Show_Call and then follows instructions until it gets to the instruction CALL Orbit. CALL Orbit causes flow to jump to SUBROUTINE Orbit, where its instructions are performed.

After finishing all the statements in subroutine Orbit, the RETURN statement tells flow to jump back to the statement immediately following CALL Orbit, which is Y1 = SQRT(178.9 - X ** 2). Then this and the rest of the program statements in the main program are executed.

The idea of calling subroutines is one of the most powerful tools in a programmer's toolbox. Subroutines make programs easy to understand, easy to modify and quicker to design with fewer initial errors. As programming with subroutines will shorten the time it takes a programmer to solve a given problem on the computer, all useful subroutines should be accumulated in libraries of subroutines. Besides creating your own library of personally written subroutines, libraries can be inherited from other programmers or purchased.

To understand why libraries are important, consider writing a program that needs a factorial routine ($n! = 1 \cdot 2 \cdot 3 \cdot 4 \cdot \ldots \cdot (n-1) \cdot n$) when there is an existing one in an available library. The factorial routine can easily be incorporated into the program being written. It won't be necessary to figure out what Fortran statements are required to calculate factorials, and the programmer won't need to go through the process of resolving any errors accidentally introduced into the program by writing the factorial routine from scratch. All of this saves time. Powerful programs can be written quickly by invoking many known-to-work library subroutines.

1.4 Top-Down Design

Creating a **top-down design** is the process of *taking a large task and breaking it down* into its component parts. It is difficult to overstate the importance of this idea. Most often, when a real-world problem is approached, the solution isn't immediately obvious. The project must be taken apart into its main components. These main ideas may then need to be subdivided further and those subideas divided yet further, and so on. Through this successive subdivision of ideas, all the pieces will soon become manageable sizes and can then be programmed as a set of subroutines linked together with CALL statements.

```
ENORMOUS JOB
CAN'T SEE THE SOLUTION
```

Take the "enormous job" arbitrarily called "TASK" in Figure 1.10, and break it down into its component parts.

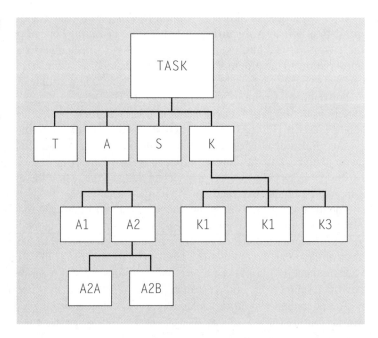

Figure 1.10

In this abstract example, problem "TASK" is divided into four subtasks, T, A, S and K. Subtask A is then divided into sub-subtasks A1 and A2. K is broken down into K1, K2 and K3. A2 is divided into sub-sub-subtasks A2A and A2B.

The diagram in Figure 1.10 is called a **structure chart;** it is a *graphic outline* that represents the *main programming ideas in the program.* Although it isn't clear what this program does because the purposes of T, A, S and K and the other subboxes aren't described, a program (Figure 1.11) can be written to reflect the Fortran version of the above structure chart.

Figure 1.11

```
PROGRAM TASK
        CALL T
        CALL A
        CALL S
        CALL K
END
SUBROUTINE T
    . . .
RETURN
END
SUBROUTINE A
        CALL A1
        CALL A2
RETURN
END
SUBROUTINE S
    . . .
RETURN
END
SUBROUTINE K
        CALL K1
        CALL K2
        CALL K3
RETURN
END
SUBROUTINE A1
    . . .
RETURN
END
        etc.
```

Important Concepts Review

- **Selection structures** allow a programmer to selectively execute groups of instructions.
- `IF (question) THEN ... ELSE ... ENDIF` is one of Fortran's selection structures.
- **Looping structures** repeat program instructions over and over in a controlled way.
- `DO WHILE ... END DO` is Fortran's basic looping structure.
- **Subroutines** are mini Fortran programs that lead to top-down designs and shorten programming efforts.
- **Libraries** are collections of subroutines that save time when writing programs by providing already-written, working subroutines.
- A **prompting message** tells the user to take some action, such as entering values from the keyboard.
- `READ *`, temporarily stops the execution (performing of instructions) of a program so the user can enter values from the keyboard.
- A **structure chart** is a graphic outline of how subroutines are called.

1.5 High-Level Languages

Throughout the preceding programming examples, Fortran seems to use fairly sensible words and symbols to describe how things work in a Fortran program. "`Print *, ...`" writes information on the computer screen, "`READ *, ...`" gets values from the keyboard, the plus symbol (+) adds values, etc. This is no accident. Fortran is a **high-level language** meant for use by anyone needing to solve a scientific or engineering problem on a computer. The elements of the Fortran language are recognizable and easily understood. However, programming wasn't always like this.

Back in the 1950s when Fortran was developed by John Backus, much programming was done in **machine language**. There are many different machine languages, and any given computer only understands its own and other computers that use the same machine language. Machine language is defined by a computer's hardware.

Hardware is composed of the physical parts of a computer. This is at the opposite end of the scale from **software,** which is only a collection of electronic pulses representing a computer program. Part of every computer's hardware is its Central Processing Unit (**CPU**), which analyzes incoming machine language instructions, performs arithmetic and logical operations, etc. Software running in a computer can be likened to its mind, and the CPU can be thought of as the computer's brain.

One CPU can't directly understand a different *type* of CPU's machine language because they probably don't "speak" the *same* machine language. It would be like a person who only speaks Chinese trying to have a conversation with a Brazilian who only understands Portuguese. The different machine languages in computers' CPUs caused many problems in the early days of computing.

When programs needed to be moved from one computer to another and the two computers understood different machine languages (had incompatible CPUs), entire programs had to be *rewritten*. Only then could the CPU of the new computer understand how to correctly perform the same job as the previous computer. This process was very costly.

High-level languages such as Fortran emerged to avoid the rewriting process that occurs when moving a program from one computer to another and the tedious, error-prone efforts involved in programming in machine language. High-level languages help programmers quickly create working programs through powerful instructions such as `IF`, `DO`, `PRINT`, `READ`, etc. High-level languages have English words and fairly recognizable mathematical symbols, but remember, **computers only understand *their own* machine language.** Somehow Fortran programs, which we understand, must be translated into a computer's native machine language, which is all the computer understands. There are a couple of methods for doing this, one of which is to use a **compiler.**

A Fortran compiler "reads" a Fortran program and systematically translates each Fortran instruction into as many machine language instructions as is necessary to accomplish the Fortran instruction. Many machine language instructions may be needed to carry out a single Fortran instruction.

Fortran program >> Fortran compiler >> Machine language program

Once a Fortran program is translated by a compiler, the computer has machine language instructions that it can understand and can then **run** the program. (Running a program means to have a computer perform the program's instructions.) Fortran compilers act as translators that can change a Fortran program into its machine language counterpart, similar to the way a human language translator might help a Brazilian and Taiwanese person communicate by understanding both Portuguese and Chinese.

With high-level languages, a programmer is saved the aggravation of working in machine languages and *programs can be moved to different computers with different CPUs by simply compiling those programs on the new computer with ITS Fortran compiler.* The compilation process will generate a whole new set of machine language instructions that the new computer can understand. The ability to move programs from one computer to another is called **portability**.

1.6 Applications

In addition to writing customized programs in Fortran to solve problems, there are many categories of purchasable **applications.** Applications are extremely useful programs written by one or more program designers to apply the computer in a specific way. Three of the most common are **spreadsheets** (detailed later in Special Sections 1 and 2), **word processors** and **d**atabase **m**anagers (**DBM**s).

Spreadsheets

Spreadsheets offer a convenient method for evaluating and graphing tabular data. Often, it is faster to solve a problem with a spreadsheet than it is to write a customized computer program that does the same job. Most spreadsheets have an appearance similar to the one illustrated in Figure 1.12.

Each rectangle is called a **cell**, and is identified by an intersecting row and column. In Figure 1.13, spreadsheet cell **D6** is shaded.

A spreadsheet cell may contain any one of three different types of entries:

1. Labels, which are words or symbols used to clarify entries in the sheet.
2. Numeric values.
3. Formulas that enable a spreadsheet user to perform calculations based on the contents of other cells or groups of cells.

Earlier in this chapter, an example program was shown that calculated the squares and square roots of the numbers from `1.0` to `4.0`:

Figure 1.12

Figure 1.13

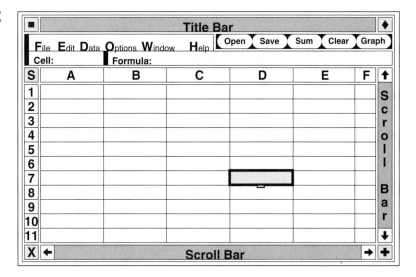

```
PROGRAM Try_DO
    REAL     ::      Base
    Base = 1.0
    DO WHILE ( Base <= 4.0 )
        PRINT *, Base, Base ** 2, SQRT ( Base )
        Base = Base + 1.0
    END DO
END PROGRAM Try_DO
```

Once basic spreadsheet skills are mastered, the results from the above program would be MUCH faster to render on a spreadsheet (Figure 1.14).

Figure 1.14

	A	B	C	D	E	F
1	NUMBER	SQUARE	SQ ROOT			
2	1	1	1.00000			
3	2	4	1.41421			
4	3	9	1.73221			
5	4	16	2.00000			

Each occupied cell in the spreadsheet from Figure 1.14 contains the following:

A1: NUMBER	A label: the word "NUMBER"
B1: SQUARE	A label: the word "SQUARE"
C1: SQ ROOT	A label: "SQ ROOT"
A2: 1	Cell A2 has the number 1 in it.
A3: 2	Cell A3 has the number 2 in it.
A4: 3	.
A5: 4	.
B2: A2 * A2	Cell B2 contains the formula A2 * A2, which multiplies the contents of cell A2 by the contents of cell A2 to calculate the square of it. A "1" is displayed in cell B2.
B3: A3 * A3	Cell B3 contains the formula: A3 * A3, which displays a 4 in cell B3.
B4: A4 * A4	.
B5: A5 * A5	.
C2: SQRT(A2)	Cell C2 contains the formula SQRT (A2), which will calculate and display the square root of the value in cell A2.
C3: SQRT(A3)	.
C4: SQRT(A4)	.
C5: SQRT(A5)	.

The formulas in a spreadsheet are *easy* to create and replicate so it is almost never necessary to type all the values and formulas in a spreadsheet.

As mentioned previously, once a basic knowledge of spreadsheet skills is mastered, many projects can be completed faster with a spreadsheet than by writing a computer program to do the same task. This is particularly true when it comes to graphing data.

Word Processors

Word processing is another type of application that is extensively useful. A word processor allows a **user** (person using the word processing computer program) to create and efficiently edit text. "Text" can be a letter of reference, a data file, a resumé, a Fortran program, a grant proposal, etc. The real power of word processing is that it allows **documents** (anything created by a word processor) to be **saved** and **changed** easily.

Imagine typing a 50-page report on a typewriter. After the job is done it is proofread, and it turns out that it is necessary to insert a paragraph on page 11. Since the document is typed on the typewriter, all subsequent pages need to be retyped! If the document is created on a word processor, however, the paragraph can be typed (inserted) on page 11 and *the computer* reformats the entire document, pushing the rest of the text down. The document is then reprinted on the computer's printer. Making this correction in a word processor only costs the user the time it takes to compose and type the paragraph and wait for the printer to reprint pages 11–50.

Figure 1.15

ELEMENT NAME	LAST MEASURED	SYM	MASS	ORGAN DAMGE	HALF-LIFE
CESIUM	11/24/04	Cs	132.9	...	...
STRONTIUM	07/18/04	Sr	87.6	...	...
URANIUM	07/22/04	U	238.0		
PLUTONIUM	03/02/04	Pu	242.0		
...	...				

It is well worth learning how to use a major word processor to present ideas professionally, and keep documents flexible and stored safely.

Database Managers

Database managers (DBMs) create a convenient, flexible filing and look-up system for data. The information in a database is commonly stored in rows of a table, each of which is usually broken down into several fields (Figure 1.15).

This DBM can be searched in a variety of ways using the DBM's **SQL** (**S**tructured **Q**uery **L**anguage). SQLs enable applications to be created that can search a database and display data that conform to a variety of criteria. For instance, one could perform a search on the database in Figure 1.15 such as, "find and print all radioactive elements that have been measured since July 1, 2004 and have an atomic weight greater than 220." The SQL search responds with "Uranium" and any other elements in the table that were measured since 04/01/04 and have a mass greater than 220. A user can also ask, "Which elements have a half-life greater than 100 years?" This would yield a list of all elements contained in the table that met the condition of having a half-life greater than 100 years. The major strength of DBMs is the flexible way large amounts of data can be stored and searched.

1.7 Networks

Networks have become commonplace and create remarkable possibilities. Networks interconnect computers and allow for rapid electronic communication and shared resources (Figure 1.16). Using networks, many computers can share resources such as printers, databases

Figure 1.16

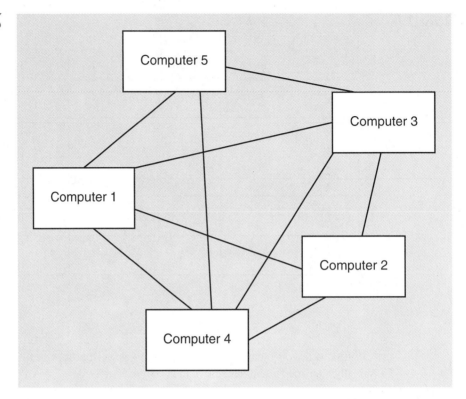

and software applications. Networks also provide the means for electronic mail (**E-mail**), a postal system in which all letters are in an electronic form (no dead trees) and the networks and computers attached to them act as the postmaster, mail carriers and mail boxes.

In Figure 1.16, a user on computer #2 might send a message to someone who uses computer #5. The e-mail would need to be routed through computer #3 or computers #4 and #1 to get to computer #5, as there is no direct network connection between #2 and #5. But that's what networks are all about—passing information from computer to computer until it reaches its destination. Another interesting point is that computer #2 might be in Chile, South America, and computer #5 in Israel. It really doesn't matter; the network will move the information on until it gets to where it belongs. Electronic messages can be sent between two users or to a group of users, and there is no practical limit to where e-mail can be sent; the planet truly is shrinking.

One other use of networks is that they also enable computer users to exchange computer files. A person might create some software and want to give it to someone else. With networks, there is no need to send the software through the physical mail system. The software can be sent electronically with a **file transfer program** . . . Amazing!

Important Concepts Review

- **High-level language**—a computer language that uses recognizable words and symbols and is used to create computer programs quickly with powerful statements such as IF, READ, etc. High-level languages require translation into machine language.

- **Machine language** is the only language a computer understands.

- **Compilers** translate programs written in high-level languages into machine language.

- **Spreadsheets** provide a convenient way to evaluate and graph tabular data. Often, spreadsheet solutions are quicker than writing customized computer programs.

- **Word processing** is a convenient way to create, format, save and modify text.

- **Database managers** (DBMs) store large amounts of data and allow easy, flexible searching of the data.

- A **network** is an interconnection of computers that makes sharing resources and electronic communication possible and convenient.

- **E-mail** is an efficient way to communicate with other computer users.

- **File transfer programs** make it easy to transfer information from one person to another through computer networks.

EXERCISES

1.1 What is program flow?

1.2 What are program instructions?

1.3 A computer program makes a computer "smart." Describe the elements of a computer program.

1.4 Name three things a computer can't do.

1.5 In the selection structure:
```
IF ( X <Testit ) THEN
    PRINT *, 'DILUTION NORMAL'
ELSE
    PRINT *, 'ISOTONIC'
ENDIF
```
    a) If `X` contains the value `3` and `Testit` has a value of `7`, what gets printed?
    b) `X` is `-22` and `Testit` is `-23`, what gets printed?
    c) `X` is `-91` and `Testit` is `-91`, what gets printed?

1.6 What do looping structures accomplish?

1.7 Subroutines are mini-_____.

1.8 A top-down design involves taking a large _____ and breaking it down into its _____.

1.9 A graphic outline of a program is called a _____.

1.10 Finish program TASK (Figure 1.11) by writing the remaining subroutines.

1.11 What is a prompting message?

1.12 What is the difference between high-level languages and machine languages?

1.13 What translates a Fortran program into machine language?

1.14 What is a (a) spreadsheet? (b) word processor? (c) DBM?

1.15 Networks allow computers to _____ _____.

1.16 E-mail means what?

1.17 A computer user can send a file to another user using a ____ ____ ____.

1.18 If a computer's CPU can be considered its brain, software would be considered the computer's _____.

Learn to Design First

Introduction

It is extremely difficult to persuade beginning programmers to take the time to think through, organize and write down a **design** for a computer program. Usually, when confronted with a programming project, neoprogrammers run to computers and begin to enter their sketchy ideas in a frenzy of typing and expect that with some trial and error and a little luck, a working program will spring forth. This is a bad plan. It is essential to use your mind *first* and create designs for your programs. It will save you time by helping you foresee most of the pitfalls that are not obvious initially. It can be very difficult to untangle a poorly designed program after most of the program has already been entered into the computer. Enter your programs as the last step of the programming process. The difficult part is having the patience to do this.

Solutions to simple programs may or may not come easily to you, but be assured that programming tasks quickly become too complicated to solve off the top of your head. When these more complicated endeavors arise, it is very important that you have developed a methodical approach to design and the **discipline to use it**. If this "think-before-you-act" approach is followed, your programs will be designed as well and simply as possible, and will take the least amount of time to complete.

2.1 Program design

Important Concepts Review

2.2 Improve the program by running it

Important Concepts Review

2.3 Program form—for each structure chart block

2.4 Program format

2.5 Identifiers

Important Concepts Review

Exercises

Computer programs should work right, be easy to understand and modify, be completed in as short a time as possible and be designed flexibly to grow and expand with time. That is the Golden Rule of program design. Programming is a means, not an end; it is a tool for data analysis that should serve you, not consume all of your time.

This chapter introduces the program design strategy **the Five-Paragraph Essay Approach to Program Design**. This title is borrowed, in part, from the Humanities. In the field of writing, the *five-paragraph essay* is a scheme for formulizing essay writing where a writer says what she's going to say in paragraph one, says it in paragraphs two, three and four and says what she said in paragraph five. Although formulized approaches to writing rarely advance a writer's skill or lead to brilliant essays, a formula-like approach to program design works very well.

Please be patient and think about the way this chapter is explaining how to design computer programs. Whenever the question, "Where do I start?" crosses your mind, come back to this chapter and start here. • • •

2.1 Program Design

Tables 2.1a, b and c summarize the main steps required to design and code a well-organized working computer program. Follow the steps outlined in Tables 2.1a, b and c every time a program needs to be written. By doing the design steps every time, a programmer's mind will slowly be brought around to "Right Thinking" (*1984*, George Orwell). As in all things, the more one practices a technique, the better one becomes.

*(Note: Table 2.1a uses the word **algorithm**. Algorithm is a great word to impress your friends with. It means a step-by-step solution to a problem or a method of solution.)*

The design steps are best illustrated by using them. Consider the following problem: Figure 2.1 (p. 26) shows a graph of a general linear equation of the form $y = ax + b$. The curve varies depending on the values of *a* and *b*. The shaded area is a triangle formed by the axes and a linear equation. Design a program that calculates the area of that triangle.

Grasp the Problem

Step 1 of the Five-Paragraph Essay Approach to Program Design says, "Grasp the problem" first. Obviously, this is a wise plan because if what's being asked isn't understood,

Table 2.1

	The Five-Paragraph Essay Approach to Program Design a. Thinking Phase		
1. Grasp the problem.	State the problem in DETAIL.		. . . a couple of paragraphs.
2. Break the problem down.	Write a structure chart.		Organize the main ideas.
3. Shape the solution for each main idea in the structure chart.	Briefly describe block's purpose.		. . . a sentence or two.
	Write a list of what to do.		Think like the computer.
	What data are traveling to and from the calling program unit (flow of data)?		Arguments.
	Write English program statements to show how the program unit's algorithms work. This step may be skipped if the algorithm is a simple *sequential statement structure*. Use brackets to set off block statements such as `IF` and `DO`.		Basic structures: • `Call`. • Sequential. • Selection. • Looping.
	b. Action Phase		
4. Debug/Test the program.	Enter the program into the computer. Use the program stub method and only develop **ONE** program unit at a time.		Don't type the program in all at once.
	Remove initial errors.		Typos, syntax errors, etc.
	Is the program producing correct answers? Improve the program by running it.		Try varied data to shake out the bugs. Check formulas by hand. Does the program handle ALL feasible situations? Error checking? Answers readable? Limited solution?— Generalize. More modularity?

continues

Table 2.1 (continued)

	c. Form Phase (documentation)	
	Use comments.	Comment blocks. Line comments.
5. Make *each program unit* clear and understandable.	Insert indentations in block statements if not already done. Five spaces is about right for indentations.	Selection structures. Looping structures. Subroutines. Other blocks.
	Include extra blank lines where appropriate.	. . . for readability (use judgment).

Figure 2.1

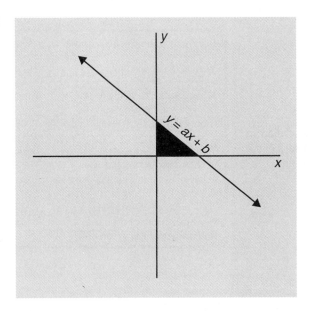

how can a computer program be written to solve the problem? Figure 2.2 shows the "State the Problem" phase of the design. The paragraph in Figure 2.2 is an effort to define the problem clearly. NOTICE THE DETAIL! When writing a "State the Problem" paragraph or two, try to include everything that comes to mind so it is as complete as possible. Also notice that "State the Problem" takes the form of the programmer discussing the problem with herself. This is a period of mental exploration and brainstorming to arrive at an understanding of what is required to design a computer program solution.

It is important to recognize at this point that the programmer may be wrong about some of the ideas given in the paragraph shown in Figure 2.2. Mistakes will probably be made at each step of the design process, but that doesn't matter. What is of concern is that the programmer makes her best effort at each step.

The paragraph in Figure 2.2 is missing some essential parts that will show up when the program is run. That's the beauty of the design process; it will discover all the programmer's mistakes if ALL steps in Table 2.1a, b and c are followed.

Figure 2.2

> **State the Problem**
>
> In this problem I want to calculate the area of a triangle formed by the axes and a linear equation. I can figure out the area of any triangle from the formula *Area* = ½ • *Base* • *Height* if I know what *Base* and *Height* are. *Base* would be the *x* intercept of the equation, which is found by solving it when *y* is zero. ∴ the *x* intercept would be 0 = *ax* + *b* (substituting zero for *y*) –> *x* = –*b*/*a*. Similarly, *Height* in *Area* = ½ • *Base* • *Height* is the *y* intercept of the equation when *x* is zero. ∴ the *y* intercept is *y* = *a* • 0 + *b* (substituting zero for *x*) –> *y* = *b*. By plugging my values for *Base* (–*b*/*a*) and *Height* (*b*) into the formula *Area* = ½ • *Base* • *Height* I will have the area of the triangle. I will need to get values for *a* and *b* and I want to print out *a*, *b* and *Area*.

Break the Problem Down

Writing a **structure chart** is step two. A structure chart is a graphic outline that shows a breakdown of the main programming issues that need to be designed to solve a given problem. Going back to the "writing" metaphor will help explain structure charts. When writing a research paper, the first sensible step is to outline the main ideas. This is usually done by putting each idea on a 3 × 5 index card or by writing an outline. By doing this, the research paper is broken down into manageable pieces. The main ideas can be broken down further, and those ideas yet further.

To write a research paper on a brewery, for example, an outline such as the one in Figure 2.3 (p. 28) might be created to help organize the main ideas. Figure 2.4 shows how to represent this organization of the brewery research paper with a structure chart.

It is clear from the structure chart how the whole paper fits together: where and when specific information is presented. The diagram in Figure 2.4 shows the *structure* of the paper. Notice that the presentation of ideas goes from left to right and top to bottom. Breaking down a program design problem into its main ideas by making a structure chart is the programmer's way of outlining the important programming issues that need to be addressed in a computer program. The connecting lines in the structure chart represent the backbone of the design; the lines show the flow of control to subprograms.

Getting back to the specific program that is being designed . . . Figure 2.5 (p. 29) shows a structure chart for the triangle program. The triangle program is broken down into three main tasks: getting values for *a* and *b*, calculating the triangle's area and displaying the answer on the computer's monitor. The three blocks under "TRIANGLE PROGRAM" are in order (from a program flow point of view) from left to right; get the values first, calculate the area second and finally, write the results.

As programs become more complicated, the concept of breaking down problems into smaller pieces and organizing them with a structure chart will become indispensable.

Shape the Solution for Each Main Idea

Each block in a structure chart is coded as a subroutine (**program unit** is a more general term that includes subroutine) except the top block, which is called the **main program**.

Figure 2.3

BREWERY RESEARCH PAPER

I. What is a brewery?
   A. Makes beer.
   B. Sells beer.
II. Environmental impact.
   A. Creates waste.
      1. Pollutes.
      2. Expensive to remove.
   B. Environmentally active?
III. Consumers.
   A. College students.
   B. Sports fans.
      . . .
      etc.

Figure 2.4

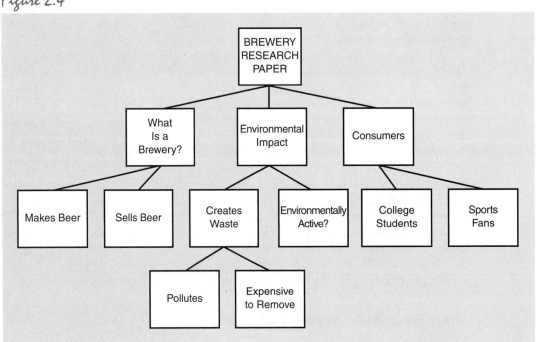

Figure 2.5

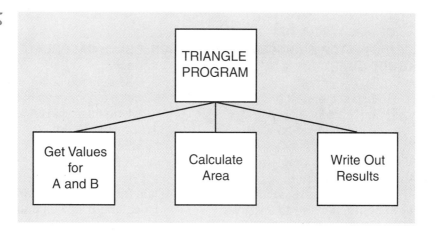

The primary purpose of a main program is to call subroutines that do the actual work; main programs are traditionally short and are used to "drive" the program. Step three in Table 2.1a titled, "Shape the solution for each main idea in the structure chart" describes how to mold each main idea (each block in the structure chart) into a subroutine. The four parts of step three are set off in Table 2.1a by a thick blue line. Use these instructions to develop each block of the structure chart including the main/top block.

Although subroutines haven't been explained in detail, the general way they work was shown in Chapter 1, Figure 1.9. It is not expected that everything about subroutines will be understood from studying the following development of the structure chart blocks shown in Figure 2.6 (p. 30). That is not the point. Grasp HOW the main ideas are developed. Starting from a brief description of what a block is supposed to do (the "purpose" step), the block is transformed into English Fortran statements that closely represent the necessary Fortran program statements.

*(Note: **arguments** are mentioned below. They are the means of moving data in and out of a subroutine . . . more on this in Chapter 3.)*

At this point in the design process, the programmer can take the written design from Figure 2.6 to the computer and translate the English Fortran parts into actual Fortran statements AS the program is typed into the computer. *(Hint: when you're not sure how to translate your English Fortran into actual Fortran, look through the text for examples that are similar to what you're trying to do and at the "Quick Reference Section" at the end of the text.)* As the English Fortran parts of each main idea are translated into true Fortran statements and typed into the computer, the program shapes up as seen in Figure 2.7 (p. 31). Program `Triang` in Figure 2.7 is written using correct Fortran syntax; Fortran will understand the way the algorithms are typed. Square brackets set off the program units and do not appear in the actual program.

A **modular programming solution** (the BEST kind), such as the one in Figure 2.7, is where each structure chart block is turned into one program unit (subroutine, main program, etc.). A finished modular program consists of a collection of program units—modules of code

Figure 2.6

MAIN PROGRAM — TOP BLOCK OF THE STRUCTURE CHART
Purpose:
   The purpose of the main program is to call the three subroutines and "drive" the program.
List of what to do:
   CALL "get values" subroutine
   CALL "calc area" subroutine
   CALL "print results" subroutine
Data (arguments) moving to and from calling program unit:
   NONE
English Fortran
   CALL Read_A_and_B
   CALL Calc_Area
   CALL Print_Answers

MAIN IDEA 1: GET VALUES FOR A AND B
Purpose:
   The purpose of this block is to get values for A and B.
List of what to do:
   GET A VALUE FOR A
   GET A VALUE FOR B
Data (arguments) moving to and from calling program unit:
   From calling program unit: NONE
   To calling program unit: A, B
English Fortran:
   READ A
   READ B

Note: the "English Fortran" in this example could have been omitted as it's a simple sequential statement structure and is essentially the same as "List of what to do." See the design examples in Chapters 6–11 for examples of more intricate English Fortran.

MAIN IDEA 2: CALCULATE AREA
Purpose:
   The purpose of this block is to calculate the area of the triangle where base = –B/A and height = B.
List of what to do:
   SET BASE TO –B/A
   SET HEIGHT TO B
   SET AREA TO ½ • BASE • HEIGHT
Data (arguments) moving to and from calling program unit:
   From calling program unit: A, B
   To calling program unit: BASE, HEIGHT, AREA
English Fortran:
   BASE = B/A
   HEIGHT = B
   AREA = ½•BASE•HEIGHT

MAIN IDEA 3: PRINT RESULTS
Purpose:
   The purpose of this block is to print out the answers.
List of what to do:
   WRITE A and B in an equation format
   WRITE BASE
   WRITE HEIGHT
   WRITE AREA
Data (arguments) moving to and from calling program unit:
   From calling program unit: A, B, BASE, HEIGHT, AREA
   To calling program unit: NONE
English Fortran:
   WRITE Y = AX+B
   WRITE "BASE =",BASE
   WRITE "HEIGHT=",HEIGHT
   WRITE "AREA =",AREA

Note: the variables listed in "Data (arguments) moving to and from calling program unit" should appear in the subroutine's argument list!! See Chapter 3 for an explanation of arguments.

Figure 2.7

```
PROGRAM Triang     ! The main program
        REAL       ::      A, B
        REAL       ::      Base, Height, Area
    CALL Get_Values_For_A_and_B ( A, B )
    CALL Calc_Triangle_Area ( A, B, Base, Height, Area )
    CALL Print_Answers ( A, B, Base, Height, Area )
END PROGRAM Triang

SUBROUTINE Get_Values_For_A_and_B ( A, B )
        REAL, INTENT ( OUT )      ::      A, B
    READ ( UNIT = *, FMT = * ) A, B
RETURN
END SUBROUTINE Get_Values_For_A_and_B

SUBROUTINE Calc_Triangle_Area ( A, B, Base, Height, Area )
        REAL, INTENT ( IN )       ::      A, B
        REAL, INTENT ( OUT )      ::      Area
        REAL, INTENT ( OUT )      ::      Base, Height
    Base   = - B / A
    Height = B
    Area   = 1.0 / 2.0 * Base * Height
RETURN
END SUBROUTINE Calc_Triangle_Area

SUBROUTINE Print_Answers ( A, B, Base, Height, Area )
        REAL, INTENT ( IN )       ::      A, B, Area
        REAL, INTENT ( IN )       ::      Base, Height
    WRITE ( UNIT = *, FMT = * ) "y=", A, "x", "+", B
    WRITE ( UNIT = *, FMT = * ) "Base=", Base
    WRITE ( UNIT = *, FMT = * ) "Height=", Height
    WRITE ( UNIT = *, FMT = * ) "Area=", Area
RETURN
END SUBROUTINE Print_Answers
```

(the Fortran statements are also called **source code**). In program Triang of Figure 2.7, each bracketed chunk of program corresponds to a structure chart block in the triangle program.

Designing a computer program is a dynamic process and mistakes will probably be made at many points, but completing the ENTIRE process will lead, most directly, to a well-designed working program.

Important Concepts Review

- **Take the time to design.**

- **Designs** aren't expected to be perfect, but the design process as a whole leads, most directly, to a well-designed working program.

- When designing a program, **take things one step at a time.**

- The **Five-Paragraph Essay Approach to Program Design** is a formulized, methodical approach to breaking down a problem into its component parts so a modular solution can be written.

- A **structure chart** is a graphic outline showing the main ideas of a program.

- **Inputs** are the data coming into a program.

- **Outputs** are the answers generated by a program.

- **Answers** are the whole reason for computer programs.

- **Modular program designs** use subroutines to implement each block of a structure chart.

- The **main program** "drives" the program by calling subroutines.

2.2 Improve the Program by Running It

The triangle program in the previous section was written in a modular way, and that is the way we'll write programs in this text. But since subroutines haven't been formally presented, the examples discussed in the rest of this chapter will NOT involve subroutines.

Once a design is written up in its final form, the program is 75% done and is ready to be entered into the computer and tested. The **debug and testing** phase (Table 2.1b) will reveal any problems the *written* design phase (Table 2.1a) didn't turn up, and will lead to a working program.

Remove Initial Errors

When the triangle program is entered into the computer, typographical errors will probably be made. These errors show up when the program is **run**. Typing errors can confuse the Fortran compiler and result in **error messages** and a program that doesn't work. Error messages are designed to guide a programmer to the problems in her program.

Another way that errors can be introduced into a program is by translating **pseudocode** incorrectly into Fortran. Pseudocode is a fancy phrase for the "English Fortran" statements developed according to Table 2.1a. Appendix C, "The Art of Debugging," thoroughly discusses how to attack program errors by using **debugging techniques** (debugging means to remove errors). Assume the triangle program is translated from its English Fortran form (pseudocode) into syntactically correct Fortran statements that are typed correctly. The rest of the design process and how to improve the program are described in the following sections.

Make the Program Friendly

The "action phase" (Table 2.1b) says to debug and test the program. It is an integral part of program design because it quickly shows where the program does not perform as the written design predicted. Often it's "back to the drawing board" when run-time problems are discovered. Keep in mind that program `Triang` in Figure 2.8 (p. 34) is a prototype; it's a rough draft that represents a best first attempt from the written design phase and will need to go through revisions, just as one would revise a research paper several times. Again, subroutines have not been completely explained yet, so a stripped-down, subroutine-free version of the triangle program will be used for discussion. Figure 2.8 depicts the first time the triangle program is run. **"Make the program 'friendly' "** is a cutesy computer-industry buzz phrase that simply means that whoever uses a "friendly program" will never feel confused or uncomfortable while operating the program.

Note that the written part of the design did not produce the idea of using a prompting message. But next time a program is designed by the designer of program `Triang`, it is more likely that she will think of prompting messages. Each mistake makes you a better program designer. Every design a programmer does will make her better at future designs, but it is important to recognize that the prompting message problem WAS discovered. It is unimportant that it was uncovered as late as the testing stage. The design process *as a whole* will uncover ALL the problems with the program. Figure 2.9 (p. 35) shows the revised program.

Make Answers Readable—Use Formats

The answers in Figure 2.9 are numerically correct but they are U–G–L–Y. The equation doesn't look much like an equation, there are too many decimal places in the coefficients and the spacing in all the answers is poor. This points out the next step of program refinement: `Triang`'s answers need to be made more readable.

```
y=   -2.3000000 x+    4.9000001
Base=    2.1304348
Height=   4.9000001
Area=    5.2195654
```

Two **formats** and some modifications to the values being written will solve this problem (Figure 2.10, p. 36). Output formats are Fortran's way of helping programmers edit values in a program *as they are sent* to an output device such as a computer screen, printer, etc. Output formats make answers look readable and professional. Appendix D explains

Figure 2.8

```
PROGRAM Triang
      REAL     ::     A, B
      REAL     ::     Base, Height, Area
      READ (UNIT = *, FMT = *) A, B
        Base   = - B / A
        Height = B
        Area   = 1.0 / 2.0 * Base * Height
        WRITE ( UNIT = *, FMT = * )  "y=", A, "x", "+", B
        WRITE ( UNIT = *, FMT = * )  "Base=", Base
        WRITE ( UNIT = *, FMT = * )  "Height=", Height
        WRITE ( UNIT = *, FMT = * )  "Area=", Area
END PROGRAM
*********************PROGRAM RESULTS***************
```

■   <------------   A blinking cursor. What gives? What is the program doing?

The program has suspended execution on the fourth line:

                READ (UNIT = *, FMT = *) A, B

The READ statement is waiting for the user to enter values for A and B. This is the first point of program refinement: always include a prompting message when asking a user to enter values at the keyboard. A prompting message prompts the user, letting her know what to enter and helps make the program more friendly.

formats; refer to that appendix whenever neater and more readable answers are needed. Specific instances of formats will be explained as necessary throughout the text.

Here's how the first format works in concert with the WRITE statement that prints out the equation:

```
        WRITE ( UNIT = *, FMT = 25 ) "Y = ", A, "X", B
   25   FORMAT ( 1X, "The equation is: ", A4, F5.2, A1, SP, F5.2, S )
```

The original WRITE statement in Figure 2.9 was changed in the following ways:

1. "y = " was changed to "Y = ".
2. "x" was changed to "X".

Figure 2.9

```
PROGRAM Triang
    REAL      ::      A, B
    REAL      ::      Base, Height, Area
    WRITE ( UNIT = *, FMT = * ) "Please Enter Values for A and B"
    READ  ( UNIT = *, FMT = * ) A, B
        Base   = - B / A
        Height = B
        Area   = 1.0 / 2.0 * Base * Height
        WRITE ( UNIT = *, FMT = * )  "y=", A, "x", "+", B
        WRITE ( UNIT = *, FMT = * )  "Base=", Base
        WRITE ( UNIT = *, FMT = * )  "Height=", Height
        WRITE ( UNIT = *, FMT = * )  "Area=", Area
END PROGRAM

******************* PROGRAM RESULTS****************

Please Enter Values for A and B     <---  Prompting message.
-2.3, 4.9                           <---  Values entered for
                                          variables A and B at the
                                          keyboard.

y=  -2.3000000 x+   4.9000001
Base=   2.1304348
Height= 4.9000001
Area=   5.2195654
```

3. "+" was omitted.

4. The WRITE statement now has a reference to FORMAT 25:

$$\text{WRITE (} \ldots \text{ FMT = 25) } \ldots$$

When a format, designated by a **label number** (25 above), is used in a WRITE statement, the values that are printed are displayed according to the editing specifications (**format descriptors**) in the FORMAT statement. Format descriptors provide exact rules for how values are written. The WRITE statement in program Triang tells Fortran to write "Y = ", A, "X", and B according to FORMAT 25. FORMAT 25 has seven parts:

1. 1X — The 1X causes column 1 on the screen to be skipped, leaving a blank.
2. "The equation is: " — This part of the format simply prints the string of characters between the quotation marks.
3. A4 — This descriptor allows four characters to be printed. The corresponding characters in the WRITE statement are: "Y∆=∆" (where ∆

Figure 2.10

```
PROGRAM Triang
    REAL    ::    A, B
    REAL    ::    Base, Height, Area
    WRITE ( UNIT = *, FMT = * ) "Please Enter Values for A and B"
    READ  ( UNIT = *, FMT = * ) A, B
        Base   = - B / A
        Height = B
        Area   = 1.0 / 2.0 * Base * Height
        WRITE ( UNIT = *, FMT = 25 ) "Y = ", A, "X", B
        WRITE ( UNIT = *, FMT = 37 ) "Base   =", Base
        WRITE ( UNIT = *, FMT = 37 ) "Height =", Height
        WRITE ( UNIT = *, FMT = 37 ) "Area   =", Area

 25     FORMAT ( 1X, "The equation is:  ", A4, F5.2, A1, SP, F5.2 )
 37     FORMAT ( 1X, A8, 2X, F10.7 )

END PROGRAM Triang

************************PROGRAM RESULTS*****************

Please Enter Values for A and B
-2.3 4.9
The equation is:  Y = -2.30X+4.90
Base    =    2.1304348
Height  =    4.9000001
Area    =    5.2195654
```

4. F5.2

5. A1

represents a blank). Therefore, Y∆=∆ is displayed on the screen.

This is a description of how the value contained in A, which is -2.3, will be written from the WRITE statement. The F means a value of type REAL, such as A, is to be written. The 5 means the entire number will take five spaces on the screen, and the .2 means there will be two places to the right of the decimal. That's why A's value is written as -2.30 (a total of five characters on the screen).

A1 is similar to A4 above. The difference is that A4 will accommodate four characters and A1 allows only *one* character to be sent to the output device. The corresponding element that is printed from the WRITE statement is "X".

6. SP Normally, positive numbers are printed without a + sign. The SP descriptor causes a + to be printed in front of positive numbers.
7. F5.2 See number 4 above. This format descriptor results in +4.90 being printed.

The final result of the WRITE statement and its associated FORMAT is shown below:

The equation is: Y = -2.30X+4.90

Three WRITE statements each use the second format, FORMAT 37:

```
WRITE ( UNIT = *, FMT = * )  "Base   =", Base
WRITE ( UNIT = *, FMT = * )  "Height =", Height
WRITE ( UNIT = *, FMT = * )  "Area   =", Area
  . . .
37   FORMAT ( 1X, A8, 2X, F10.7 )
```

FORMAT 37 does four things:

1. 1X   This skips a space.
2. A8   This allows eight characters to be printed (e.g., "BaseΔΔΔ="— again, Δ represents a blank).
3. 2X   Allows two spaces to be skipped.
4. F10.7   This displays the value of the variable in the corresponding WRITE statement (e.g., Height, which has a value of 4.9000001). In the format, the 10 dictates that the entire number will take ten spaces on the screen and the .7 puts seven places to the right of the decimal.

Use Varied Data to Shake Out the Bugs

It is always wise to test a program with a variety of values to make sure it works under all conditions. Examine the results from program Triang in Figure 2.11 (p. 38). Program results show a negative Base and Area. The magnitudes of the answers are correct, but the slope of the equation is backwards from the original interpretation in Figure 2.2. With a positive coefficient for x and a positive y intercept, the curve would graph as seen in Figure 2.12 (p. 38). The equation has a negative x intercept and consequently produces a negative value for the base of the triangle. No problem. One of Fortran's built-in functions, the ABS() function, is used to force an absolute value. The next generation of program Triang is illustrated in Figure 2.13 (p. 39).

Does the Program Handle All Cases?

In the last section, a was positive, which led to flawed results. It is important to continue testing the program with different values to turn up any other conditions that produce wrong answers or cause the program to crash. When certain data cause a program to crash, it is necessary to write program statements to avoid the problem. Consider the program results for program Triang in Figure 2.14 (p. 39). The program crashed because the

Figure 2.11

```
************** PROGRAM RESULTS ***************

 Please Enter Values for A and B
 1.3 2.6
  The equation is:  Y =   1.30X+2.60
  Base    =   -2.0000000
  Height  =    2.5999999
  Area    =   -2.5999999
```

Figure 2.12

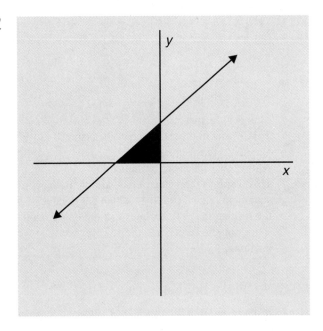

read-in value for A was 0.0, and line 6 of the program tried to perform the statement, Base = -B / A. Computers cannot divide by zero any more than humans can.

One solution to the problem lies in using a selection structure to do **error checking.** Error checking is the process of looking for potential problems that could interfere with the proper operation of a program because of invalid incoming data. An IF statement such as the one in Figure 1.4 of Chapter 1 demonstrates this:

```
IF ( question ) THEN
      block1
ELSE
      block2
ENDIF
```

Selection structures are covered in detail in Chapter 6, but for now it is enough to understand that if the answer to question is yes, the program will do whatever Fortran statements are in block1 and NOT the statements in block2. Likewise, if the answer

Figure 2.13

```
PROGRAM Triang
    REAL     ::      A, B
    REAL     ::      Base, Height, Area
    WRITE ( UNIT = *, FMT = * ) "Please Enter Values for A and B"
    READ  ( UNIT = *, FMT = * ) A, B
        Base   = ABS (- B / A )
        Height = ABS ( B )
        Area   = 1.0 / 2.0 * Base * Height
        WRITE ( UNIT = *, FMT = 25 )  "Y = ", A, "X", B
        WRITE ( UNIT = *, FMT = 37 )  "Base   =", Base
        WRITE ( UNIT = *, FMT = 37 )  "Height =", Height
        WRITE ( UNIT. = *, FMT = 37 ) "Area   =", Area

    25    FORMAT ( 1X, "The equation is:  ", A4, F5.2, A1, SP, F5.2 )
    37    FORMAT ( 1X, A8, 2X, F10.7 )

END PROGRAM Triang
```

Notice the ABS() absolute value function usage in the Base and Height calculations.

```
*************** PROGRAM RESULTS **************

 Please Enter Values for A and B
1.3 2.6
 The equation is:  Y = 1.30X+2.60
 Base   =    2.0000000
 Height =    2.5999999
 Area   =    2.5999999
```

Figure 2.14

```
*************** PROGRAM RESULTS ***************

Please Enter Values for A and B
0.0, 17.3       <--- Values typed at the keyboard for A and B.

ERROR*** PROGRAM Triang ABORTED - Division by zero in line 6.
```

Figure 2.15

```
PROGRAM Triang
      REAL      ::      A, B
      REAL      ::      Base, Height, Area
      WRITE ( UNIT = *, FMT = * ) "Please Enter Values for A and B"
      READ  ( UNIT = *, FMT = * ) A, B
      IF( A == 0.0 ) THEN
            WRITE ( UNIT = *, FMT = * ) "There is no Triangle"
      ELSE
            Base   = ABS ( - B / A )
            Height = ABS ( B )
            Area   = 1.0 / 2.0 * Base * Height
            WRITE ( UNIT = *, FMT = 25 ) "Y = ", A, "X", B
            WRITE ( UNIT = *, FMT = 37 ) "Base   =", Base
            WRITE ( UNIT = *, FMT = 37 ) "Height =", Height
            WRITE ( UNIT = *, FMT = 37 ) "Area   =", Area
      ENDIF

  25  FORMAT ( 1X, "The equation is:  ", A4, F5.2, A1, SP, F5.2 )
  37  FORMAT ( 1X, A8, 2X, F10.7 )

END PROGRAM Triang

************** PROGRAM RESULTS ***************

 Please Enter Values for A and B
 0.0 17.3
 There is no Triangle
```

The IF statement avoids the division-by-zero program crash.

to `question` is no, the program will do `block2` and not `block1`. This can solve our division-by-zero problem by checking for the error before it happens (Figure 2.15).

Always Hand Check Formulas with a Calculator

A final phase of program testing that ensures the program is producing correct results is to hand check formulas to be sure they are working correctly and generating accurate values. This can be done easily with a calculator (Figure 2.16). The numbers check, so it can be assumed that the formulas in the program are working correctly.

Generalize the Solution

Program `Triang` has come a long way from the stated goal: "Write a computer program to find the area of a triangle formed by axes and a linear equation." The process of

Figure 2.16

```
************* PROGRAM RESULTS **************

Please Enter Values for A and B
-2.3  4.9
The equation is:   Y = -2.30X+4.90
Base   = 2.1304348
Height = 4.9000001
Area   = 5.2195654
```

Using a calculator, the program's results are checked:

```
base   ==> -B/A  ==> -4.9/(-2.3) ==> 4.9/2.3 ==> 2.1304
height ==>  B    ==> 4.9
area   ==>  ½ • base • height ==> ½•2.1304•4.9 ==> 5.2195
```

The hand-checked values match so the program is producing correct answers.

creating a computer program from conception to implementation has been shown through testing the program. Triang could be taken further by making it more general. **Generalized programs** handle more situations, and for Triang, this might include putting the body of the program in a loop so a user could calculate triangle areas for several linear equations. There are many subtle ways that programs can and should be generalized, and they will be explained as they come up.

Important Concepts Review

- Once a program is typed into the computer, it must be **thoroughly tested and refined**.
- **Debugging** is the act of removing errors from a program.
- **Friendly computer programs** use prompting messages and generate readable answers to minimize confusion for the user.
- FORMATs in WRITE statements edit values as they are displayed to make answers more readable.
- **Use varied data** when testing a program to be sure it works under all conditions.

- **Error checking** is any programming technique used to check incoming values to prevent bad data from corrupting the program's performance.

- **Hand check** programs' answers with a calculator to ensure that the formulas are working correctly.

2.3 Program Form—For Each Structure Chart Block

When a program is written, it is essential that a programmer uses techniques to make the main program and all of its subroutines as understandable as possible. This is accomplished with comments, blank lines and indentations to improve readability and visually set off structures.

One of the main reasons it is necessary to make programs very clear is the **software life cycle,** which is the process where programs must be modified to meet new requirements. After a program is completed and is working correctly, it may remain in service for a year or two. Then, perhaps because of changes in data or the necessity to produce different answers, the program has to be changed. If the original programmer used poor documentation techniques by not choosing **meaningful identifiers** (names supplied by the programmer), or didn't use indentations or comments, the programmer doing the modifications will spend excessive time trying to decipher exactly what the program is doing. This is an **expensive waste of time** that can be avoided by good documentation methods and program style.

Comments

Comments are whole or partial lines that are embedded in a program as brief notes to help explain important parts of the program. These notes are ignored by the Fortran compiler and don't affect the way the program works. Two styles of commenting are recommended:

1. Comment blocks
2. Line commenting

Comment lines are started with an exclamation point. Anything that comes after the "!" to the end of the line is ignored. A **comment block** is a group of comment lines at the beginning of a main program, subroutine or other type of program unit that briefly explains the purpose of the program unit, including an explanation of each variable used. The main program's comment block should also include the author's name, date written, location, etc. **Line commenting** is used to explain any part of a program that might be confusing. Examine Figure 2.17's commenting techniques.

Figure 2.17

```
                PROGRAM Triang

!---------------------------------------------------------------
!   PROGRAMMER :     C. FORSYTHE
!   DATE WRITTEN:    11/15/2004
!---------------------------------------------------------------
!   The purpose of this program is to calculate the area
!   of a triangle formed by a linear equation and axes
!
!     A and B - real variables that contain the
!               values of the coefficients for Y=AX+B
!
!     Base    - is a real variable representing the length
!               of the base of the triangle
!
!     Height  - is a real variable representing the length
!               of the height of the triangle
!
!     Area    - is a real variable representing the area
!               of the triangle: 1/2*Base*Height
!---------------------------------------------------------------
!
      REAL     ::    A, B
      REAL     ::    Base, Height, Area
      WRITE ( UNIT = *, FMT = * ) "Please Enter Values for A and B"
      READ  ( UNIT = *, FMT = * ) A, B
      IF( A == 0.0 ) THEN   ! don't want to divide by zero
           WRITE ( UNIT = *, FMT = * ) "There is no Triangle"
      ELSE
           Base   = ABS ( - B / A )    ! ABS() prevents neg Area
           Height = ABS ( B )          ! ABS() prevents neg Area
           Area   = 1.0 / 2.0 * Base * Height
           WRITE ( UNIT = *, FMT = 25 ) "Y = ", A, "X", B
           WRITE ( UNIT = *, FMT = 37 ) "Base   =", Base
           WRITE ( UNIT = *, FMT = 37 ) "Height =", Height
           WRITE ( UNIT = *, FMT = 37 ) "Area   =", Area
      ENDIF
!
  25  FORMAT ( 1X, "The equation is: ", A4, F5.2, A1, SP, F5.2 )
  37  FORMAT ( 1X, A8, 2X, F10.7 )

END PROGRAM Triang
```

Figure 2.18

```
PROGRAM Triang
REAL::A, B
REAL::Base, Height, Area
WRITE ( UNIT = *, FMT = * ) "Please Enter Values for A and B"
READ  ( UNIT = *, FMT = * ) A, B
IF ( A == 0.0 ) THEN
WRITE ( UNIT = *, FMT = * ) "There is no Triangle"
ELSE
Base   = ABS ( - B / A )
Height = ABS ( B )
Area   = 1.0 / 2.0 * Base * Height
WRITE ( UNIT = *, FMT = 25 ) "Y = ", A, "X", B
WRITE ( UNIT = *, FMT = 37 ) "Base   =", Base
WRITE ( UNIT = *, FMT = 37 ) "Height =", Height
WRITE ( UNIT = *, FMT = 37 ) "Area   =", Area
ENDIF
25 FORMAT ( 1X, "The equation is:  ", A4, F5.2, A1, SP, F5.2 )
37 FORMAT ( 1X, A8, 2X, F10.7 )
END PROGRAM Triang
```

Indentations and Blank Lines

Indentations should be used to improve the readability of program structures such as IF statements and loops. The reason is obvious if the program in Figure 2.17 is compared to the A–W–F–U–L program in Figure 2.18 that works exactly the same way. See how unreadable program Triang has become without comments and indentations? Figure 2.18's program demonstrates very poor **program style.** The IF statement has faded away into a mass of Fortran statements.

2.4 Program Format

The general format of a Fortran program is illustrated in Figure 2.19. Fortran allows programmers to place program statements anywhere on a program line. Each line is limited to a maximum length of 132 characters. Occasionally it may be necessary or desirable to continue a line onto the next line. This is accomplished with the ampersand (&).

```
READ &
A, B
```

The two lines above are identical to the following one line:

```
READ A, B
```

Figure 1.19
```
PROGRAM prog-name-identifier
      comment block
      variable declarations
      Fortran statements
END   PROGRAM prog-name-identifier
SUBROUTINE identifier1()
   ...
END SUBROUTINE identifier1
SUBROUTINE identifier2()
      etc.
```

The `&` must be the last nonblank character on a line. It also may be necessary to split a statement in the middle of a **character constant** (any series of characters enclosed in apostrophes or quotes). These situations require ampersands on both lines so that unwanted blanks are not accidentally introduced between characters:

```
PRINT * , "This is th&
            &e way to split&
            & a character constant."
```

The `PRINT` statement above is equivalent to the following:

```
PRINT * , "This is the way to split a character constant."
```

2.5 Identifiers

In Figure 2.19 the word *identifier* comes up a lot. An **identifier** is a name created by a programmer to identify a Fortran entity, such as a variable or a subroutine. Identifiers must **start with a letter** and that first letter may be followed by **any combination of letters, digits, and underscores** ("_"). Identifiers may be up to **31 characters** long, and Table 2.2 shows some variations on identifiers.

Table 2.2

Identifiers	Valid?
A_____J_____	YES
3_SHOULDNT_START_WITH_A_NUMBER	NO
CANT LEAVE SPACES	NO
TRY_THE_ROOT	YES
TOO_LONG_12345678901234567890123	NO
JUST_LONG_ENOUGH_ABCDEFGHIJKLM7	YES
_CANT_START_WITH_AN_UNDERSCORE	NO

Choosing meaningful identifiers (**mnemonic** identifiers) is vital. Program designers must ALWAYS shoot for clarity because the software life cycle implies that useful programs will probably be modified at some point in the future. Whoever does a modification ("who" may be you modifying your own program) can do the revisions much more swiftly if the program's identifiers were well chosen in the first place. Figure 2.18 demonstrated how a lack of comments and indentations (poor program style) made program `Triang` harder to understand. Couple that with poor identifiers (additionally poor program style), and a programmer can create a real disaster by trying to make modifications. Examine Figure 2.20 as an example of unacceptable program style. This program does, however, function perfectly.

Figure 2.20

```
PROGRAM Potato
REAL::Toast, Frog_Legs
REAL::Frost_Bite, Freddy, Rug_Burn
WRITE ( UNIT = *, FMT = * ) "Please Enter Values for A and B"
READ  ( UNIT = *, FMT = * ) Toast, Frog_Legs
IF ( Toast == 0.0 ) THEN
WRITE ( UNIT = *, FMT = * ) "There is no Triangle"
ELSE
Frost_Bite   = ABS ( - Frog_Legs / Toast )
Freddy = ABS (Frog_Legs )
Rug_Burn   = 1.0 / 2.0 * Frost_Bite * Freddy
WRITE ( UNIT = *, FMT = 25 )  "Y = ", Toast, "X", Frog_Legs
WRITE ( UNIT = *, FMT = 37 )  "Base   =", Frog_Legs
WRITE ( UNIT = *, FMT = 37 )  "Height =", Freddy
WRITE ( UNIT = *, FMT = 37 )  "Area   =", Rug_Burn
ENDIF
25 FORMAT ( 1X, "The equation is:  ", A4, F5.2, A1, SP, F5.2 )
37 FORMAT ( 1X, A8, 2X, F10.7 )
END PROGRAM Potato
```

Important Concepts Review

- **Software life cycle** implies that most programs WILL be modified at some point, so programs should be made as clear and easy to modify as possible.

- **Comments** are notes that clarify programs. Comments start with exclamation points and are ignored by the compiler.

- **Comment blocks** are at the beginning of the main program and every other program unit. A comment block provides a brief description of the purpose of the program unit and descriptions for all variables. In main programs, include the author's name, date written (or modified), etc.

- **Line commenting** clarifies individual lines of a program.

- **!** starts a comment line.

- **Blank lines and indentations** make structures more obvious and the program easier to understand.

- **Program style** is the way a programmer presents her program.

- **&** enables a line to be continued on the next line. Use ampersands on both lines if internal spaces are undesirable, such as in the middle of a string of characters.

- **Identifiers** should have well-chosen names that imply their use. An identifier must start with a letter and may be followed by any combination of letters, digits, or underscores and can be up to 31 characters long.

EXERCISES

In Exercises 2.1 through 2.3, follow the design steps in Table 2.1a.

2.1 Design a program to evaluate the resistance in a circuit containing three resistors connected in parallel.

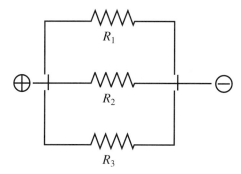

The resistance of parallel resistors in a circuit is given by the following formula:

$$1/R_T = 1/R_1 + 1/R_2 + \cdots + 1/R_N$$

where R_T is the total resistance.

2.2 Design a program to solve a system of linear equations:
$$ax + by = c$$
$$dx + ey = f$$

2.3 Design the following three programs.
   a. Design a program to convert Celsius temperatures to Fahrenheit.
   $$Fahrenheit = 9.0/5.0 * Celsius + 32.0$$
   b. Design a program to convert Fahrenheit to Celsius.
   c. Design a program that will present a two-item menu:

   1. Convert Fahrenheit to Celsius.
   2. Convert Celsius to Fahrenheit.

   Allow the user to choose either conversion. Use two subroutines, one for each conversion.

2.4 Describe how identifiers are constructed and give six examples.

2.5 What purpose do comment blocks serve?

2.6 Why do programmers use varied data to test programs?

2.7 What is error checking?

2.8 Write a set of Fortran statements to error check the menu selection in Exercise 2.3c.

2.9 What are the elements of good program style?

2.10 Design a program to calculate the half-life of a radioactive substance. Half-life is defined by the following formula:
$$\frac{a}{a_o} = \frac{1}{2^N}$$

The original concentration is a_o, current concentration is a and N is the half-life.

Subroutines

3

Introduction

Subroutines are the road to minimum-time-spent program designs. Subroutines make programs easier to design and code, and make them much closer to error-free when they are first entered into a computer. This is VERY desirable as programming is a means, not an end, for scientists and engineers; you want convenient and reliable data analysis without spending your lives writing computer code. Sometimes the most efficient analysis is not accomplished by designing a computer program at all. Answers can also be rendered with software packages such as powerful spreadsheets, where rapid replication of formulas and robust graphing facilities can outspeed designing and coding a computer program.

Frequently, however, designing a customized computer program is the only possible method of solving a problem. When this unfortunate circumstance occurs, GET LAZY! Reuse programs or parts of programs whenever possible. One way to accomplish this is by training yourself to design programs with subroutines in the top-down style explained in Chapter 2. Then you can save any created subroutines that might be of general use in future designs. Reusing working subroutines is made easy by creating collections of them called "libraries."

Also, take advantage of existing software libraries such as *LA Pack*, which is a public domain linear algebra library. There are many such libraries in existence; some are free and others must be

- 3.1 Writing a subroutine
- 3.2 Subroutine definition
- 3.3 Scope of subroutines
- 3.4 Arguments

   Important Concepts Review

- 3.5 Flow of data through arguments

   Important Concepts Review

- 3.6 Local variables
- 3.7 Entering a modular program into a computer
- 3.8 Subroutine libraries

   Important Concepts Review

   Exercises

purchased. It is well worth having libraries available that relate to your research because it will save you a lot of time designing programs. Fortran 90 itself has a host of built-in routines that can shorten programming efforts. Use existing subroutines; don't reinvent the wheel, and remember to **recognize when a subroutine might have future use**. Once a subroutine works, it will always work and won't have to be recreated.

Writing customized subroutines is the focus of this chapter because it's more difficult to create a subroutine than to simply use an existing one. As explained before, many situations will arise where a necessary subroutine doesn't exist in any of your software libraries and you will be called upon to create one.

• • •

3.1 Writing a Subroutine

Subroutines are mini Fortran programs that are either called from the **main program** (the top block of a structure chart) or from **another program unit**. A Fortran program may only have ONE main program. The distinction between a main program and subroutines will become fuzzier and fuzzier, as it should, because the only difference is that the main program can't be called. When a program's tasks are outlined in a **structure chart**, each block on the chart is designed to do one well-defined task. Program designers should code each block's **algorithm** (method of solution) as a subroutine. This will simplify the calling program, which can be a main program or another program unit. It is possible to write programs without subroutines, but the result is usually extremely difficult to understand, and it breaks the **Golden Rule of Program Design**: programs should work right, be easy to understand and modify, be completed in as short a time as possible and be designed flexibly to grow and expand with time.

Flow of control for subroutine calls works as shown in Figure 3.1 When a `CALL` is met, program flow is transferred to the called subroutine, and its instructions are performed. After the subroutine's job is done, the `RETURN` statement at the end of the subroutine transfers program flow back to the statement immediately following the `CALL` that invoked the subroutine in the calling program.

In the previous couple of paragraphs, main programs were referred to as **calling programs** and subroutines as **called programs**. This is a general method of identifying a program unit by its relationship to any **parent** or **children** program units (Figure 3.2). Program unit 2 in Figure 3.2 is called by MAIN so program unit 2 is a **called** program unit.

Figure 3.1

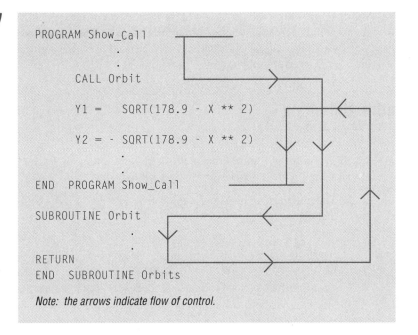

Note: the arrows indicate flow of control.

Figure 3.2

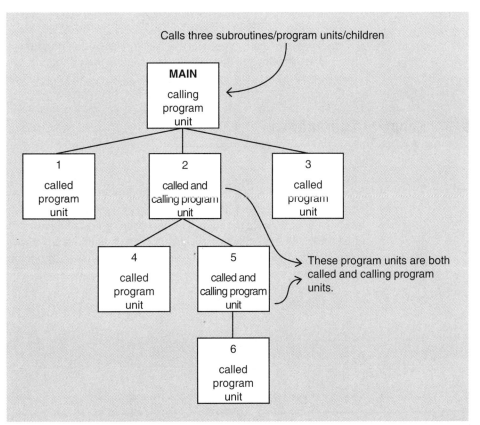

Program unit 2 also calls program units 4 and 5, so program unit 2 is a **calling** program unit as well. Program unit 5 has similar properties in that it is called by unit 2 and calls 6. The expressions "called" and "calling" will be used to define a program unit's relative position in the structure of a program from now on.

3.2 Subroutine Definition

The general form of a subroutine is illustrated in Figure 3.3. Each element of the generic subroutine in Figure 3.3 is defined in the following list:

1. The **subroutine heading** is the word `SUBROUTINE` followed by an identifier that names it.
2. `IMPLICIT NONE` should be included in EVERY program unit. `IMPLICIT NONE` helps minimize subtle program errors by forcing all variables to be explicitly declared (more on this in Chapter 4).
3. **Comment blocks** should be included in EVERY subroutine. A comment block briefly describes the purpose of a subroutine and the variables used in it.
4. **Declarations** are a set of special Fortran 90 specification statements placed at the beginning of a program unit where variables (and other data items) are defined.
5. **Program statements** accomplish the subroutine's purpose.
6. `RETURN` sends program flow back to the calling program to the statement immediately following the `CALL` that invoked the subroutine.
7. `END SUBROUTINE` is the last physical statement of the subroutine.

3.3 Scope of Subroutines

The **scope** of a subroutine is defined by which other program units in a program have access to the subroutine's internal data and definitions. A Fortran subroutine's scope is itself only. Subroutines' internal data and definitions are NOT "known" to any other program units that comprise the program. This **isolation** of program units is a key feature of structured programming languages such as Fortran. Isolation helps stop data from accidentally being modified in an inappropriate way. Figure 3.4 shows a program example that demon-

Figure 3.3

```
1. SUBROUTINE identifier_1
2.        IMPLICIT NONE
3.        comment block
4.        declarations
5.        program statements
6. RETURN
7. END SUBROUTINE identifier_1
```

Figure 3.4

```
      PROGRAM Demonstrate_The_Scope_Of_Subs
                IMPLICIT    NONE
      !
      !-------------------------------------------------------------------
      ! Written by      :   C. Forsythe
      ! Date Written    :   3/25/2004
      !-------------------------------------------------------------------
      ! The purpose of this program is to show how program units are
      !   isolated from each other.
      !
      ! Test - An integer variable
      !-------------------------------------------------------------------
          INTEGER   ::   Test
      !
          Test = 2
          PRINT *, "In main, Test =       ", Test
          CALL For_Testing_Scope
          PRINT *, "Back from sub, Test =", Test
      !
      END PROGRAM Demonstrate_The_Scope_Of_Subs
!====================================================================
      SUBROUTINE For_Testing_Scope
                IMPLICIT    NONE
      !
      !-------------------------------------------------------------------
      !  This subroutine creates a local variable called Test.
      !
      ! Test  -  an INTEGER (no decimal part allowed) variable
      !             that is set to 97 and then printed
      !-------------------------------------------------------------------
      !
          INTEGER :: Test    ! a local variable
      !
          Test = 97
          PRINT *, "I'm in sub, Test =", Test
      !
      RETURN
      END SUBROUTINE For_Testing_Scope

***************PROGRAM RESULTS***************

In main, Test =        2
I'm in sub, Test = 97
Back from sub, Test = 2
```

strates the isolation of subroutines. Data and definitions that are created in a subroutine are said to be **local** to it.

In Figure 3.4, `Test` is given a value of `2` and printed (displayed on the screen) in the main program. Then in subroutine `For_Testing_Scope`, a variable named `Test` is given a value of `97` and printed. But when `Test` is printed a final time in the main program, a `2` is displayed, just as it was in the first `PRINT` statement. How could `Test` have two different values? The reason for the different printed values of `Test` is that data in one program unit aren't known to any other program units. Variable `Test` in subroutine `For_Testing_Scope` is a *completely different* variable from `Test` in the main program. In subroutine `For_Testing_Scope`, variable `Test` is created by the following declaration statement:

```
INTEGER :: Test ! a local variable
```

`Test` is **local** to subroutine `For_Testing_Scope`. After the subroutine is left via the `RETURN` statement, the subroutine's local variable `Test` is discarded. But variable `Test` in the main program *still* has a value of `2` because it is a completely different variable than subroutine `For_Testing_Scope`'s local variable `Test`.

Isolation of program units is critical to structured programming. Without program unit isolation, two program units that each contain a variable with the **same name** are inadvertently referring to the same value. Accidental and inappropriate modifications could be made to that variable causing the program to work incorrectly. On the other hand, there MUST be a **method of getting data in and out of subroutines** or the useful results of a given subroutine would be lost inside the event horizon of its isolation.

3.4 Arguments

Arguments solve the problem of how to move data in and out of program units and make the **flow of data** possible (Figure 3.5). The generic subroutine in Figure 3.5 differs from the one in Figure 3.3 in a couple of ways:

1. Figure 3.5's subroutine statement includes a set of parentheses containing **arguments**:

```
SUBROUTINE identifier_1 (arg1, arg2, ... )
```

Figure 3.5
```
1. SUBROUTINE identifier_1 ( arg1, arg2, ... )
2.      IMPLICIT NONE
3.      comment block
4.      argument declarations
5.      local declarations
6.      program statements
7. RETURN
8. END SUBROUTINE identifier_1
```

2. Figure 3.5 suggests that there are two kinds of declarations:

   a. Argument declarations

   b. Local declarations

Figure 3.6 shows an example of a program that uses an argument to move a `9` into a subroutine. In the example from Figure 3.6 below, the `CALL` to `SUBROUTINE Demonstrate_Arguments` specifies an **actual argument** named `Test_Value_1`. Actual arguments supply data to subroutines and/or receive data back from subroutines; they are the conduit through which data flows in and out of program units.

Actual arguments must, however, "connect" with their counterparts, called **formal arguments**. Formal arguments are listed between the parentheses in a subroutine's heading, and represent actual arguments symbolically within the subroutine. In `SUBROUTINE Demonstrate_Arguments`, formal argument `Test_Value` represents actual argument `Test_Value_1`. That is why when `Test_Value` is printed, `Test_Value_1`'s `9` is displayed; `Test_Value` adopts `Test_Value_1`'s value.

Formal arguments become one-and-the-same with their corresponding actual arguments, making the actual arguments **available to the called program**; formal arguments impersonate the actual argument correlatives. Actual arguments (which are supplied to a subroutine by an associated `CALL` statement) and the formal arguments that represent them are united until control is returned to the calling program when the subroutine is finished.

It is interesting that in Figure 3.6, the actual argument's name, `Test_Value_1`, is different from the formal argument's name, `Test_Value`. This happens frequently because subroutines are often used and reused at several different places in a program and in *different* programs. Formal arguments *become* the actual arguments that are passed to

Figure 3.6

```
PROGRAM ...
   ...
Test_Value_1 = 9
CALL Demonstrate_Arguments ( Test_Value_1 )
   ...
END PROGRAM ...
SUBROUTINE Demonstrate_Arguments ( Test_Value )
   ...
     PRINT *, Test_Value

RETURN
END  SUBROUTINE

*************** PROGRAM RESULTS ***************

      9.000000
```

In the figure, `Test_Value_1` in the `CALL` statement is labeled **Actual Argument**, and `Test_Value` in the `SUBROUTINE` heading is labeled **Formal Argument**.

Figure 3.7

```
PROGRAM ...
   ...
Test_Value_1 = 9
CALL Demonstrate_Arguments ( Test_Value_1 )
   ...
Another_Value = -4190
CALL Demonstrate_Arguments ( Another_Value )
   ...
CALL Demonstrate_Arguments ( -7.31 )
END PROGRAM ...
SUBROUTINE Demonstrate_Arguments ( To_Be_Printed )
   PRINT *, To_Be_Printed
RETURN
END   SUBROUTINE

************** PROGRAM RESULTS ***************

      9.0000000
  -4190.0000000
     -7.3100000
```

Annotations: **Actual Argument** (→ Test_Value_1), **Different Actual Argument** (→ Another_Value), **Another Actual Argument** (→ -7.31), **Formal Argument** (→ To_Be_Printed)

them, so formal arguments' names are not so terribly important other than that they should be named in a meaningful way. More significant than formal arguments' names is the way that they are declared. They must reflect the corresponding actual arguments' declarations. Figure 3.7 shows an example of different actual arguments being passed to the same formal argument.

In Figure 3.7, the first `CALL` is below:

`CALL Demonstrate_Arguments ( Test_Value_1 )`

This `CALL` associates actual argument `Test_Value_1` (its value is `9.0000000`) with the subroutine's formal argument, `To_Be_Printed`. `Test_Value_1` and `To_Be_Printed` become one (zen). `Test_Value_1` is now represented within the subroutine by the formal argument `To_Be_Printed`. When `To_Be_Printed` is printed, it displays `9.0000000` (`Test_Value_1`'s value). Then control returns to the calling program and `Test_Value_1` and `To_Be_Printed` are disassociated. Later in the program, another `CALL` to subroutine `Demonstrate_Arguments` is performed, but this time a different actual argument, `Another_Value` (`-4190`), forms a relationship with formal argument `To_Be_Printed`. When `To_Be_Printed` is printed this time, `-4190.0000000` is displayed. Finally, a numeric constant, `-7.31`, is sent to the subroutine as an actual argument by the third `CALL` and is printed. The ability to connect different actual arguments to a subroutine's formal arguments is one of the most powerful features of subroutines—it permits *generalized subroutines to be created and reused.*

It is also reasonable to have more than one argument to move data in and out of a subroutine. Often a subroutine will require several values to perform its required task. This doesn't present a problem if two rules are observed:

1. There must be the same number of formal arguments as actual arguments.
2. Formal and actual arguments must correspond in type (`REAL`, `INTEGER`, etc.)

Figure 3.8A shows a program segment that uses arguments correctly. Figure 3.8B has an **argument number mismatch**; the `CALL` has four actual arguments, but the subroutine

Figure 3.8

```
PROGRAM
   ...
   INTEGER                 :: a
   CHARACTER (LEN = 7)     :: b
   LOGICAL                 :: c
   ...
   CALL Hyperbolic ( a, b, c )
   ...
SUBROUTINE Hyperbolic ( x, y, z )
   INTEGER                 :: x
   CHARACTER (LEN = 7 )    :: y
   LOGICAL                 :: z
   ...
             (A)
-------------------------------
PROGRAM
   ...
   INTEGER                 :: d
   CHARACTER (LEN = 7)     :: b
   LOGICAL                 :: c, d
   ...
   CALL Hyperbolic ( a, b, c, d )
   ...
SUBROUTINE Hyperbolic ( x, y, z )
   INTEGER                 :: x
   CHARACTER (LEN = 7 )    :: y
   LOGICAL                 :: z
   ...
             (B)
-------------------------------
```

continues

```
PROGRAM
   ...
   REAL                     :: a
   CHARACTER (LEN - 7)      :: b
   LOGICAL                  :: c
   ...
   CALL Hyperbolic ( a, b, c )
   ...
SUBROUTINE Hyperbolic ( x, y, z )
   INTEGER                  :: x
   CHARACTER (LEN = 7 )     :: y
   LOGICAL                  :: z
   ...
             (C)
```

Figure 3.8 (continued)

definition only has three formal arguments. Figure 3.8C has a **type mismatch**. The actual argument `a` is of type `REAL` in the `CALL`, but it is associated with formal argument `x` in the subroutine that is type `INTEGER`—no good.

Important Concepts Review

- Subroutines are **mini Fortran** programs.

- When a `CALL` statement is met, program flow is transferred to the called subroutine, and its instructions are performed. After the subroutine's job is completed, the `RETURN` statement at the end of the subroutine transfers flow back to the statement immediately following the `CALL` that invoked the subroutine in the calling program.

- A **calling program unit** is the program unit that called the currently active subroutine.

- A **called program unit** is a program unit that is called.

- A **program unit** is a subroutine, main program, etc.

- Subroutines are **isolated**—values in one subroutine aren't known in another.

- **Arguments** are the way data flows in and out of subroutines.

- **Actual arguments** occur in a `CALL` statement.

- **Formal arguments** occur in a subroutine header. Formal arguments represent actual arguments within the subroutine.

- Actual and formal arguments must **correspond** in number and type.

3.5 Flow of Data Through Arguments

When arguments are used with subroutines, it is important to define the direction that the *data* flows through the arguments. It is analogous to getting a clear vision of the various ways flow of *control* can be altered. Having well-defined **flow of data** is another way to prevent data from being modified or set inappropriately, resulting in programs that fail. The three different ways that data can move through actual/formal argument conduits are listed below. Does an actual/formal argument relationship:

1. supply data to a subroutine?   . . . in
2. get data back from a subroutine?   . . . out
3. supply data to and get data from a subroutine?   . . . inout

Classifying arguments is done by declaring *formal* arguments with an `INTENT` attribute. Examine the program in Figure 3.9 (pp. 60–62) to see `INTENT` in action. Program `Classify_Arguments` in Figure 3.9 calls three subroutines:

1. `Get_Values_For_E_and_R`.
2. `Figure_The_Current`.
3. `Print_The_Results`.

When `Get_Values_For_E_and_R` is called, actual arguments, `E1` and `R1`, are united with formal arguments, `Voltage` and `Resistance`. Since `Voltage` and `Resistance` are `INTENT ( OUT )`, their read values will become `E1` and `R1`'s in the calling program unit once the subroutine is finished.

Actual arguments' contents entering `Get_Values_For_E_and_R` are *irrelevant* to the subroutine because the arguments' values are assigned in `Get_Values_For_E_and_R`. These types of arguments are said to be `INTENT ( OUT )`, meaning that it is intended that the arguments *only* allow data to travel *out* of the subroutine to the calling program's actual arguments. Actual arguments sent to formal arguments that have the `INTENT ( OUT )` attribute *become undefined* upon entering a subroutine. This means that if the actual arguments contain values when they arrive in a subroutine as formal arguments, those values are destroyed, making the formal arguments undefined and unusable until they are given values *in the subroutine*.

Figure 3.9

```fortran
      PROGRAM Classify_Arguments
               IMPLICIT   NONE
      !
      !-------------------------------------------------------------------
      ! Written by:   C. Forsythe
      ! Date      :   09/30/04
      !-------------------------------------------------------------------
      !  The purpose of this program is to show the difference
      !     between arguments that:
      !
      !   a) carry information (data) into a called program unit.
      !
      !   b) arguments that send information out of the called
      !      program unit back to the calling program unit.
      !
      ! E1      - type REAL, represent voltage
      !
      ! R1      - type REAL, represent resistance
      !
      ! I1      - type REAL, represent current
      !-------------------------------------------------------------------
      !
           REAL           ::    E1, I1, R1
      !
           CALL Get_Values_For_E_and_R ( E1, R1 )
           CALL Figure_The_Current     ( E1, R1, I1 )
           CALL Print_The_Results      ( E1, R1, I1 )
      !
      END PROGRAM Classify_Arguments
!===================================================================
      SUBROUTINE Get_Values_For_E_and_R ( Voltage, Resistance )
               IMPLICIT   NONE
      !
      !-------------------------------------------------------------------
      ! This subroutine is designed to get values for voltage and
      !    resistance from the keyboard and send them back
      !    to the calling program unit.
      !
      ! Voltage    -  type REAL, represents voltage -
      !                  sent back to the calling program.
      !
      ! Resistance - type REAL, represents resistance -
      !                  sent back to the calling program.
      !-------------------------------------------------------------------
      !
```

Figure 3.9 (continued)

```
      REAL,       INTENT ( OUT )  ::   Voltage, Resistance
!
      PRINT *,   "Please enter values for E & R"
      READ *,     Voltage, Resistance
!
      RETURN
      END SUBROUTINE Get_Values_For_E_and_R
!=================================================================
      SUBROUTINE Figure_The_Current ( Voltage, Resistance, Current )
               IMPLICIT   NONE
!
!-----------------------------------------------------------------
! This subroutine is designed to calculate current given
!    resistance and voltage from the formula:
!
!        current = voltage / resistance
!
! Voltage  -   type REAL, represents voltage -
!                 imported from the calling program.
!
! Resistance - type REAL, represents resistance -
!                 imported from the calling program.
!
! Current  -   type REAL, represents current —
!                 sent back to the calling program.
!-----------------------------------------------------------------
!
      REAL,       INTENT ( IN )   ::   Voltage, Resistance
      REAL,       INTENT ( OUT )  ::   Current
!
      Current = Voltage / Resistance
!
      RETURN
      END SUBROUTINE Figure_The_Current
!=================================================================
```

continues

Figure 3.9 (continued)

```
      SUBROUTINE Print_The_Results ( Voltage, Resistance, Current )
            IMPLICIT    NONE
!
!---------------------------------------------------------------
!    This subroutine is designed to display the results of the
!       calculation for current.
!
!  Voltage   -    type REAL, represents voltage -
!                     imported from the calling program.
!
!  Resistance - type REAL, represents resistance -
!                     imported from the calling program.
!
!  Current   -    type REAL, represents current -
!                     imported from calling program.
!---------------------------------------------------------------
!
      REAL,        INTENT ( IN )   ::   Voltage, Resistance
      REAL,        INTENT ( IN )   ::   Current
!
      PRINT *, "Voltage    =", Voltage
      PRINT *, "Resistance =", Resistance
      PRINT *, "Current    =", Current
!
      RETURN
      END SUBROUTINE Print_The_Results

*************** PROGRAM RESULTS ***************

 Please enter values for E & R
23.71 3.78
 Voltage    =    23.7099991
 Resistance =     3.7800000
 Current    =     6.2724867
```

It is a wonderful feature of Fortran to "undefine" arguments that are intended to only carry data out of a subroutine. It helps make errors easier to spot. If, for instance, an INTENT (OUT) formal argument is not given a value inside the subroutine, the associated actual argument in the calling program becomes undefined. Any attempt to use that undefined variable in the calling program will result in a program crash. That may sound bad, but it actually works to the programmer's advantage: when the program crashes, the problem can be immediately diagnosed because the INTENT (OUT) formal argument SHOULD have returned a usable value to its actual argument analog, and not rendered it undefined. The question then becomes, why doesn't the actual argument in the calling program have a value? The answer is obvious—it wasn't set properly in the subroutine—the problem is isolated.

In subroutine `Figure_The_Current`, a different situation is brewing. `Voltage` and `Resistance` get their values from actual arguments, `E1` and `R1`, in the calling program. The formal arguments, `Voltage` and `Resistance`, are `INTENT ( IN )` because they are receiving essential information from the calling program and their values should NOT be changed; `Voltage` and `Resistance` have their appropriate values and are only needed in the calculation for `Current`. `INTENT ( IN )` attributes protect formal arguments, and therefore the associated actual arguments, from being modified by not allowing those formal arguments to be in any explicit statement that *can change* their values.

Formal argument `Current` in subroutine `Figure_The_Current` of Figure 3.9 is given its value in the subroutine and is sent back to the calling program to actual argument `I1`, so it must be an `INTENT ( OUT )` argument. The third subroutine, `Print_The_Results`, has all `INTENT ( IN )` arguments because the subroutine is only displaying the incoming values and shouldn't allow any arguments to be modified. An argument may also be `INTENT ( INOUT )` if it is supposed to both supply a value to a subroutine and return a modified value to its actual argument.

Again, the main purpose of `INTENT` attributes is to help program designers create and debug programs quickly and efficiently. ALWAYS use `INTENT` attributes when declaring formal arguments!

Important Concepts Review

- **Arguments** are classified as `INTENT ( IN )`, `INTENT ( OUT )` or `INTENT ( INOUT )`.

- `INTENT ( IN )` arguments are used for data flowing into a program unit and must not be modified within the program unit.

- `INTENT ( OUT )` arguments become undefined upon entering a subroutine. This makes unset `OUT` arguments easy to spot because any attempt to use them in the calling program will result in a program crash.

- `INTENT ( INOUT )` arguments supply useful values to a subroutine and may be modified by it—`INTENT ( INOUT )` arguments provide a two-way street for data flow.

- **Flow of data** is the movement of data in and out of program units.

3.6 Local Variables

When writing a subroutine, variables (*remember: for now, just think of a variable as a container that can hold a value*) other than the arguments are often needed to help with calculations. These variables' values are set inside the subroutine and are not usually needed after leaving the subroutine. Such variables are called **local variables.** Subroutine `Local_Variables` in Figure 3.10 uses the variable `Product` to hold the results of the calculation: `num1 * num2`. `Product` is then used in the `PRINT` statement to display the arithmetic product of `num1` and `num2`. Variable `Product` doesn't come from the calling program and isn't sent back to it; `Product` is "local" to the subroutine.

On occasion, it is desirable to keep the value of a local variable so that when the subroutine is called again, its previous value is available. When this necessity occurs, the local variable(s) should be listed in a `SAVE` statement, or have `SAVE` as one of its attributes. In a subroutine such as the one in Figure 3.11, the value of `Count` would

Figure 3.10

```
SUBROUTINE Local_Variables ( num1, num2 )
          IMPLICIT NONE
!
      INTEGER, INTENT ( IN )   :: num1, num2
      INTEGER     :: Product    ! a local variable
!
   Product = num 1 * num2
   PRINT *, Product
!
RETURN
END SUBROUTINE Local_Variables
```

Figure 3.11

```
   ...
SUBROUTINE How_Many_Times_Has_It_Been_Done
          IMPLICIT NONE
    INTEGER       ::   Count = 0 ! initialize to 0
    SAVE Count
    Count = Count + 1    ! add 1 to Count
    PRINT *, "I've done it", Count, " times"
RETURN
END
```

normally be lost every time the `RETURN` statement is executed because it is a local variable. Since `Count` has been put in a `SAVE` statement, its value won't be lost. `Count` will be saved so its previous value will be available each time the subroutine is called. This allows `Count` to be incremented by `1` every time subroutine `How_Many_Times_Has_It_Been_Done` is called. Also, putting a `SAVE` statement in a program unit's declarations with no variables listed will save ALL local variables in the subroutine.

3.7 Entering a Modular Program into a Computer

When a program is entered into a computer, it is unwise to first type in the entire program and then run it. Typing itself can generate many errors, and there are usually inherent errors in program designs that will further confuse a Fortran compiler. Running a several-hundred-line program for the first time after having typed it all in will lead to pages and pages of errors. There is a better way.

Once a program design is written and the English Fortran is ready to be translated and typed into a computer, begin by entering the main program. The main program, if well designed, will mostly consist of calls to subroutines. Create a program stub for each of the called subroutines in the main program. **Program stubs** are yet-to-be-completed subroutines that each contain a single `PRINT` statement that displays the subroutine's name whenever flow of control enters it. A program won't run if it's missing any of its subroutines, but these program stubs are legitimate subroutines. They allow the main program to execute even though the program is very incomplete. The main program can then be rid of errors. Each of the called subroutines (currently stubs) in the main program is then developed in exactly the same way as the main program—one at a time. Create program stubs for any called program units and work on them later.

Figure 3.9 showed a program that calculated *current* given *resistance* and *voltage*. Examine how the program stub approach to program development simplifies the enter-the-program-into-the-computer phase for program `Classify_Arguments`. An example of a first typing effort at a computer is displayed in Figure 3.12 (p. 66).

Notice how the strategy in Figure 3.12 keeps things simple. It is clear from the program results that the main program is working correctly and "driving" the program. Once the program is working as in Figure 3.12, each of the subroutines can be developed one at a time, adding declarations and arguments as necessary.

At this point in the development, no comment blocks have been included. Each subroutine must have a comment block similar to the one in the main program, but NOT NOW. Add line comments as needed for clarity WHILE developing subroutines, but WAIT until the program unit is finished before creating its comment block.

Figure 3.12

```
      PROGRAM Classify_Arguments
         CALL Get_Values_For_E_and_R
         CALL Figure_The_Current
         CALL Print_The_Results
      END PROGRAM Classify_Arguments
!================================================
      SUBROUTINE Get_Values_For_E_and_R
         PRINT *, "I am in Get Values"
      RETURN
      END SUBROUTINE Get_Values_For_E_and_R
!================================================
      SUBROUTINE Figure_The_Current
         PRINT *, "I am in calculate"
      RETURN
      END SUBROUTINE Figure_The_Current
!================================================
      SUBROUTINE Print_The_Results
         PRINT *, "I am in the print sub"
      RETURN
      END SUBROUTINE Print_The_Results

*************** PROGRAM RESULTS ***************

I am in Get Values
I am in calculate
I am in the print sub
```

3.8 Subroutine Libraries

Subroutines are mini Fortran programs that accomplish one well-defined task. One subroutine might read in values from a disk file and another subroutine may perform calculations on those values. Often, the same kinds of jobs need to be done in different programs. When the same work needs to be accomplished in various programs, DON'T design the same subroutines over and over; it is an obvious waste of time. Whenever a useful subroutine is written, save it in a collection of program units called a **library**. Programmers can create their own libraries or use existing libraries.

The Black Box

To understand libraries one first needs to understand the black box concept. A black box is a program unit that has the following features:

1. One well-defined purpose.
2. One entrance at the top.

3. One exit through the bottom.
4. A description of the exact purpose, including required and generated values (arguments). This description, together with the black box's name, is called an **interface**.

A black box can be visualized as in Figure 3.13. Flow of control enters a black box, and the black box performs some useful function. It is not important *how* the black box does its job. It is only relevant for the programmer to understand how to *use* the black box.

Consider the black box defined in Figure 3.14, which shows a black box that might be a subroutine inherited from a colleague or another source. The point is that it is *unnec-*

Figure 3.13

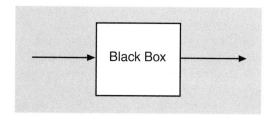

Figure 3.14

```
            SUBROUTINE Get_Value ( Prompt, Value, Low, High )
!------------------------------------------------------------
! The purpose of this subroutine is to:
!
!    a) Prompt the user for a value
!    b) Get the value from the keyboard
!    c) Store the input value in variable "Value"
!    d) Verify that the value is between Low and
!       High inclusively.
!
!  Prompt    - a CHARACTER type variable INTENT ( IN )
!              that contains a prompting message.
!
!  Value     - a DOUBLE PRECISION variable INTENT ( OUT )
!              used to hold the input value and
!              send it back to the calling program to
!              its corresponding actual argument.
!
!  Low       - a DOUBLE PRECISION variable INTENT ( IN )
!              used as the acceptable lower limit for
!              variable Value.
!
!  High      - a DOUBLE PRECISION variable INTENT ( IN )
!              used as the acceptable upper limit for
!              variable Value.
!------------------------------------------------------------
```

essary to understand *how* the routine works. A programmer can simply read the description of the subroutine's arguments and purpose and use the subroutine at will; the interface is completely defined.

The method of programming that uses existing black boxes is THE WAY to go. It allows the user to create large programs where a few hundred lines of program statements invoke thousands of lines of existing, working subroutines. Never write subroutines that have been written before; reuse them whenever possible.

Creating and Accessing Libraries

To create a Fortran library, type the subroutines one after another like a Fortran program WITHOUT a *main program*. As time progresses and useful subroutines are created, they can be added to the library. This collection of subroutines is called the **source library** because it contains the actual Fortran source statements (source code) for the subroutines. The source library is then compiled to create an **object library**, which is the machine language version of the source library (Table 3.1). Object libraries have machine language versions of every program unit contained in corresponding source libraries. An object library can be made available to any program that requires one or more of its subroutines.

When a Fortran program is compiled, a machine language version of the program is created, but the program isn't quite ready to run. A second process called **linking** incorporates any additional required library subroutines into the program to finish it. Figure 3.15 shows the whole process of getting a Fortran program into an executable form, including linking, where library subroutines are added to the program that is being created.

Table 3.1

Source Library	Compiler ==>	Object Library
Subroutine Traj (...) ... End Subroutine Traj		1001001010011101110101000000 1010011101000101110011001110
**********************		0100010010011101000000011010
Subroutine I_O (...) ... End I_O		0000110100100010010101110100 0010100111011001101010101010010
**********************		0010011010010000110101101111
Subroutine Random ... End Random		1111101010010100110101010 etc.
********************** etc.		

Figure 3.15

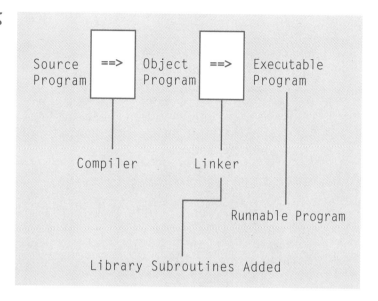

Important Concepts Review

- **Local variables** are variables that are needed within a subroutine and aren't supplied or needed by the calling program.

- The `SAVE` statement or `SAVE` attribute preserves the values of local variables so that the next time a subroutine is called, the local variables will have their previous values.

- **Program stubs** are simple, incomplete subroutines that allow the development of the calling program unit without worrying about the called programs' units.

- **Libraries** are collections of useful subroutines.

- The **black box** concept is when a programmer invokes a useful subroutine by understanding the description of its purpose and arguments. It is NOT necessary to understand *how* it works.

- A **source library** contains the actual Fortran statements that comprise the library.

- An **object library** is the compiled machine language version of a source library.

- **Linking** is the process where required library subroutines are added to a program to make it ready to run.

EXERCISES

```
1              PROGRAM Pendulum
2                  IMPLICIT NONE
3                      DOUBLE PRECISION  ::  Length, Period
4                      CALL Read_Length ( Length )
5                      CALL Calc_Period ( Length, Period )
6                      CALL Print_Pendulums_Period ( Period )
7              END PROGRAM Pendulum
8              !-----------------------------------------------------
9              SUBROUTINE Read_Length ( L )
10                 IMPLICIT NONE
11                     DOUBLE PRECISION, INTENT(????????)    :: L
12                 PRINT *, "Please enter pendulum length in cm"
13                 READ *, L
14             RETURN
15             END SUBROUTINE Read_Length
16             !-----------------------------------------------------
17             SUBROUTINE Calc_Period ( L, Per )
18                 IMPLICIT NONE
19                     DOUBLE PRECISION, INTENT(????????)    :: L
20                     DOUBLE PRECISION, INTENT(????????)    :: Per
21                     DOUBLE PRECISION              :: The_Period
22                 The_Period = 2.0 * 3.141593 * SQRT ( L / 9.8 )
23                 Per = The_Period
24             RETURN
25             END SUBROUTINE Calc_Period
26             !-----------------------------------------------------
27             SUBROUTINE Print_Pendulums_Period ( Period )
28                 IMPLICIT NONE
29                     DOUBLE PRECISION, INTENT (????????) :: Period
30                 PRINT *, "The pendulum's period is", Period
31             RETURN
32             END SUBROUTINE Print_Pendulums_Period
33             !-----------------------------------------------------
```

The lines in the program above have been numbered for reference. Exercises 3.1 through 3.8 refer to the above program.

3.1 List the line numbers in the order that they will be executed.

3.2 Which program units are (a) called? (b) calling?

3.3 Identify actual arguments by name and line number.

3.4 Identify formal arguments by name and line number.

3.5 Which variable(s) in the program are local variables?

3.6 What are the INTENT attributes for each of the formal arguments?

3.7 Comment the entire program in an appropriate way.

3.8 Identify any poorly named identifiers and rename them.

3.9 What is the purpose of the SAVE statement?

3.10 Write a "first-attempt" version of the program in Exercise 3.1 using the program stub technique.

3.11 Why is the stub technique a good one, and how is it consistent with the design approach described in Chapter 2?

3.12 What is the purpose of program libraries?

3.13 What is the difference between a source library and an object library?

3.14 What is unimportant to know about a black box?

3.15 The linking process collects _____ and creates _____.

Data, Variables, Constants and Expressions

Introduction

The basic building blocks of program data need to be understood before you get deeply involved with program flow-altering techniques, such as selection structures (Chapter 6) and looping structures (Chapter 7). **Variables** and **constants** comprise program data. You need to know what a variable can and cannot do, and what a constant is and how to construct one correctly. Letting any of the information about program data "slip through the net" will lead to wrong answers or screens full of error messages.

It is equally important to understand the way Fortran evaluates arithmetic operations involving variables and constants, such as:

```
a / b + n ** .037
```

Often, beginning programmers assume that if more digits of accuracy are required for a calculation, all that needs to be done is simply type in the extra digits, and voilà, more accuracy. Not so. Fortran 90 offers two basic kinds of "REAL" variables with associated constants—**single precision** and **double precision**. Single precision REAL values have about 7 to 8 significant digits as a rule, while double precision can handle about 14–15 digits.

4.1 What are data?

4.2 Variables

4.3 Constants

Important Concepts Review

4.4 Defining and using variables

4.5 Variable declarations

Important Concepts Review

4.6 Arithmetic expressions

Important Concepts Review

Exercises

As far as `INTEGER` values are concerned, Fortran is only required to supply one kind of accuracy (single precision). Single precision `INTEGER`s can represent about 9 digits. Some versions of Fortran 90 do support "long" `INTEGER`s and more than two `REAL` precisions, but compilers are not required to do so. See "Kind Functions" in Appendix B to find out what kinds of accuracy are available for `REAL` and `INTEGER` values in a given Fortran 90 compiler.

In addition to `INTEGER` and `REAL` data, Fortran supplies ways to represent `CHARACTER`, `LOGICAL` and `COMPLEX` data. It is even possible for you to create original data types to fit your own, specific computing needs. These custom "types" are called **derived data types**. Take the time to understand how data are (*data* IS a plural noun) represented and think about how they move in and out of program units; the whole idea behind programs is that they take data and manipulate them to produce useful answers. • • •

4.1 What Are Data?

Data can take several forms:
1. as required information coming into a program from one of the following places:
   a) keyboard
   b) file of values
   c) another source
2. as information created within a program for use by the same program:
   a) explicitly created within a program
   b) created as a combination of external and internal values
3. as information generated by a computer program to provide:
   a) answers for humans
   b) data for other computer programs

Data must be stored and represented in a program so that they can be used and modified at different places in the program while it's running. Data are stored in **variables**. Besides variables that store program data, *specific* data values are often required. Such data might include e, the natural base of logarithms (≈2.718281828459045 . . .), the character string "Answer =" or simply the number 1. Unchanging instances of data are aptly called **constants** because they don't change; they remain constant.

4.2 Variables

A **variable** is a named place in a computer's memory that can be thought of as a container to hold a value. A variable has four significant features:

1. A **name**—a variable's name should imply its use; it should be mnemonic. If a variable holds a value for the angle of deflection of some beam, don't name the variable Banana or Fried_Egg_Plant; name it Angle_Of_Beam_Deflection.
2. A **value**—the whole point of variables is to store values so they can be used at various points in a program. Until a variable is given a value, it is considered **undefined** and MUST NOT be used in any capacity except in a statement that gives the variable a value.
3. A **type**—what kind of data does it hold?
4. **Its value may be changed**—when a variable's value is changed, its old value, if there was one, is lost.

When one thinks of computers from a scientific or engineering point of view, numbers come to mind. In reality, numbers are not the only form of data that is necessary in computer programs. Variables and the corresponding kind of data they may hold are classified into two categories: **numeric** and **nonnumeric** (Table 4.1).

Numeric Types

1. INTEGER variables can only contain positive or negative whole numbers . . . **no decimal parts**.

Table 4.1

	Variables	
Classification	Variable Type	Values Stored
Numeric	INTEGER REAL DOUBLE PRECISION COMPLEX	Whole numbers, about 9 digits Values for decimals, 7–8 digits Decimal values, 14–15 digits Imaginary numbers
Nonnumeric	LOGICAL CHARACTER DERIVED	.TRUE. or .FALSE. Strings of characters Derived type

2. REAL type variables store values that *may* have decimal parts. Variables of type REAL represent numbers in **scientific notation**. They have a positive or negative mantissa of about 7 to 8 digits times a power of 10 that ranges in magnitude from approximately 10^{-37} to 10^{37}. A number such as 1.92007×10^{-18} can conveniently be stored in a variable of type REAL.

3. The DOUBLE PRECISION type is the same as REAL except that DOUBLE PRECISION variables can hold approximately twice as many significant digits (14 to 15) in the mantissa. Their magnitude generally ranges from about 10^{-307} to 10^{307}. This extra accuracy proves invaluable in many calculations.

4. COMPLEX variables can store values that have imaginary parts. The imaginary number, $a + bi$, would be represented as (a, b) in Fortran. When using the COMPLEX data type, a and b may be REAL, DOUBLE PRECISION or INTEGER.

Nonnumeric Types

1. LOGICAL variables can only contain a value of true or false.
2. CHARACTER type variables store strings of characters such as: "The Fahrenheit temperature is:" or "$-$ &".
3. Derived data types are created by a programmer. Derived data types are constructed from:
   a) Elementary data types (INTEGER, LOGICAL, etc.).
   b) Elementary and other already-defined derived data types.

4.3 Constants

As mentioned before, **constants** are specific values that **don't change**. An example of an INTEGER constant is: -3012. Constants are defined for EACH data type. They are limited by the way a compiler handles the variables of the constant's type. A constant must be able to be stored in a variable of the constant's type. Constants are defined by three features:

1. **Form**—the way they need to be typed in a Fortran program.
2. **Precision**—the number of significant digits in the mantissa and exponent (numeric only).
3. **Range of values**—the acceptable values.

Integer Constants

Form: An INTEGER constant is a series of digits preceded by a sign indicating whether the constant is positive or negative. If the constant is positive, the "+" is optional. There can be NO decimal point or commas among the digits.

Precision: INTEGER constants can contain up to about 9 or 10 digits (more if your compiler supports long integers).

Range: -2147483648 to 2147483647

Examples:

```
           0
         -22
 -2147483648
730092549811     NG (No Good)—too many digits
        7800
     144,870     NG—no commas allowed
        22.7    NG—no decimal points
```

Real Constants

Form: REAL type constants can take one of two forms:

    a) Signed decimal constant.
    b) Signed scientific notation constant.

Signed decimal constants have up to 7 or 8 significant digits plus a decimal point. A minus sign must precede the constant if it's negative. As with integers, no sign is required if the constant is positive.

Scientific notation constants can have up to 7 or 8 digits with an *optional* decimal point. This positive or negative mantissa is followed by Esxx, which means "times 10 to the power of sxx." In the sxx part, s is a sign for the exponent (again, only essential if the exponent is negative), and xx represents one or more digits in the exponent.

Precision: REAL constants can have up to 7 or 8 digits in either of the forms described above. It is important to understand that supplying more than 7 or 8 digits WILL NOT increase the accuracy of a constant.

3.14159265358979323846264338327950288841971693993

will be maintained by the computer as:

    3.141593

Range: macro: -3.402823E+38 to 3.402823E+38

and

micro: -3.402823E-38 to 3.402823E-38

Again, these ranges may vary from compiler to compiler.

Examples:

```
   7.0338
   -.22
   -2.2E-1    (identical to -.22)
   E29        NG—MUST have at least one digit in the mantissa
              ... E29 would actually be interpreted as a variable name.
   1E29
4,500.00      NG—no commas allowed
 4500.00
```

Double Precision Constants

Form: Same as type REAL except this type of constant may have up to 14 or 15 digits in the mantissa and the power of 10 is denoted by a D rather than E.

Precision: 14 to 15 digits. When using DOUBLE PRECISION constants, the *scientific notation* form should ALWAYS be used to ensure DOUBLE PRECISION accuracy in the mantissa.

Range: macro: -1.7976931348623D+308 to 1.7976931348623D+308

and

micro: -1.7976931348623D-308 to 1.7976931348623D-308

Examples: 3.1415926535898   NG—will be considered the REAL constant 3.141593 because, as stated above, DOUBLE PRECISION constants will only be maintained with full accuracy when used in scientific notation. The next example shows the correct way to represent a double precision version of π.

3.1415926535898D0
31008975630987.D+3
2D-231

Complex Constants

Form: Complex numbers are of the form $a + bi$ from a mathematical point of view. Fortran represents $a + bi$ by enclosing a and b in parentheses and separating them by a comma: (a, b).

Precision: The components of a complex constant, a and b, can be any combination of REAL, DOUBLE PRECISION or INTEGER. Precision is the same as each component's type.

Range: Same as REAL, DOUBLE PRECISION or INTEGER.

Examples: (7.3, -9.8)
(-4, 3.7E-35)
(0, 0.0)
(-8D-7, 11)

Logical Constants

Form: There are only two LOGICAL constants: .TRUE.
.FALSE.

Periods on each side of TRUE and FALSE are required.

Character Constants

Form: A character constant is any sequence of characters (called a **string** of characters) enclosed in either *apostrophes* or *quotation marks*.

Examples:
```
'abcdefghijklmnopqrstuvwxyz'
"ABCDEFGHIJKLMNOPQRSTUVWXYZ"
'A12"                                   NG—must be either
                                        apostrophes or quota-
                                        tion marks, not one of
                                        each.
'412'
'9 h      "& -9, hello'
"-75"
"~!@#$%^&*()_-=+|\][{}:;,./'1234567890"
```

In the examples above it should be noted that the character constant "-75" is different from the number -75. The character constant "-75" is a string of three characters: a minus sign, the character 7 and the character 5. Arithmetic operations cannot be performed on "-75". This distinction is explained by collating sequences; see Chapter 5.

Derived Data Type Constants

There is no real concept of constants for derived data types. Each **subobject** in a derived data type is ultimately a basic Fortran type. Each of the individual components can accommodate constants as described in the previous sections. Derived data types are thoroughly explained in Chapter 5.

Important Concepts Review

- **Data** is information that is required by a program, generated by it, or both.

- A **variable** is a named container that can hold a value of a specific type.

- A variable **must** receive a value before the variable is used. It is considered **undefined** until it receives its first value.

- The contents of variables may **change**.

- When a new value is put into a variable, **its old value is lost**.

- **Numeric variables** allow numeric data to be stored and used in a program.

- **Nonnumeric variables** allow other types of data to be stored and used in programs (LOGICAL, CHARACTER, etc.).

- A **constant** is an unchanging instance of a Fortran type; it remains constant.

- Constants have **form** (what they look like), **precision** (digits of accuracy—numeric only) and **range** (possible values).

- **Scientific notation** constants have a mantissa and a power of 10 (e.g., 6.02E+23 or 3.509341232209D-102).

- **Always use the scientific notation form for** DOUBLE PRECISION constants, or they will actually be single precision REAL constants.

- **Adding extra digits** beyond a constant's precision WON'T make it more accurate.

4.4 Defining and Using Variables

As mentioned before, a variable must be given a value in a program before it is used. If this isn't done, the variable is considered **undefined** and any attempt to use it will crash the program. This is a GOOD feature of Fortran ... it forces programmers to think carefully about program data and stops spurious program results from cropping up if all variables are given some arbitrary initial value, such as zero.

The next sections explain how to get values into variables.

Variable Assignment

A variable is a container of a specific type (LOGICAL, INTEGER, etc.) that can hold a value. How do values get into variables? There are a couple of different ways. First, there is **variable assignment**. This is when a value is "assigned" to a variable. The general form of an assignment statement is:

```
variable = value
```

Assignment statements should be thought of as, "*variable* **is set equal to** *value*." The value on the right-hand side of the equal sign is moved into the variable (Figure 4.1, p. 80). Box variable in Figure 4.1 represents a variable. For now, it doesn't matter how the computer actually stores the value. It is only important that a value can be put into a variable and whenever the variable's name is used, the variable's current value is used.

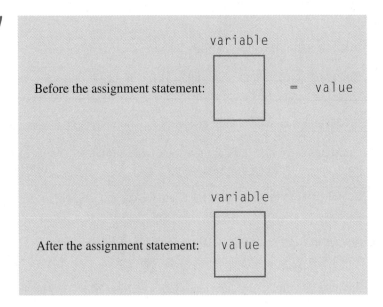

Figure 4.1

Look at the statements in the Fortran program segment in Figure 4.2 to get a better idea of how variable assignment statements work. In the third assignment statement, the contents of variable Mass_Of_Object, which is 200, is multiplied by the contents of variable Acceleration, 9.8. The resulting value is stored in variable Force. The mechanics of any assignment statement are precisely defined as follows: *the expression on the right side of the equal sign is evaluated and that value is then stored into the variable on the left side of the equal sign.*

Incremental Assignment Statement

The **incremental assignment statement** is a special case of the assignment statement where a value (often 1) is added to a variable to increase its contents by some number. This is frequently used to count things or accumulate a series of values. For instance, every time an event happens, a programmer can add 1 to a Counter variable to keep track of how many times the event happens.

Observe how the incremental assignment statement works in Figure 4.3 (p. 81), as it is confusing to see Counter_3 on both sides of the assignment statement's equal sign. In the second assignment statement in Figure 4.3, 1 is added to Counter_3:

Counter_3 = Counter_3 + 1

The expression, Counter_3 + 1, is evaluated (1 + 1) to 2, and that 2 is then stored into Counter_3, destroying its previous value of 1. Effectively, Counter_3 has been incremented by 1.

Figure 4.2

```
Acceleration   = 9.8
Mass_Of_Object = 200
Force          = Mass_Of_Object * Acceleration
```

                                Acceleration

Before the assignment: [   ] = 9.8

                                Acceleration

After the assignment: [9.8]

                                Mass_Of_Object

Before the assignment: [   ] = 200

                                Mass_Of_Object

After the assignment: [200]

                              Force = Mass_Of_Object * Acceleration

Before the assignment: [   ] = [200] X [9.8]

                                Force

After the assignment: [1960]

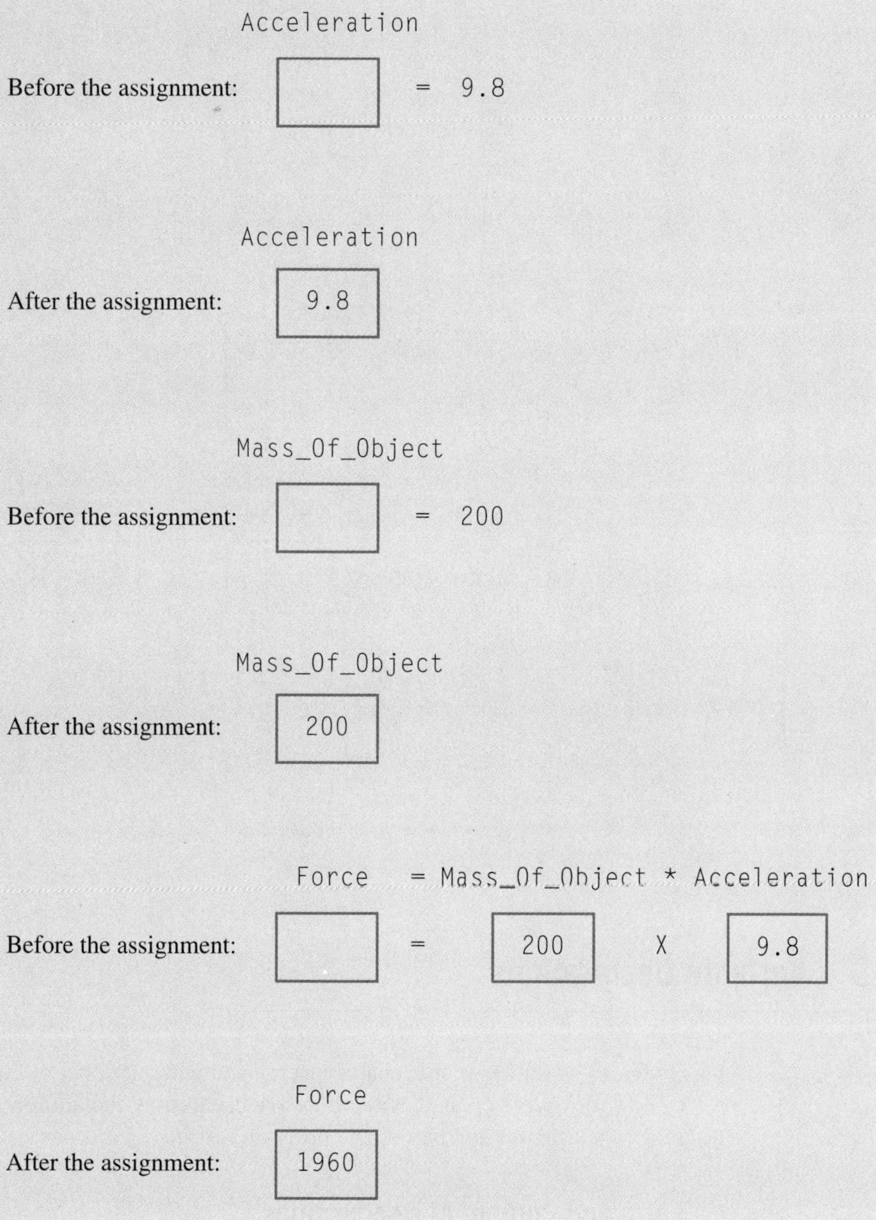

Figure 4.3

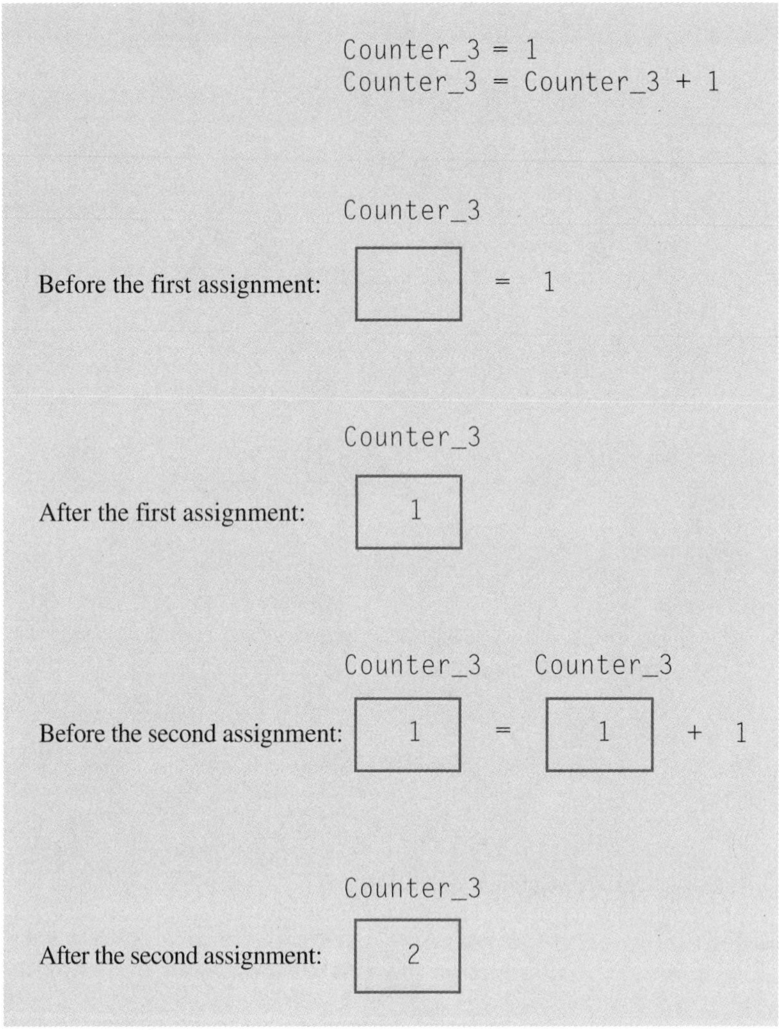

```
                              Counter_3 = 1
                              Counter_3 = Counter_3 + 1
```

                                     Counter_3
Before the first assignment: [   ]   = 1

                                     Counter_3
After the first assignment: [1   ]

                                     Counter_3   Counter_3
Before the second assignment: [1   ]   =   [1   ] + 1

                                     Counter_3
After the second assignment: [2   ]

4.5 Variable Declarations

So far, all variables in this chapter were used without concern for where they came from or how they were created. **Variables are created** in declaration statements at the beginning of a program unit before any program instructions.

Form and Content of Declarations

All variables used in a program unit MUST be **declared** in well-designed programs. Declarations define all variables and other kinds of program data that are used in a program

4.5 Variable Declarations

Figure 4.4

```
program unit header
comment block
    declarations
        statement 1
        statement 2
            .
            .
            .
        statement n
end of program unit
```

unit. To be clear on where this variable genesis takes place, remember the general organization of a Fortran program unit (Figure 4.4).

To declare a variable (which creates it) in a program unit, it is necessary to do the following:

1. Supply various essential pieces of information, such as the variable's type (CHARACTER, COMPLEX, REAL, etc.), the variable's name, etc.
2. Put the information in correct Fortran form; make the declaration syntactically correct.

The information and form combined are shown in Figure 4.5.

1. type is any of the Fortran types: REAL, INTEGER, LOGICAL, etc.
2. (kind) is an optional suffix that is used with some types to further define type. For example, in the following (LEN = 10) is a *kind* that is used to define CHARACTER type variables that can hold up to 10 characters. LEN means *len*gth in characters.

    CHARACTER (LEN = 10) :: . . .

3. [,attribute]... is a list of Fortran **attributes** that define special characteristics of the variables in var-list. Attributes are separated from type and each other by commas. Attributes and their purposes are introduced as needed throughout this text.

    (Note: the square brackets, [], and ellipsis, . . . , are NOT included in an actual declaration. When defining the syntax of a part of the Fortran language, anything

Figure 4.5

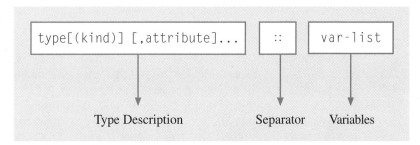

between square brackets is optional, and an ellipsis means that there may be as many as desired. See the Quick Reference Section *for additional explanation and examples of these conventions.)*

4. `::` separates the type description (Figure 4.5) from the variables that will conform to that description.
5. `var-list` is a list of variable names separated by commas that will have the properties defined on the left side of the `::`.

When choosing a variable name, choose one that is **mnemonic** (nih-maw′-nik). Mnemonic essentially means that a name is chosen to imply the purpose of what it names. If a variable is named `aaaaaa_bbbbbb`, what is it used for? Alternatively, a variable named `Melting_Point_Of_ZrO2` will probably contain a value for the melting point of zirconium dioxide.

Names may be up to 31 characters long. Take advantage of this and invent descriptive names; make all programmer-chosen names mnemonic. (Programmer-chosen names are usually called **identifiers**.)

`IMPLICIT NONE`, Declare All Variables

When Fortran was originally designed by John Backus in the 1950s, an unfortunate decision was made. For the convenience of programmers (what irony), it was decided that, unless otherwise declared, a variable starting with any of the letters `I`, `J`, `K`, `L`, `M`, or `N`, would be considered type `INTEGER`. All others would be type `REAL` (again, unless otherwise declared). So, if a variable, `IVAL1`, was not declared in a type statement (**explicitly typed**), it would be **implicitly typed** as `INTEGER`. This convention was carried throughout each new version of Fortran.

Implicit typing is a poor idea because it is very easy to get it into your mind that a variable is to be of, say, type `REAL`, and then forget to declare it as such, and find it implicitly typed as `INTEGER`. An example of such a variable is `Mean_Of_N_Numbers`. A variable named `Mean_Of_N_Numbers` would probably be created to hold the average of *n* numbers. `Mean_Of_N_Numbers`'s purpose being so defined would probably need to be of type `REAL` because averages aren't generally integral. If variable `Mean_Of_N_Numbers` isn't declared in a type statement (explicitly typed), it would be *implicitly typed* as `INTEGER` because the first letter of the variable is `M`. This causes decimal parts to be lost whenever the variable is used, because `INTEGER` variables can't hold decimal parts.

The most compelling reason to prevent implicit typing is that it makes the process of creating programs more efficient. One common cause of a high frustration-meter reading for beginning programmers is subtle typographical errors. Consider the program segment in Figure 4.6.

In line 3, there is an assignment statement whose purpose is to add the value in `Next_Trapezoid` to variable `Trapezoidal_Method` and store the result into `Trapezoidal_Method`. This increases `Trapezoidal_Method` by the value contained in `Next_Trapezoid`; it is an incremental assignment statement similar to the one seen earlier in this chapter: `Count = Count + 1`. Look carefully at the assignment statement. The word *trapezoidal* in variable `Trapezoidal_Method` is misspelled

Figure 4.6

```
(1)     PROGRAM Numerical_Integration
            ...
(2)         REAL    ::   Trapezoidal_Method, Definite_Integral
            ...
(3)         Trapzoidal_Method = Trapezoidal_Method + Next_Trapezoid
            ...
(4)     END PROGRAM Numerical_Integration
```

on the left side of the equal sign. The program will not work, and figuring out why isn't trivial. It is EXTREMELY EASY to overlook this sort of error.

There is, however, a way to get Fortran to discover the spelling error: include the statement IMPLICIT NONE at the beginning of **EVERY** program unit. IMPLICIT NONE disables the implicit typing trait of Fortran and forces ALL variables to be explicitly declared.

Without IMPLICIT NONE, Trapzoidal_Method (the misspelled variable in Figure 4.6), would be implicitly typed, allowing the program to run, but generating wrong answers. Including IMPLICIT NONE in the program unit causes the compiler to say, "Hey, what's this variable, Trapzoidal_Method?" The programmer then takes a careful look and realizes that she left out the "e" after "Trap". IMPLICIT NONE caused the error to be discovered! Fortunately, implicit typing will not be a part of future releases of Fortran.

Data Initializations

All variables in Fortran must be given values before they can be used. Any variable that hasn't received a value is considered **undefined**. Variables are given values in four different ways:

1. In an **assignment statement**.
2. In a READ **statement** where values come from an external source, such as the keyboard or a data file.
3. By **Data initializations in declarations**.
4. In DATA statements.

Items 1 and 2 above have already been discussed, so we'll move on to data initializations in declaration statements. This is a very convenient method for giving variables first-time (initial) values.

```
            INTEGER     ::      Count_1 = 0, Count_2 = 1
```

In the above declaration, Count_1 and Count_2 are created as type INTEGER variables and are given their first values, 0 and 1 respectively, at the time the program is *compiled*. Be careful when using this method of initialization because there is a trap.

Beginning program designers often use data initializations in declarations in subroutines. They assume that each time the subroutine is called, that variables will be **reinitialized** by the declaration statements. This is not true.

A variable is given its first value in a declaration statement when the program is **compiled**, which is done only **once**. Reentering a subroutine by calling it again does not cause the subroutine to be recompiled, so the variables won't be reinitialized to the values specified in the declaration statements. Any variables in program units other than those initialized in the main program's declaration statements are automatically treated as though they were listed in a SAVE statement.

The DATA statement (method 4 from above) is another way to initialize variables at compile time. Giving variables their first values in a DATA statement allows a set of variables to be initialized all at once.

```
DATA var-list / constants /
```

Here is an example of how a DATA statement works:

```
INTEGER      ::   Val1, Val2, Val3, Val4
DATA              Val1, Val2, Val3, Val4 / 5, 3 * 0 /
```

In the above example, Val1 to set to 5 at compile time while Val2, Val3 and Val4 are each set to 0 (3 * 0 means three individual zeros, not three multiplied by zero). Since DATA statements are also acted on at compile time, all the warnings regarding data initializations in subroutines apply.

DATA statements are essential for initializing variables with BOZ constants. BOZ constants are binary, octal or hexadecimal constants. They may be used in DATA statements to initialize INTEGER variables with binary, octal or hexadecimal values.

```
DATA    I1, I2, I3 / B"10011010", O"326" , Z"E5" /
```

The value of I1 would be 154 (10011010_2), I2 would be 214 (326_8), and I3 would be 229 ($E5_{16}$). BOZ constants may **only** be used in DATA statements or read using BOZ FORMATs.

Parameters: Named Constants

A **parameter** is a constant that has been given a name. To create a parameter, use the parameter attribute (Figure 4.7).

PI is a PARAMETER; it is a **named constant**. Whenever the value for π is required, the programmer simply uses PI.

So what's the difference between the named constant PI and variable PIE in Figure 4.7? The answer is that PI CAN'T be changed; it is a constant in the true sense of the word. Any attempt to explicitly modify PI's value will result in a syntax error. Named constants ARE truly constants and can't be changed, but the contents of a variable such as PIE can be altered at will.

PARAMETERs can also be created using the PARAMETER statement. PARAMETER statements have the general form seen below:

```
PARAMETER ( named_const_1 = constant_1 [, named_const_2 = constant_2] . . . )
```

Figure 4.7

```
PROGRAM Trig_Stuff
    ...
              IMPLICIT NONE
    ...
    DOUBLE PRECISION,    PARAMETER :: PI=3.1415926535898D0
    ...
    DOUBLE PRECISION                :: PIE=3.1415926535898D0
    ...
END PROGRAM Trig_Stuff
```

`constant_1`, `constant_2`, etc., may be simple constants or defined in terms of other constants. An example of this is shown below:

```
REAL     ::   PI, PI_Over_2
PARAMETER ( PI = 3.1415927, PI_Over_2 = PI / 2.0 )
```

Using named constants is another method of preventing errors from getting into programs. If a program is, for example, using a 15-digit approximation of π nine or ten times, the probability of typing π incorrectly at least once is high. Instead of typing the 15 digits repeatedly, create a named constant, `PI`, that requires the approximation for π to be typed **only once**. Use `PI` instead of `3.1415926535898D0` the nine or ten times it occurs in the program.

Named constants coupled with `IMPLICIT NONE` (to ensure that their names are typed correctly), will reduce the possibility of making an error with constants to nearly zero. There are other very good reasons for using parameters. The reasons will be pointed out as they come up in this text.

Important Concepts Review

- In an **assignment statement**, the expression on the right side of the equal sign is evaluated and that **value** is then **stored into the variable** on the left side of the equal sign.

- The **incremental assignment statement** adds a value to a numeric variable.

$$j = j + 2$$

  The statement above adds 2 to j's contents. If j is -9, for example, it would be -7 after the assignment.

- **Declarations** (Figure 4.5) create and define program variables and named constants, and allow data initializations.

- `type` defines whether a variable or named constant is `INTEGER`, `REAL`, `DOUBLE PRECISION`, `LOGICAL`, `COMPLEX`, `CHARACTER`, or a derived data type.

- `(kind)` is a suffix that is used with some types to further define `type`.

- `[,attribute] ...` is a list of Fortran **attributes** that define any additional special characteristics for `type`.

- `::` separates the type description from the variables that will conform to the description.

- `var-list` is a list of variable names separated by commas that have the properties defined on the left side of the `::`.

- **Mnemonic names (identifiers)** imply the purpose of the named object. A variable named `Square_Root` that holds the square root of some number has a mnemonic name.

- **Implicit typing** is a language feature (a B-A-D feature) that gives undeclared variables a default type. Variables starting with `I, J, K, L, M` or `N` are of type `INTEGER` and all others are `REAL`.

- `IMPLICIT NONE` turns implicit typing off, which helps catch mistyped identifier names by forcing all variables to be explicitly declared.

- **Data initializations** can be performed in declaration statements, but they are only done **once** at **compile time**. Beware of its use in program units other than a main program!

- `DATA` **statements** allow multiple variables to be initialized in one statement at compile time.

- **BOZ constants** may be used in `DATA` statements or read with BOZ formats to assign binary, octal or hexadecimal values to `INTEGER` variables.

- A declaration statement containing the **parameter** attribute is used to create named constants.

4.6 Arithmetic Expressions

Arithmetic expressions are made of meaningful combinations of **operands** and **operators**. The following arithmetic expression is an example of a meaningful expression where the contents of variable A (an *operand*) is added (+ is an *operator* that performs addition) to 5 (another *operand*):

$$A + 5$$

An example of a nonmeaningful arithmetic expression is seen below:

z+-+*-G//-+7***gt

Operands are operated on by **operators**. An operand can be any of the following:
1. Variable.
2. Constant.
3. Arithmetic expression in parentheses.

Arithmetic operators are shown in Table 4.2.

Arithmetic Order of Precedence

When arithmetic expressions are created, you must understand how the computer will evaluate them. If this is not understood, formulas can give incorrect answers. Expressions are evaluated according to the rules of **arithmetic order of precedence**—the order that arithmetic operations are done. To get a feel for arithmetic order of precedence, consider how the following expression could be evaluated:

3 + 6 / 3 - 3 * 3

You might use the method in Figure 4.8. Unfortunately, the result in Figure 4.8 is wrong. That way of evaluating expressions **is not** the way Fortran does them. Fortran evaluates expressions according to the rules of arithmetic order of precedence listed in Table 4.3 (p. 90).

Table 4.2

Arithmetic Operators	
Operator	Purpose
**	Exponentiation, ($BASE^{POWER}$)
*	Multiplication
/	Division
+	Addition
−	Subtraction

Figure 4.8

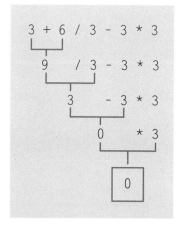

Table 4.3

Arithmetic Order of Precedence		
Order	Operator	Direction
first	**	from right to left
second	*, /	from left to right
third	+, −	from left to right

Arithmetic expressions are searched first for exponentiation operators, and if there are more than one, they are evaluated from **right to left**. When all the exponentiation operators are resolved, all multiplication and division operations will be evaluated from **left to right** as each multiplication **OR** division operator is met. Finally, addition and subtraction will be performed from **left to right**. Armed with this understanding of arithmetic order of precedence, the example expression of Figure 4.8 is correctly evaluated in Figure 4.9.

Parentheses Override Order of Precedence

It is impossible to construct certain expressions with only the rules of arithmetic order of precedence. For example, try writing a Fortran version of the simple expression below using only arithmetic operators and a knowledge of how arithmetic order of precedence works.

$$\frac{a+b}{c+d}$$

Use parentheses to group terms to avoid the awkward brainstrain of trying to figure out if and how an expression can be written so it only uses arithmetic order of precedence. Expressions contained in parentheses will be evaluated first. The expression above can easily be written in Fortran with parentheses:

$$(a + b) / (c + d)$$

Within any set of parentheses, all the rules of arithmetic order of precedence apply. It's also reasonable to use parentheses simply to clarify an expression, even when the parentheses aren't necessary.

Figure 4.9

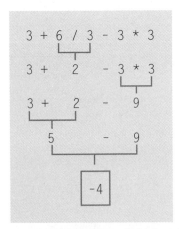

Nesting Parentheses in Expressions

Parentheses may be **nested** as needed. Nesting parentheses simply means having matching sets of parentheses within other sets of parentheses. When parentheses are nested, the contents of the innermost set of parentheses are evaluated first (Figure 4.10).

Notice that the expression inside the innermost set of parentheses, 6 + 2, is evaluated first. And in the fourth line, the expression, 6 - 4, is evaluated before 2 ** 8; the parentheses override the precedence of the exponentiation operator.

Arithmetic Operator Modes

Arithmetic operators will produce different results depending on the types of their operands. Fortran doesn't operate on INTEGER values the same as it does on REALs, even if the actual numbers are the same. For example 9 and 9.0 **are not** the same. Nine is nine is nine, it's true, but in expressions, REALs and INTEGERs can behave very differently. These evaluation differences are a result of the **operator modes** summarized in Table 4.4.

Table 4.4 simply means that if there are integer values as the left and right operands in an expression, the result of the arithmetic operation will be an integer. Any other combination of operands will result in a value of type REAL.

There are two important implications of the operator modes that must be carefully considered:

1. **Integer division.**
2. **Mixed-mode expressions.**

Integer Division   **Integer division** is one integer operand divided by another integer operand. The result of integer division is the whole part of the quotient; any decimal part

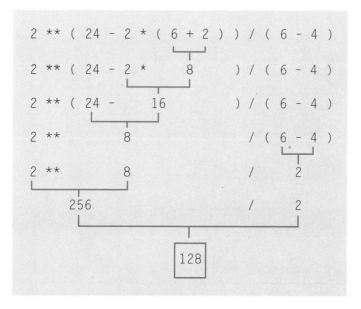

Figure 4.10

Table 4.4

	Operator Modes	
Left Operand	Right Operand	Operation Result
INTEGER	INTEGER	INTEGER
REAL	INTEGER	REAL
INTEGER	REAL	REAL
REAL	REAL	REAL

is **truncated** (not rounded). *Truncate* means to chop off. Truncation of decimals in integer division is not bad. In fact, it can be useful in some contexts, but not understanding integer division can lead to incorrect formula evaluation. To understand how integer division and truncation work, see the examples in Figure 4.11.

In each case, the result of the division is the whole part of the quotient. 5 / 9 has a calculator value of 0.555555555. However, in Fortran, since the operands are both integers, the result of the division is 0; the whole part of the quotient (.555555555... is truncated). Note that there is NO ROUNDING of values, just truncation. This obviously affects formula evaluation. Figure 4.12 illustrates the impact of integer division on expression evaluation. If the expression in Figure 4.12 is worked out on a calculator, the result is 2.916666666667.

Figure 4.11

```
5   / 9   = 0   ...   0.5555 ...
9   / 5   = 1   ...   1.8
22  / 7   = 3   ...   3.142857142857
99  / 100 = 0   ...   0.99
100 / 99  = 1   ...   1.010101010101 ...
```

Figure 4.12

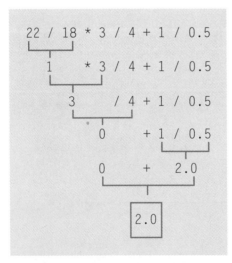

Mixed-Mode Expressions   In a **mixed-mode expression**, one operand is INTEGER and the other is REAL. The result of these kinds of operations is type REAL. This knowledge can be used to a programmer's advantage. If, for instance, the expression in Figure 4.12 does not produce the correct answer because of unexpected truncation from integer division, the problem can be solved by forcing a mixed-mode expression. For example, introduce a REAL term that doesn't change the value of the expression, such as multiplying by 1.0, but does change its mode (Figure 4.13). The 1.0 inserted at the beginning of the expression changes 22 / 18 to 22.0 / 18. It becomes a mixed-mode expression that will be evaluated as REAL and preserves decimal accuracy.

Mixed-Type Assignments for Arithmetic Types   When a value of one arithmetic type is assigned to a variable of another arithmetic type, it is called a **mixed-type assignment**. An example is assigning a COMPLEX value to an INTEGER variable. Mixed-type assignment is permitted in Fortran, but **only among numeric types!** The type of the expression on the right-hand side of the assignment statement will be converted to the receiving variable's type. This kind of assignment must be done with great care.

There are two concerns when performing mixed-type assignments:

1. Assigning a REAL value to an INTEGER variable.
2. Assigning a REAL value to a DOUBLE PRECISION variable.

When a REAL value is assigned to an INTEGER variable, the problem is, again, truncation. Moving a REAL value into an INTEGER variable will cause any decimal part to be lost. When a REAL value is assigned to a DOUBLE PRECISION variable, only the first 7 to 8 digits can be trusted in the DOUBLE PRECISION variable (Figure 4.14, p. 94).

In Figure 4.14, the value 3.1415927410125732 is incorrect. As everyone knows, the first 17 digits of π are 3.1415926535897932. Only the first 7 digits of π are

Figure 4.13

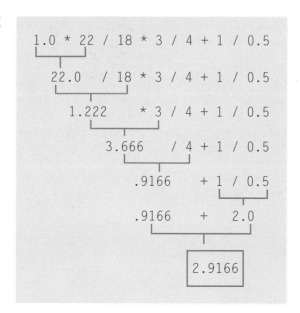

Figure 4.14

```
           ...
REAL                 ::      PI = 3.1415927
DOUBLE PRECISION     ::      DPI
           ...
DPI = PI
PRINT *, DPI
           ...

*************** PROGRAM RESULTS ***************

3.1415927410125732
```

represented correctly after the REAL variable PI (single precision) is assigned to the DOUBLE PRECISION variable DPI.

Important Concepts Review

- **Operands** are operated on by operators. An operand can be a constant, a variable or an expression contained in parentheses.

- An **operator** performs some operation on operands. In the expression a / b, the "/" is a division operator that performs division on operands a and b.

- **Arithmetic order of precedence** is the order that arithmetic operators are performed in an expression. Refer to Table 4.3.

- **Parentheses** override arithmetic order of precedence. Parentheses may be nested as necessary, and the innermost set is evaluated first. Within any set of parentheses, all the rules of arithmetic order of precedence apply. It is also reasonable to use parentheses simply to clarify an expression.

- **Operator modes** are the different results operators will generate based on the *types* of operands involved in an expression. Refer to Table 4.4.

- **Integer division** is when one integer operand is divided by another; any decimal part is truncated.

- **Mixed-mode expressions** occur when one operand is type INTEGER and the other operand is one of the real types (REAL or DOUBLE PRECISION). The result of such expressions is type REAL.

EXERCISES

4.1 Design a program to multiply two complex numbers using two different methods. Method 1 should use variables of type COMPLEX. Method 2 should use four REAL variables to represent the two complex multiplicands. Use the design techniques explained in Chapter 2.

4.2 Write a program that will store the value 2.7182818 into a variable named e (use the PARAMETER attribute) and PRINT out e to the power of 1, e to the power of 3, e to the power of 5 and e to the power of 7. Be sure to use mnemonic identifiers and subroutines. (*Hint: only one subroutine is required.*)

4.3 Generalize the program from Exercise 4.2 by making it read a value from the keyboard into a REAL variable called x, and then print out the value of e^x. Make your answers readable, and don't forget a prompting message for the READ statement and IMPLICIT NONE in ALL subroutines and the main program. Use subroutines. (*Hint: use two subroutines, one to obtain a value for x and a second to display e^x.*)

4.4 Explain why the following program will not work.

```
PROGRAM    Pendulum_Period
           IMPLICIT NONE
!-------------------------------------------------
!
!  The purpose of this program is to calculate
!     the period, P, of a pendulum of length,
!     L, in a gravitational field.
!     G, using the formula:
!         P = 2 * pi * ( L / G ) ** 0.5
!
!------------ Declarations ------------------
!
!   Length   - a REAL variable representing the
!              length of the pendulum
!   Period   - a REAL variable representing the
!              period of the pendulum
!   PI       - a REAL named constant containing
!              an approximation of pi
!   Gravity  - a REAL named constant containing
!              the force of gravity for earth
!
!-------------------------------------------------
!
     REAL                ::  Length, Period
     REAL, PARAMETER     ::  PI      = 3.1415927
     REAL, PARAMETER     ::  Gravity = 9.8
```

```
            !
                        Period = 2*PI * ( Length / Gravity )**0.5
                        PRINT *, "Period is:", Period
            !
            END PROGRAM   Pendulum_Period
```

4.5 What will the contents be for the following variables after the assignment statement or sequence of assignments?

```
            INTEGER                              ::      Itest
            REAL                                 ::      Rtest
            DOUBLE PRECISION       ::      Dtest
```
  a) Itest = 6
     Itest = 7
  b) Rtest = 2.718281828459045
  c) Rtest = 2.718281828459045D+00
  d) Dtest = 2.7182818
  e) Dtest = 2.7182818
  f) Dtest = 2.7182818D+00
  g) Rtest = 2.7182818
  h) Rtest = 2.4

4.6 Write a Fortran program that prints out the values: 5, 10, 15, 20, 25, 30, 35, 40, 45 and 50. Use only one `INTEGER` variable and the incremental assignment statement for the computations. Don't forget `IMPLICIT NONE`.

4.7 Explain the purpose of `IMPLICIT NONE` and why it should be included in *every* program unit.

4.8 Use `DATA` statements to initialize ten integer variables, V1, V2, . . . , V10 to the following values:
  a) All zeros—there are two methods; show both.
  b) 1, 2, 3, 4, 5, 6, 7, 8, 9, 10.
  c) -1, -1, -1, -1, -1, 1, 1, 1, 1, 1—again, there are two methods; show the one that requires the least typing.

4.9 Explain why you should use caution when using the `DATA` statement in a subroutine.

4.10 Why won't the following program work?

```
PROGRAM Whats_Wrong
!
            IMPLICIT NONE
!
            REAL, PARAMETER   ::  Celsius_To_Fahrenheit_Ratio, Thirty_2 = 16.0
            REAL              ::  Celsius = 100, Fahrenheit
!
```

```
      Thirty_2 = Thirty_2 * 2.0
      Celsius_To_Fahrenheit_Ratio = 9.0 / 5.0
      Fahrenheit = Celsius_To_Fahrenheit_Ratio * Celsius + Thirty_2
      PRINT *, Celsius, "Celsius is", Fahrenheit, "Fahrenheit"
!
END PROGRAM Whats_Wrong
```

4.11 Rewrite the program in Exercise 4.10 so that it will work.

4.12 Arithmetic order of precedence is the order that arithmetic operators are evaluated. Arithmetic operator modes define the results of an operation (such as / or **) based on the types of the operands involved. Using this information, evaluate the following expressions:

    a) `255/2**2**3`
    b) `1057/1058/13.7+11-(22*3/11)*2`
    c) `29.0/4+(2**4**(3-2.5))`
    d) `(((19/8/2)/4)**4*4/2.0+5)**3/10`
    e) `23/6/(1.0/3.0)/2/2/0.5/0.5/2`
    f) `17.0/2+(2**2**3+4.5)`

Derived Data Types, MODULEs and Data Representations

5

Introduction

In Chapter 4, you learned about variables, constants and expressions. Fortran has a rich set of built-in data types including INTEGER, REAL, DOUBLE PRECISION, COMPLEX, LOGICAL and CHARACTER. Unfortunately, however, the data we need to represent the world around us is often composed of aggregates of numbers, text and other types of data. Consider designing a program that needs to represent and manipulate information about trees. Trees aren't simple organisms and you can't represent them with any single, fundamental Fortran variable with a type such as INTEGER or CHARACTER. Through combining elemental Fortran types, however, an aggregate type can be created, and variables of that type can be declared to more closely represent the features of trees:

5.1 Derived data types and MODULEs

Important Concepts Review

5.2 Special operations for CHARACTER type data

Important Concepts Review

5.3 How numeric values are stored in memory

Important Concepts Review

Exercises

```
 TYPE Tree
      CHARACTER ( LEN = 15 )      ::      Trees_Name
      LOGICAL                     ::      Conifer_Deciduous
      ! .TRUE. = Conifer, .FALSE. = Deciduous
      REAL                        ::      Growing_Rate
 END TYPE Tree
 !
 TYPE ( Tree )                    ::      Birch, Maple, Blue_Spruce
```

TYPE Tree has three subobjects—Trees_Name, Conifer_Deciduous and Growing_Rate—that are defined as elementary Fortran types. This declaration is called a **derived data type**, which is then used to declare variables, Birch, Maple and Blue_Spruce. *Each* of these variables contains the three subobjects Trees_Name, Conifer_Deciduous and Growing_Rate that can hold values to more accurately describe the tree they represent. Birch, for example, might contain "Betula", .FALSE. and 4.7, while Blue_Spruce might contain "Picea pungens", .TRUE. and 3.9.

Derived data types are inherently complicated because there is no practical limit to how many subobjects you can include in a derived type. In larger programs that use these types in various program units, it's cumbersome to redeclare the derived types in each unit.

To remedy this problem, Fortran offers a specialized, general purpose program unit called a MODULE. MODULEs are used to package derived data types so that they can be easily accessed by any program unit. This has another subtle and very desirable advantage in that the TYPE definitions in a program only need to be entered *once* in the MODULE. This minimizes the chance of accidentally introducing typographical errors throughout the program. MODULEs can also be stored in libraries, just like subroutines.

Derived data types allow you to represent complex data structures in a natural, clear way that makes sense. As you read this chapter, understand that with derived data types, there is no limit to the types you can create and the data you can represent.

There are two additional data-related topics that are covered toward the end of this chapter. First, a detailed explanation of CHARACTER data and the ways that you can manipulate them is presented. CHARACTER data require special attention because of the various operations that are possible: you can attach/chain any combination of CHARACTER variables or constants together to form new strings. Individual characters or sequences of characters within a CHARACTER variable or constant can also be manipulated.

Second, the last few sections in this chapter are dedicated to offering a basic understanding of how fundamental numeric types are actually represented in a computer. Binary and hexadecimal arithmetic are explained, as well as bits, bytes and two's complement arithmetic.

● ● ●

5.1 Derived Data Types and MODULEs

Derived data types are customized types created by a programmer. They group useful characteristics of data into aggregate data types. Derived data types are made of one of the following:

1. Simple Fortran data types (LOGICAL, CHARACTER, INTEGER, etc.)
2. Simple Fortran data types and/or other derived data types.

To create a derived data type, do the following:

1. Define the TYPE.
2. Use the newly created TYPE to declare variables of that TYPE.

Figure 5.1

```
(1)      TYPE type-kind
(2)          declarations
(3)      END TYPE type-kind
         ...
(4)      TYPE ( type-kind ) :: var-list
         ...
```

Define and use derived data types as seen in the general form shown in Figure 5.1.

In Figure 5.1, line (1) begins the derived data type definition. The derived data type will be called $type\text{-}kind$, which should be a mnemonic name for the kind of data type being created. Line (2) contains the declarations that define $type\text{-}kind$. Use as many lines of declarations as necessary to define the derived TYPE. Line (3) ends the TYPE definition. Now that the TYPE is defined, it may be used to define variables of type TYPE ($type\text{-}kind$) as seen in line (4) of Figure 5.1.

To see a concrete example of derived data types, assume information about chemical elements is required for a program. The program needs an element name, its chemical symbol and its atomic number (Figure 5.2).

In the TYPE definition for Element, a CHARACTER (LEN = 15) variable called Element_Name holds the element's name. The *kind*, (LEN = 15), means variable Element_Name can hold up to 15 characters. A second variable, Symbol, is CHARACTER (LEN = 2) and holds the element's symbol. Finally, INTEGER variable, Atomic_Number, holds the element's atomic number.

After the TYPE definition for Element is completed, three variables, Hydrogen, Helium and Lithium are defined as TYPE(Element). This means that variables Hydrogen, Helium and Lithium *each* contain *three* subordinate variables:

Figure 5.2

```
PROGRAM Element_Derived_Data_Type
        IMPLICIT NONE
    ...
!
    TYPE Element
        CHARACTER ( LEN = 15 )   ::   Element_Name
        CHARACTER ( LEN = 2  )   ::   Symbol
        INTEGER                  ::   Atomic_Number
    END TYPE Element
!
    TYPE ( Element )  ::  Hydrogen, Helium, Lithium
!
    ...
```

Element_Name, Symbol and Atomic_Number. These subordinate variables are called **subobjects**. There are two ways to assign values to a derived type variable:

1. **Structure constructors**.
2. Individual element assignment; the **% operator**.

Structure constructors are used to create specific instances of a derived type and can give **all** of a derived data type variable's subordinate variables their values in **one** assignment statement. The general form of a structure constructor is shown below:

type-kind (value1, value2, ...)

When a structure constructor is used in an assignment statement, value1, value2, etc., will be assigned to the individual subobject variables contained in the derived data type variable on the left side of the equal sign. Program Derived_Data_Types shows how this works in Figure 5.3.

As in Figure 5.2, a derived data type for chemical elements is defined in program Derived_Data_Types of Figure 5.3. Once the type is defined, four variables (Hydrogen, Helium, Lithium and Carbon) are declared. Each of the four variables is then assigned a set of values with structure constructors. The assignment statement for Helium from Figure 5.3 follows:

Helium = Element ("Helium", "He", Atomic_Number_Of_Hydrogen * 2)

All three variables (subobjects) in variable Helium, which are Element_Name, Symbol and Atomic_Number, receive values. It is important that each of the fields in a structure constructor is an appropriate type that corresponds to the type of the subobject in its assigned derived type variable. It is also allowed to use expressions in any field of a structure constructor. Variable Helium's Atomic_Number is set to 2, which is the result of the following expression in the structure constructor:

Atomic_Number_Of_Hydrogen * 2

Structure constructors are great when a programmer wants to assign values to ALL the subobjects contained in a derived data type variable. It is, however, often necessary to access individual parts of a derived data type variable. To accomplish this, Fortran supplies the % (percent) operator. The general form for using the % operator is shown below:

derived-type-variable % subobject-variable

As an example of the use of the % operator, assume the programmer of PROGRAM Derived_Data_Types in Figure 5.3 needs to print variable Carbon's chemical symbol. The following PRINT statement does the job:

PRINT *, Carbon % Symbol

Carbon % Symbol contains a value of "C" and may be used in any CHARACTER expression. Carbon % Symbol is simply Carbon's Symbol variable, which is type CHARACTER.

Figure 5.3

```
PROGRAM    Derived_Data_Types
                               IMPLICIT NONE
!
!--------------------------------------------------------------------
! Written by      :    C. Forsythe
! Date Written    :    3/25/2004
!--------------------------------------------------------------------
! The purpose of this program is to demonstrate the creation of
!   a derived data type and the use of structure constructors to
!   assign values to variables of that derived type.
!
!  Hydrogen, Helium, Lithium, Carbon  --   are all variables of type
!                                             "Element"
!--------------------------------------------------------------------
!
        INTEGER                          :: Atomic_Number_Of_Hydrogen = 1
!
        TYPE Element           ! Define a derived type called Element
            CHARACTER ( LEN = 15 ) :: Element_Name
            CHARACTER ( LEN = 2  ) :: Symbol
            INTEGER                :: Atomic_Number
        END TYPE Element
!
! Define variables of TYPE with a kind of Element
!
        TYPE ( Element )   ::   Hydrogen, Helium, Lithium, Carbon
!
! Assign the variables' values using structure constructors
!
    Hydrogen = Element ( "Hydrogen", "H",  Atomic_Number_Of_Hydrogen )
    Helium   = Element ( "Helium",   "He", Atomic_Number_Of_Hydrogen * 2 )
    Lithium  = Element ( "Lithium",  "Li", Atomic_Number_Of_Hydrogen * 3 )
    Carbon   = Element ( "Carbon",   "C",  6 )
!
    PRINT *, Hydrogen
    PRINT *, Helium
    PRINT *, Lithium
    PRINT *, Carbon
!
END PROGRAM Derived_Data_Types

*************** PROGRAM RESULTS ***************

Hydrogen         H   1
Helium           He  2
Lithium          Li  3
Carbon           C   6
```

MODULEs

When a derived data type is created, it may be useful or even necessary for other program units within the program. In fact, the derived type may be useful in different programs altogether! In Figure 5.2, a derived data type was created:

```
TYPE Element
    CHARACTER ( LEN = 15 )    ::    Element_Name
    CHARACTER ( LEN = 2  )    ::    Symbol
    INTEGER                   ::    Atomic_Number
END TYPE Element
```

Type Element could be useful in other programs that deal with chemical data. Following the "GET LAZY" principles of reusing existing code, complicated types such as Element can be saved in MODULE program units. These useful MODULEs can then be added to a programmer's library, making the MODULE's data definitions available to any program unit in any program.

There are two steps for using MODULEs:

1. **Define** the MODULE.
2. **Make its contents available** to a given program unit.

The general form of a MODULE used to hold data definitions is seen in Figure 5.4. Figure 5.5 shows how putting type Element in a MODULE works.

Once a module is created and either included in a program or library, it can be accessed by the USE statement. The general form of a USE statement, which makes all of a module's definitions available to a program unit, is seen below:

```
USE module-name
```

USE statements *must* be placed between the program unit header and IMPLICIT NONE. As an example of the USE statement, assume that the MODULE in Figure 5.5 currently exists in an accessible library of program units, and study the program in Figure 5.6 (p. 106). This is a modified version of program Derived_Data_Types from Figure 5.3. Examine the comments in program Derived_Data_Types from Figure 5.6; they explain how module Chemistry_Derived_Data_Types is accessed.

It is very desirable to have all TYPE derived data definitions neatly tucked away in modules. It reduces the likelihood of typographical errors by eliminating the necessity of

Figure 5.4

```
MODULE  module-name
     data-definition-1
     data-definition-2
          ...
     data-definition-n
END MODULE  module-name
```

Figure 5.5

```
MODULE  Chemistry_Derived_Data_Types
!
!-----------------------------------------------------------------
!  The purpose of this module is to provide derived data types
!     for chemistry related programs.
!
!  Element    - A derived data type that contains an element's name,
!               its symbol and its atomic number
!
!  MORE DATA DEFINITIONS CAN BE ADDED TO THIS
!  MODULE AS THEY ARE CREATED
!-----------------------------------------------------------------
!
      TYPE Element
           CHARACTER ( LEN = 15 )   ::    Element_Name
           CHARACTER ( LEN = 2  )   ::    Symbol
           INTEGER                  ::    Atomic_Number
      END TYPE Element
!
END  MODULE Chemistry_Derived_Data_Types
```

typing and retyping the TYPE definitions, and it simplifies programs by making them less cluttered, more readable and therefore more understandable.

Derived Data Type Arguments

Passing a derived data type variable to a program unit is the same as passing any other type of variable as an argument, except that there are special concerns about the TYPE definitions. Packaging derived data type definitions into MODULEs is particularly important when a programmer passes a derived data type variable to another program unit. This is because a derived data type definition in a calling program unit is only considered the *same* as a derived data type definition in a called program unit if the following rules are strictly adhered to:

1. Both TYPE definitions must contain the SEQUENCE property.
2. All components of the TYPE definition must be identical in *every* respect (except spacing and comments), including identifier names, order of declarations, etc.

Examine the program in Figure 5.7 (p. 107), which demonstrates how to declare a derived data type in the main program and another derived type in a subroutine that is considered the *same* derived type. The TYPEs must be the same so that the argument is properly passed. Doesn't it make more sense to store the TYPE definition in a MODULE and access that TYPE with a USE statement in the program and subroutine, rather than being fettered by all the precise TYPE definition reproduction?

Figure 5.6

```
PROGRAM    Derived_Data_Types
                    USE Chemistry_Derived_Data_Types  ! This USE gives this
                                                      ! program unit access
                                                      ! to the definitions
                                                      ! in the module.
                                                      ! Notice that the USE
                                                      ! is the first
                                                      ! statement after the
                                                      ! PROGRAM statement.
                    IMPLICIT NONE
!
!-------------------------------------------------------------------------
! Written by      :   C. Forsythe
! Date Written    :   3/25/2004
!-------------------------------------------------------------------------
! The purpose of this program is to demonstrate the creation of
!    a derived data type and the use of structure constructors to
!    assign values to variables of that derived type.
!
! Hydrogen, Helium, Lithium, Carbon  - are all variables of type
!                                           "Element"
!
! Atomic_Number_Of_Hydrogen          - INTEGER variable
!-------------------------------------------------------------------------
!
        INTEGER                   :: Atomic_Number_Of_Hydrogen = 1
!
! IT IS UNNECESSARY TO DEFINE TYPE ( Element ) BECAUSE IT IS SUPPLIED
! BY MODULE Chemistry_Derived_Data_Types
!
! Define variables of TYPE with a kind of Element
!
        TYPE ( Element )   ::   Hydrogen, Helium, Lithium, Carbon
!
! Assign the variables' values using structure constructors
!
     Hydrogen = Element ( "Hydrogen", "H",  Atomic_Number_Of_Hydrogen )
     Helium   = Element ( "Helium",   "He", Atomic_Number_Of_Hydrogen * 2 )
     Lithium  = Element ( "Lithium",  "Li", Atomic_Number_Of_Hydrogen * 3 )
     Carbon   = Element ( "Carbon",   "C",  6 )
!
     PRINT *, Hydrogen
     PRINT *, Helium
     PRINT *, Lithium
     PRINT *, Carbon
!
END PROGRAM Derived_Data_Types
```

Figure 5.7

```fortran
PROGRAM  Passing_A_Derived_Argument
              IMPLICIT NONE
!
!-----------------------------------------------------------------------
!   Written by     :   C.  Forsythe
!   Date Written   :   3/25/2004
!-----------------------------------------------------------------------
!   This program demonstrates creating two derived TYPE definitions.
!      One is in the main program and the other is in the subroutine.
!      Both TYPE definitions contain the SEQUENCE property and are
!      exactly physically identical, so they are considered the same
!      exact TYPE. If SEQUENCE were omitted or the physical TYPE
!      definitions varied in ANY way other than spacing or comments, the
!      TYPE in the subroutine would be local to it and different from
!      the one in the main program. The consequence of this is that the
!      argument would not be passed to the subroutine correctly.
!
!   Creature            - TYPE ( Classify_Life ) variable.
!
!   Creature_Name       - a CHARACTER ( LEN = 30 ) variable.
!-----------------------------------------------------------------------
!
      TYPE  Classify_Life
            SEQUENCE
            CHARACTER ( LEN = 25 )      ::   Kingdom
            CHARACTER ( LEN = 25 )      ::   Phylum
            CHARACTER ( LEN = 25 )      ::   Sub_Phylum
            CHARACTER ( LEN = 25 )      ::   Class
            CHARACTER ( LEN = 25 )      ::   Order
            CHARACTER ( LEN = 25 )      ::   Family
            CHARACTER ( LEN = 25 )      ::   Genus
            CHARACTER ( LEN = 25 )      ::   Species
      END TYPE Classify_Life
!
      TYPE ( Classify_Life )      ::   Creature
      CHARACTER ( LEN = 30)       ::   Creature_Name
!
      PRINT *, "Please enter the creature's name."
      READ *, Creature_Name
      PRINT *, " " ! Leave a blank line
      PRINT *, "Please enter the creature's classification."
      READ *, Creature
      CALL Print_Classification ( Creature_Name, Creature )
!
      END PROGRAM   Passing_A_Derived_Argument
!=======================================================================
```

continues

Figure 5.7 (continued)

```
      SUBROUTINE  Print_Classification ( Name, Classification )
                IMPLICIT NONE
!-------------------------------------------------------------------
!   This subroutine prints out Name and Classification.
!
!   Name          - a CHARACTER ( LEN = * ) INTENT ( IN ) argument
!                   holding a creature's name
!
!   Classification - a TYPE ( Classify_Life ) INTENT ( IN ) argument
!                    that holds a creature's classification
!-------------------------------------------------------------------
!
        TYPE   Classify_Life
            SEQUENCE
            CHARACTER ( LEN = 25 )       ::    Kingdom
            CHARACTER ( LEN = 25 )       ::    Phylum
            CHARACTER ( LEN = 25 )       ::    Sub_Phylum
            CHARACTER ( LEN = 25 )       ::    Class
            CHARACTER ( LEN = 25 )       ::    Order
            CHARACTER ( LEN = 25 )       ::    Family
            CHARACTER ( LEN = 25 )       ::    Genus
            CHARACTER ( LEN = 25 )       ::    Species
        END TYPE   Classify_Life
!
        TYPE ( Classify_Life ), INTENT ( IN )    ::   Classification
        CHARACTER ( LEN = * ), INTENT ( IN )     ::   Name
!
        PRINT *, " "      ! Leave a blank line
        PRINT *, " "
        PRINT *, "    Life form's name: ", Name
        PRINT *, " "
        PRINT *, "    Life form's classification: "
        PRINT *, " "
        PRINT *, "Kingdom     ", Classification % Kingdom
        PRINT *, "Phylum      ", Classification % Phylum
        PRINT *, "Subphylum   ", Classification % Sub_Phylum
        PRINT *, "Class       ", Classification % Class
        PRINT *, "Order       ", Classification % Order
        PRINT *, "Family      ", Classification % Family
        PRINT *, "Genus       ", Classification % Genus
        PRINT *, "Species     ", Classification % Species
!
        RETURN
        END SUBROUTINE    Print_Classification
```

continues

Figure 5.7 (continued)

```
************** PROGRAM RESULTS **************

 Please enter the creature's name.
"Colorado Blue Spruce"        <--- Enclose CHARACTER input that has embedded
                                   blanks in quotes or apostrophes.
 Please enter the creature's classification.
Embryophytes
Tracheophyta
Pterophytina
Gymnosperms
Coniferinae
none
Picea
Blue


     Life form's name: Colorado Blue Spruce

     Life form's classification:

Kingdom    Embryophytes
Phylum     Tracheophyta
Subphylum  Pterophytina
Class      Gymnosperms
Order      Coniferinae
Family     none
Genus      Picea
Species    Blue
```

Clearly, program `Passing_A_Derived Argument` would be much simpler to understand if `TYPE Classify_Life` was stored in a `MODULE` that relates to botany; the `TYPE` definitions wouldn't need to be in the program per se.

Important Concepts Review

- **Derived data types** are types that are defined by a programmer. Derived types are composed of:
    1. Simple Fortran data types (`LOGICAL`, `CHARACTER`, `INTEGER`, etc.)
    2. Simple Fortran data types and/or other derived data types.

- **Structure constructors** create specific instances of a derived type and may be used to assign all its parts in one assignment.

- **The % operator** allows a programmer to access individual elements of a derived data type.

- MODULEs are Fortran program units that contain data definitions. Modules have other abilities that are explained in Chapter 13.

- USE statements are placed immediately after a program unit's header and give program units access to the named modules.

- Passing a derived data type variable as an argument is the same as passing any other type of variable. It's best to put TYPE definitions in modules to avoid errors and simplify programs.

5.2 Special Operations for CHARACTER Type Data

There are powerful operations for CHARACTER data that need to be understood in order to exploit the full potential of this important data type. The three areas of focus for CHARACTER data in the next few sections are listed below:

1. Assignment.
2. Substring references.
3. Concatenation.

Assigning Values to CHARACTER Variables

There are some special considerations you need to understand when assigning values to CHARACTER variables. First, examine the following declaration:

```
CHARACTER ( LEN = 5 ) :: String_1, String_2
```

The declaration above creates two CHARACTER variables that can each hold five characters. Think of String_1 and String_2 as each having five little boxes, each box holding only one character (Figure 5.8). A simple example of assigning a value to String_1 is shown in Figure 5.9.

Figure 5.9's assignment worked out in a sensible way; each character in the string, "LINDA", fit sequentially into String_1's boxes. But what if the assignment was one of the following two?

5.2 Special Operations for CHARACTER Type Data

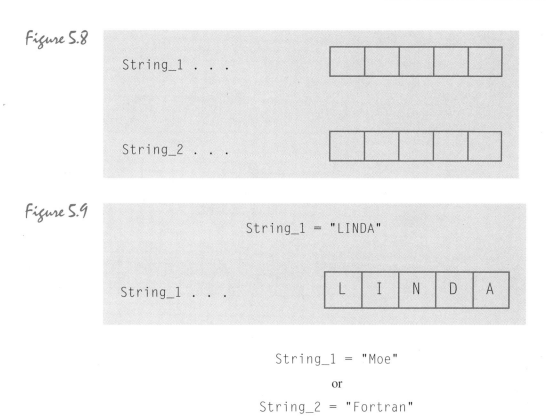

Figure 5.8

Figure 5.9

$$\text{String\_1 = "Moe"}$$
or
$$\text{String\_2 = "Fortran"}$$

The first assignment tries to put less than five characters into String_1, and the second tries to put more than five characters into String_2. What will happen? There are two rules that govern these situations:

1. **Left justified and blank filled.**
2. **Truncation.**

When a CHARACTER variable is assigned a string of characters that is shorter than the *declared* length of the variable, the string's characters are positioned in the leftmost set of boxes, and unused boxes to the right are *filled* with blanks. This is called **left justified and blank filled** (Figure 5.10, p. 112).

It is noteworthy that the left-justified-and-blank-filled action in the second assignment of Figure 5.10 wipes out the "DA" of "LINDA" from the previous assignment. When the right portion of a CHARACTER variable is filled with blanks, the blanks destroy whatever characters are there. Often, beginning programmers view blanks as being transparent, but they aren't; a blank has as much presence as any other character. When a blank is assigned to a position in a CHARACTER variable, whatever was there before is lost.

An example of **truncation**, the second rule that governs assignment to CHARACTER variables, is shown in Figure 5.11 (p. 112). Extra characters on the right side of character constant "Fortran" are truncated; variable "String_2" holds as many characters as it can and then discards the rest.

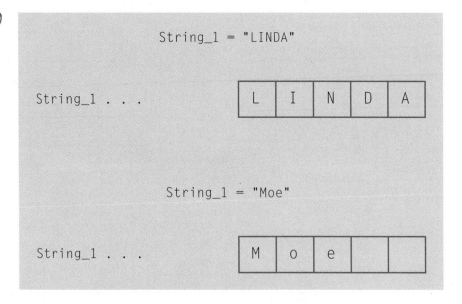

Figure 5.10

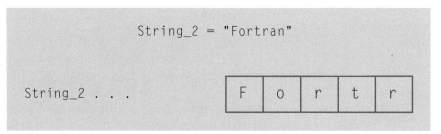

Figure 5.11

Substring References

A **substring** is a contiguous sequence of characters within a CHARACTER variable or constant. Substring references enable you to examine, extract or assign individual characters of a CHARACTER variable. A substring is specified as follows:

string-item ([start-character] : [end-character])

Limitations for start-character, end-character and string-item are listed below:

1. start-character must be less than or equal to end-character or the specified substring will contain nothing. If start-character is omitted (the square brackets indicate that start-character is optional), it is assumed to be 1.
2. start-character must be greater than zero or omitted.
3. Both start-character and end-character must be INTEGER values.

4. end-character must be less than or equal to the length of string-item or omitted. If end-character is omitted, it is assumed to be in the position of the last character in string-item.
5. string-item may be a CHARACTER variable or a CHARACTER constant.

Figure 5.12 shows a few examples of how substring references work. Assume that the assignment statements are done in order, and therefore each can be affected by the preceding

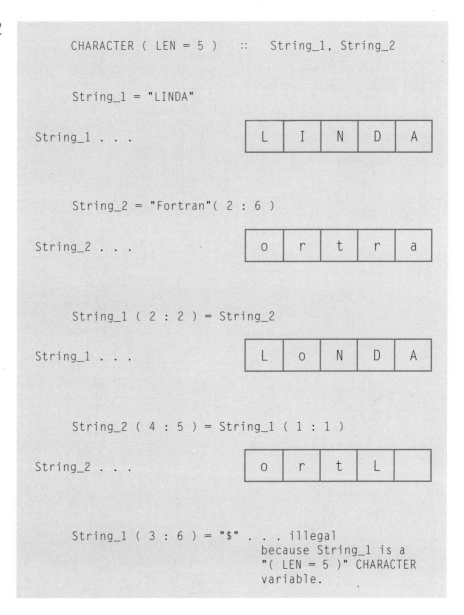

Figure 5.12

assignments. Studying the examples in Figure 5.12 reveals that the assignment statement rules—left justified and blank filled and truncation—also apply to ranges of characters specified in substring references.

Concatenation

Concatenation is the third and final special operation that can be performed on CHARACTER data. Concatenation means to "chain together." When two pieces of CHARACTER data are concatenated, they are attached to each other. Fortran's operator for concatenation is the double slash, //. Figure 5.13 shows examples of concatenation.

There are built-in Fortran functions that help manipulate CHARACTER data besides the specialized ways of working with CHARACTER data described in the last three sections. These functions are in Appendix B and include: LEN( ), which returns the length of a string,

Figure 5.13

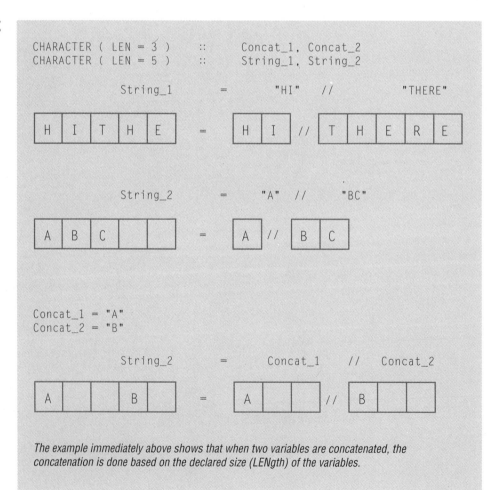

The example immediately above shows that when two variables are concatenated, the concatenation is done based on the declared size (LENgth) of the variables.

LEN_TRIM( ), which returns the number of characters in the string minus the number of trailing blanks; and INDEX( ), which tries to find one string within another string, etc.

Passing Character Data to Subroutines

When passing CHARACTER data to a subroutine, it is possible to pass actual arguments of different declared lengths to the same formal argument. One CALL could have integration as an argument, while another CALL to the same subroutine might send a variable declared as LEN = 4 to the same formal argument. This is consistent with the idea that one of the goals of subroutines is that they are reused; programmers should strive to make subroutines reusable. Formal arguments' ability to handle various string lengths is accomplished by specifying a *kind* of "(LEN = *)". Any formal argument bearing the "(LEN = *)" *kind* will accommodate any length of CHARACTER actual argument.

Important Concepts Review

- **Left justified and blank filled** means that if fewer characters are assigned to a variable than it can hold, the characters will be placed at the left of the variable, and the rest of the variable will be filled with blanks.

- **Character truncation**—more characters are assigned to a variable than it can hold; the rightmost characters that don't fit will be lost.

- **Substring reference:**

    variable (start-chr : end-chr)

- **Concatenation**—attaching two strings together. The concatenation operator is //.

- "(LEN = *)" is a *kind* for CHARACTER formal arguments that allows them to accommodate any length CHARACTER actual arguments.

5.3 How Numeric Values Are Stored in Memory

Inside a computer there are essentially two types of memory:
1. **RAM** (Random Access Memory).
2. **ROM** (Read Only Memory).

Random access memory is the high speed memory where programs are run. When a computer is turned off, everything in RAM is lost. Having to re-enter a program each time a computer is turned off and back on is obviously impractical, so programs are copied to a more permanent environment called **storage**. Storage usually takes the form of a hard disk, a floppy diskette or magnetic tape. These types of storage are made of magnetic media similar to audio cassette tapes and facilitate long-term storage of programs and data.

Read only memory (ROM) is memory in the computer that has permanently written information. When a computer is turned off, the information in ROM isn't lost and will be available the next time the computer is turned on. ROM is used for low-level running of a computer.

Binary and Hexadecimal

A computer's memory is based on **binary** (base 2) arithmetic. Binary numbers are composed of zeros and ones. This method is used because, from an electrical standpoint, it is easy to create little switches that are either **on** (1) or **off** (0). It is amazing that all the incredibly sophisticated things that computers can do come down to a series of zeros and ones! Figure 5.14 shows the eight-digit binary number 11100101, including each digit's powers of 2 and corresponding place value. To figure out the base 10 value of a binary number, simply add up the place values for each digit that is a 1.

Each binary digit (0 or 1) in the computer's memory is called a **bit**. A grouping of eight bits is called a **byte** and, generally, a grouping of four bytes is called a **word** (32 bits). Bytes are thought of as being in two groups of four bits each, as shown in Figure 5.15. The binary number 11100101 is split in two. The four most significant digits, 1110, are arranged in a vertical column, and the four least significant digits are placed similarly in a column to the right.

Figure 5.14

128	64	32	16	8	4	2	1	place value
2^7	2^6	2^5	2^4	2^3	2^2	2^1	2^0	power of 2
1	1	1	0	0	1	0	1	binary number
128 +	64 +	32 +	0 +	0 +	4 +	0 +	1	= 229 base 10

Figure 5.15

		BYTE			
2^7	128	1	0	8	2^3
2^6	64	1	1	4	2^2
2^5	32	1	0	2	2^1
2^4	16	0	1	1	2^0

Two columns of four bits that form a byte are simplified further by thinking of them as two hexadecimal (base 16) digits. Sixteen "digits" are used to represent the hexadecimal numbering system and are shown below:

```
0 1 2 3 4 5 6 7 8 9  A  B  C  D  E  F  — Hexadecimal digits
0 1 2 3 4 5 6 7 8 9 10 11 12 13 14 15  — Decimal equivalent
```

Binary number 11100101 from the byte representation in Figure 5.15 is E5 in hexadecimal. E in E5 comes from the binary number 1110 (14 decimal, which is E in hexadecimal) and the 5 comes from the binary number 0101 (5 decimal, which is also 5 in hexadecimal). **Every byte is composed of two hexadecimal digits.**

Digits in hexadecimal numbers, being base 16, have place values based on powers of 16 (Figure 5.16). Using the previously explained conversion methods, it is clear that the binary number 11100101 from Figure 5.14 is E5 in hexadecimal (229 decimal). Notice that E5, the right two digits of Figure 5.16, are 224 + 5 = 229, as well.

Collating Sequences

When a human communicates with a computer, binary isn't a very easy method to use. Imagine what it would be like if people had to communicate with computers by entering information in binary, which is all that computers understand. What would a programmer type to tell the computer to run a program? 11101001011101001001111010100, perhaps?

Obviously, binary is not very helpful from a human standpoint. To overcome this problem, collating sequences were devised. A **collating sequence** is a system of associating common symbols such as V, &, *, -, =, j, 5, etc. (in general, any symbol on a keyboard) with corresponding numbers that can then be represented in binary for the computer. There are several collating sequences such as **ASCII** (American Standard Code for Information Interchange—as'-kee), **ANSI** (American National Standards Institute—an'-see), **EBCDIC** (External Binary Coded Decimal Interchange Code—ebb'-seh-dik), etc.

A collating sequence associates a byte's worth of bits with a recognizable symbol. This way, meaningful symbols that humans understand are associated with some eight-bit (one byte) configuration. A capital A has an ASCII value of 65, for example. Its corresponding byte representation is shown in Figure 5.17 (p. 118). Both ASCII and ANSI collating sequences associate the same meaning to the numbers from 0–127. Values 0–31 and

Figure 5.16

Converting the hexadecimal number, 3AE5, to decimal:

4096	256	16	1	place value
16^3	16^2	16^1	16^0	power of 16
3	A	E	5	hexadecimal number

3*4096 + A*256 + E*16 + 5
3*4096 + 10*256 + 14*16 + 5

12288 + 2560 + 224 + 5 = **15077** base 10

Figure 5.17
The ASCII byte represention for capital A is 65 (64 + 1), 41 Hex.

128	0	0	8
64	1	0	4
32	0	0	2
16	0	1	1

127 have special meaning to the computer. Values from 32 to 126 are the symbols on a standard computer keyboard. See Appendix A for the complete ASCII collating sequence.

Representations for REAL and INTEGER

The next couple of sections deal with low-level representations of numeric types, such as REAL and INTEGER. Actual representations vary from machine to machine.

INTEGER Representations When a "standard" single precision integer value is stored in a computer, it generally occupies one word, which is four bytes. The bit in the upper-left corner is for the sign bit (Figure 5.18). If the sign bit is 1, the integer is negative. Starting with the sign bit, the bit values are 2^{31} down to 2^0. This gives integers a range of:

$$-2147483648 \quad \text{to} \quad 2147483647$$
$$-(2^{31}) \quad \text{to} \quad 2^{31} - 1$$

10000000 | 00000000 | 00000000 | 00000000   to   0111111 | 11111111 | 11111111 | 11111111

Some compilers support eight-byte integers that increase the magnitude of integer values to about 18 to 19 digits. The value of the integer in Figure 5.18 is 60143D hexadecimal; 6296637 decimal.

Two's Complement Arithmetic **Two's complement** is a way to represent *negative integer values*. Simply stated, two's complement means change every 0 bit to 1 and every

Figure 5.18
4 Bytes = 1 Word

Sign Bit

0	0	0	0	0	0	0	1
0	0	1	0	0	1	0	1
0	0	1	0	0	0	1	0
0	0	0	0	1	0	1	1

1 bit to 0 and add 1 to the result. It is simpler to examine this process in a 1-byte integer rather than a full-word integer.

Sign Bit

| 0 | 0 | 1 | 0 | 0 | 1 | 1 | 0 |

The value of this integer is 38. To create the two's complement (the negative) of this number, first change every 0 to 1 and every 1 to 0.

Sign Bit

| 1 | 1 | 0 | 1 | 1 | 0 | 0 | 1 |

Notice that the sign bit is now a 1, indicating that the value is negative. The final step in the process is to add 1 to the integer, resulting in the following bit configuration:

Sign Bit

| 1 | 1 | 0 | 1 | 1 | 0 | 1 | 0 |

If one adds the representation for 38 (00100110) and its two's complement representation (11011010), the result is 100000000 (nine bits). Since the example integer can only hold *eight bits*, the leftmost bit that was carried to the 9th bit position from the addition will be lost, leaving 00000000. This demonstrates that the two's complement representation of 38 is indeed equivalent to -38. The key to this method of representing negative integers is the *loss* of the bit that is carried beyond the leftmost bit position—the 9th bit.

REAL **Representations**   To represent a REAL number, a word of bits is broken into two parts: the **mantissa** and the **power of 10's exponent**, as seen in Figure 5.19. In Figure

Figure 5.19
Representation of REAL Numbers

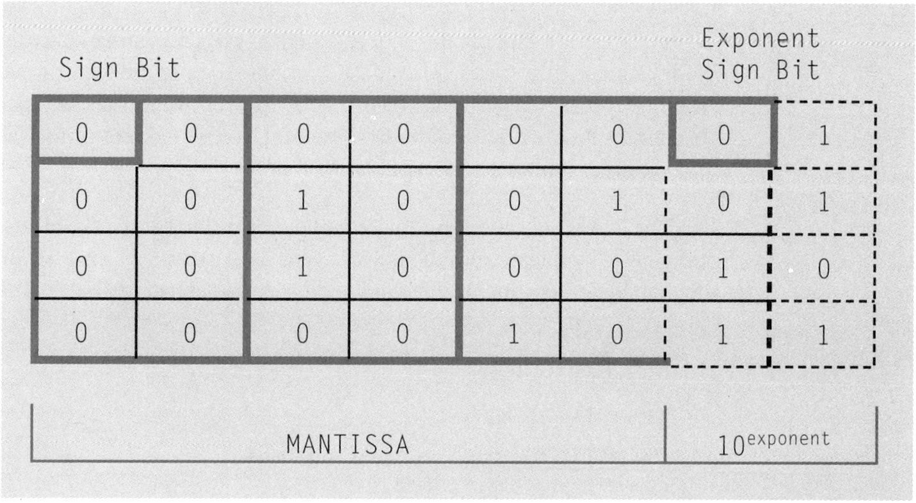

Sign Bit							Exponent Sign Bit	
0	0	0	0	0	0	0	1
0	0	1	0	0	1	0	1
0	0	1	0	0	0	1	0
0	0	0	0	1	0	1	1

| MANTISSA | $10^{exponent}$ |

5.19, the left three bytes are used for the mantissa and the rightmost byte, set off in dashed lines, is used for the power of ten. Note that the exponent byte also has a sign bit.

Important Concepts Review

- A **bit** is a binary digit that is either **on** (one) or **off** (zero).
- A **byte** is composed of eight bits.
- A **word** is a grouping of four bytes.
- The **hexadecimal** numbering system is base 16 and uses the digits 0, 1, 2, 3, 4, 5, 6, 7, 8, 9, A, B, C, D, E and F. A byte is composed of two hexadecimal digits.
- A **collating sequence** is a system where common symbols, such as those on a keyboard, are associated with binary numbers that the computer understands.
- A **sign bit** determines the sign of an integer or, in type REAL numbers, the sign of the mantissa and exponent. A sign bit of one indicates a negative value.
- **Two's complement arithmetic** is a common method for representing negative values for integers.

EXERCISES

5.1 Write a program that creates a derived data type consisting of two parts: 1) a variable to hold the name of some metal, and 2) a variable to hold the metal's melting point in degrees Kelvin. Define two variables of this TYPE to represent two metals. Use a structure constructor to assign values to the first metal. Use the % operator to assign values to the second variable. Print the two variables so that they are readable.

5.2 Declare a derived data type for computers consisting of subobjects that specify a given computer's characteristics. Name this type Computer_Type. Include the following features of computers and decide what type each subobject of Computer_Type should be:

    Brand name
    Megabytes of RAM
    Clock speed (Megahertz)
    Monitor type

CPU (Central Processing Unit) type

How much does the system cost?

Does the computer have multimedia capabilities?

5.3 Create a derived data type named Lab_Hardware to represent the hardware in a university computer lab. Assume the labs have the following characteristics:

Network type

Does the lab have a printer?

Printer type?

The computers—use the derived data type created in Exercise 5.2 for this subobject.

5.4 Declare a variable named Computer_1 of type Computer_Type (from Exercise 5.2). Assign Computer_1 appropriate values for a computer using a structure constructor.

5.5 Declare a variable, Lab_In_Wells_Hall, of type Lab_Hardware (from Exercise 5.3). Assign Computer_1 from Exercise 5.4 to the computer subobject of Lab_In_Wells_Hall.

5.6 Assign the elements of the computer subobject in Lab_In_Wells_Hall (from Exercise 5.5) using the percent, %, operator. (*Hint: the values assigned are not particularly important, it's the mechanics of the assignments that is significant.*)

5.7 Convert the following binary numbers to decimal (base 10).
   a) 11111111
   b) 100000000
   c) 11011001
   d) 1111111111111111
   e) 10000000000000000
   f) 1111

5.8 Convert the binary numbers from Exercise 5.7 to hexadecimal. (*Hint: start from the right of the binary number. Each group of four binary digits forms one hexadecimal digit.*)

5.9 What are the decimal values for the following symbols according to the ASCII collating sequence?
   a) 4
   b) *
   c) "
   d) a blank
   e) e
   f) >

5.10 What are the two's complements (negatives) of the following binary numbers?
   a) 01110100
   b) 11100011
   c) 00000000
   d) 11111111

5.11 What statement allows a given program unit to have access to the data definitions in a MODULE?

5.12 Where in a program unit should the statement from Exercise 5.11 be physically placed?

5.13 In the introduction to this chapter, a derived data type for trees was defined. Create another derived data type for fruit containing the subobjects Fruit_Name, CITRUS (LOGICAL) and Growing_Period (INTEGER-days).

5.14 Put the Tree derived data type and the Fruit derived data type into a MODULE whose name is Foliage.

5.15 What are the results of the following assignment statements? (*Hint: do not assume that the assignment statements in the questions are cumulative; each assignment does not depend on the previous ones.*)

```
CHARACTER ( LEN = 4  )    ::   String_23, Char_It
CHARACTER ( LEN = 6  )    ::   Many_C, Var_Char_1 = "MU"
CHARACTER ( LEN = 10 )    ::   Result_7

String_23 = "many"
Char_It = "target"
Many_C = "37.2aB"
   a) Result_7 = String_23 // Char_It
   b) Result_7 = Var_Char_1
   c) Result_7 = "begin" // Char_It // Many_C
   d) Result_7 = "end" ( 2 : 3 )
   e) Result_7 = Char_It ( 3 : 6 ) // String_23 ( 1 : 3 )
   f) Result_7 = Many_C // Many_C // Many_C // Many_C
```

5.16 What is the final value of Result_7 after the next assignment statements are executed? Assume these assignments are done sequentially and that each is dependent on the previous ones. The declarations and variable initializations are the same as in Exercise 5.15.

```
Result_7 = Var_Char_1
Result_7 ( 4 : 6 ) = Many_C
Result_7 = Result_7 // "de4A$"
```

Selection Structures

Introduction

Chapter 1 took you through an overview of the world of computing to give you a feel for the possibilities offered by computers. Chapter 2 said, "Think before you act" and laid out the principles of problem solving and program design by breaking down a large problem into smaller pieces. Chapter 3 provided the method for programming/coding those pieces, which is subroutines, and examined the **flow of data** in and out of program units through arguments. Finally, in Chapters 4 and 5, the foundation of program data was explained, including variables, constants, expressions, derived data types and modules. That is a significant body of information!

However, there are many problems that can't be solved with only sequential statement structures, CALLs and a knowledge of program data. For example, if you want to divide variable Top_Value by Bottom_Value, and variable Bottom_Value contains a zero, what happens?

```
Result_Value = Top_Value / Bottom_Value
```

The program crashes. You need **flow-altering structures** to detect such problems before they happen and force flow of control to swerve around them. The many other reasons to skip a program statement or a group of statements will become clear as you read this chapter.

Computers are obedient and blindly follow program instructions in sequential order. It is like someone reading an English

6.1 Conditions, blocks and block statements

6.2 Conditions are logical expressions

6.3 Logical data

Important Concepts Review

6.4 Block IF statement

6.5 Design Example

6.6 Multialternative IF

6.7 SELECT CASE statement

Important Concepts Review

Exercises

book: computers read a program's lines from left to right and top to bottom. As explained in Chapter 1, this moving from one program line to the next is called **program flow** and each line represents an **instruction** that tells the computer to do something.

Although Fortran always tries to flow through program statements sequentially, there are ways to force it to alter program flow. To solve the division-by-zero problem above, for example, a flow-altering structure called a **selection structure** comes to the rescue:

```
IF ( Bottom_Value == 0 ) THEN
        PRINT *, "Sorry, can't do the division"
ELSE
        Result_Value = Top_Value / Bottom_Value
ENDIF
```

If `Bottom_Value` is zero (== means: "is equal to ?"), an **error message** is printed and the division is not performed. If `Bottom_Value` contains something other than zero, the error message is NOT printed and the division IS performed. Program flow is appropriately altered because the selection structure checks for the error *before* it happens.

● ● ●

6.1 Conditions, Blocks and Block Statements

Selection structures execute **blocks** of statements based on the value of a **condition**. One style of a selection structure is illustrated in Figure 6.1. If `condition` in the IF statement from Figure 6.1 is true, the statements in `block-1` will be executed and those in `block-2` will not. If `condition` is false, `block-2` is executed and `block-1` isn't. There is **no condition** that will cause both `block-1` and `block-2` to be executed.

Figure 6.1

```
┌─ IF ( condition ) THEN
│         block-1
│   [ ELSE
│         block-2 ]
└─ ENDIF
```

Conditions

To clearly understand selection structures, it is necessary to understand **conditions**, **blocks** and **block statements**.

Conditions

Conditions are logical comparisons that ultimately evaluate to either true or false (would that life were so). A simple example of a condition is shown below:

$$\text{Counter\_1} > 100$$

If the value contained in variable Counter_1 is larger than 100, the condition is true. If the value in Counter_1 is 100 or less, the condition is false. Fortran evaluates such conditions and produces a true or false result. Then a selection structure uses the true or false to decide which block of statements to execute.

Blocks

A **block** is a group of one or more program statements. Any of the **flow-control structures** can be used to create any of the statements in a block. All procedural programs should be written using some combination of the four basic flow-control structures (Table 6.1). Figure 6.2 shows an example of blocks and how they can fit together.

Table 6.1

Flow-Control Structures
CALL
Sequential
Selection
Looping

Figure 6.2

```
IF ( N <= 25 ) THEN
      ┌─CALL Roots ( a, b, c )
    1 │  PRINT *, "Back from SUB Roots"
      └─N = N + 1
ELSE
      ┌─DO WHILE ( Limit > 0.0001 )
      │    ┌─CALL Get_Next ( N )
      │    │  IF ( T < 0 ) THEN
    2 │  3 │  4══PRINT *, "Error"
      │    │  END IF
      │    │  N = N + 1
      │    └─PRINT *, N
      └─END DO
END IF
```

Block 1 is the block of statements that will be executed if the value contained in variable N is 25 or less. Block 2 will be executed if N is larger than 25. There is also a block of statements inside the DO WHILE loop that will be repeated as long as Limit > 0.0001 (block 3). That block contains an IF statement that will execute block 4 if the value contained in variable T is negative.

Block Statements

A **block statement** is any statement that does something with blocks. Selection structures selectively execute blocks of statements depending on the value of a condition; selection structures are block statements. DO WHILE is a looping structure that executes a block repeatedly; it is a block statement.

Figure 6.2's structures are well organized. Some block statements are contained within other block statements. The DO WHILE loop, for example, is inside the ELSE part of the first IF statement. There is also an IF inside the DO WHILE.

One of the most important concepts to remember when putting block statements together is that they must **never straddle each other**. Figure 6.3 shows an illegal organization of block statements where they overlap. When a programmer writes up a program design, she should use drawn brackets to connect the first line of a block statement with its last line. This will set off block statements and help point out any illegal organizations because it's obvious when the brackets overlap, such as they do in Figure 6.3. Brackets that represent block statements aren't allowed to intersect! Block statements may *follow* other block statements or may be *completely contained* within a block statement ONLY.

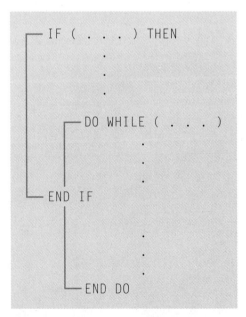

6.2 Conditions Are Logical Expressions

Chapter 4 introduced the idea of arithmetic expressions that were comprised of **operands** being operated on by **operators**, such as **, *, /, and -. In conditions, a programmer creates **logical expressions** that also have operands and operators. The operands in logical expressions MUST NOT be derived data types. Logical expressions always have a value of either **true** or **false** and, as mentioned before, can be **tested** by selection structures.

Simple Logical Expressions

Simple logical expressions are formed with **relational operators.** Fortran's relational operators are listed in Table 6.2. There are two forms for each. With the operators from Table 6.2, a programmer can create simple logical expressions. Examples of such expressions are shown in Figure 6.4. Each logical expression is either true or false ... ALWAYS. In the first example in Figure 6.4, the == **operator** compares two CHARACTER **operands**, "YES" and "yes", and produces a result of *false*. "YES" and "yes" are NOT equal because uppercase characters are not equal to their lowercase counterparts (see the ASCII collating sequence in Appendix A). All the relational operands behave the same way—they *compare their two operands and generate an appropriate true or false*.

Table 6.2

Relational Operators		
Form 1	Form 2	Definition
==	.EQ.	Equal
/=	.NE.	Not equal
>=	.GE.	Greater than or equal
<=	.LE.	Less than or equal
>	.GT.	Greater than
<	.LT.	Less than

Figure 6.4

```
Expression          Result    Explanation
"YES" == "yes"      false     uppercase is not == lowercase
  3   /=   4         true
 7+3  <=  91-2       true     10 is less than 89
-3.7 .GT. -7.3       true
-3.7  >   -7.3       true
 "P"  /=   "P"      false
```

Operands in a simple logical expression must be the same type (e.g., both type CHARACTER, both numeric, etc.). The operands must NOT be derived data types and may be any combination of:

1. Variables.
2. Constants.
3. Expressions.
4. ~~Derived data types.~~

Complex Logical Expressions

Complex logical expressions are those that use **logical operators** (Table 6.3). Logical operators have a period on each side. All operators that have a "dot" on each side are generally referred to as **dot operators**.

Logical operators are provided by Fortran so programmers can form more complicated logical expressions. The purposes of .EQV. and .NEQV. are not as obvious as .NOT., .AND. and .OR.. .EQV. and .NEQV. compare *logical* expressions to see whether they're equivalent or not. The other logical operators behave in standard truth-table fashion (Table 6.4).

Selection structures, logical expressions, relational operators, logical operators, truth tables ... it all seems very complicated when a programmer first learns about selection structures. However, complex logical expressions are not indecipherable. Most complex logical expressions make sense in a verbal context.

From a very early age, everyone has a natural sense of how to use language. Children arrive in kindergarten with a reasonable understanding of words such as *and*, *or* and *not*. These logical constructions in language are reinforced throughout our educational and social lives. When constructing a complex logical expression, a programmer is attempting to communicate the same kind of logic to a computer. One method of figuring out how to construct a logical expression is simply to say the expression out loud. If it sounds right, it probably is right. This is an excellent empirical test.

As an example of this, take the following description of a logical condition and try to construct an appropriate logical expression.

Table 6.3

Logical Operators
.NOT.
.AND.
.OR.
.EQV. and .NEQV.

Table 6.4

p	q	p.and.q
T	T	T
T	F	F
F	T	F
F	F	F

(a)

p	q	p.or.q
T	T	T
T	F	T
F	T	T
F	F	F

(b)

p	.not.p
T	F
F	T

(c)

When the value of How_Many_Times is bigger than LIMIT and the user entered a "Y" into CHARACTER variable Response to do more processing, print a warning message.

How would a programmer construct an appropriate logical expression for that set of conditions? What does the example say first? . . . "When the value of How_Many_Times is bigger than LIMIT . . ."

```
How_Many_Times > LIMIT
```

The user previously entered some value regarding whether to continue processing or not, and it is stored into variable Response. Was it "Y"?

```
Response == "Y"
```

Both things need to be true for a warning message to be printed, so the two relationships are connected with .AND..

```
How_Many_Times > LIMIT .AND. Response == "Y"
```

The complex logical expression above sounds right when said out loud and, indeed, it is right. Now a selection structure can be written to test the logical expression:

```
IF ( How_Many_Times > LIMIT .AND. Response == "Y" ) THEN
    PRINT *, "Some warning message"
END IF
```

The logical expression makes sense; if the program actions have been done until How_Many_Times exceeds LIMIT **AND** the user insists on doing more (Response == "Y"), a warning message is displayed. Logical expressions make sense.

Operator Order of Precedence

As with arithmetic operators, relational and logical operators have an **order of precedence** (Table 6.5). Relational operators are all performed from left to right and leave true and falses in their wake. Then the .NOT.s are evaluated from left to right, .AND.s from left to right, .OR.s from left to right and, finally, .EQV.s and .NEQV.s from left to right. As with arithmetic expressions, logical expressions may be grouped with parentheses as needed to override order of precedence, or simply for clarity.

Table 6.5

Order	Operator
first	Relational
second	.NOT.
third	.AND.
fourth	.OR.
fifth	.EQV. and .NEQV.

6.3 Logical Data

Creating variables of type LOGICAL is the same as creating any other variable. Construct a standard declaration statement such as:

```
LOGICAL :: Logic_1, Logic_2, Flag
```

Each of the variables, Logic_1, Logic_2 and Flag are type LOGICAL and can contain either .TRUE. or .FALSE.. This can lead to some rather interesting assignment statements. Remember the general form of an assignment statement:

```
variable = expression
```

The expression on the right of the equal sign is evaluated and the resulting value is stored into the variable on the left of the equal sign. When using LOGICAL variables, the expression must always be either true or false. Here is a simple example of an assignment to a LOGICAL variable:

```
Logic_1 = 5 > 7
```

Relational operators always produce true or false. The logical expression 5 > 7 is NOT true and therefore .FALSE. will be stored into variable Logic_1. ANY valid logical expression may be on the right-hand side of an assignment statement for Logic_1, Logic_2 and Flag.

Important Concepts Review

- A **condition** has a value of either true or false and may be tested by a selection structure.

- A **block** is a collection of program statements. The statements may include any combination of the four basic flow-control structures: 1. CALL, 2. Sequential, 3. Selection, 4. Looping.

- **Block statements**, IF, DO WHILE, etc., operate on blocks.

- Block statements must **not overlap.** Use brackets to help identify incorrectly structured block statements.

- A block statement may **follow** another block statement or be **completely contained** within a block statement. No straddling is allowed.

- **Logical expressions** are conditions created with operands, relational operators and logical operators.

- **Simple logical expressions** don't use logical operators.
- **Complex logical expressions** use logical operators.
- The **relational operators** are: ==, /=, >, <, >=, <=.
- The **logical operators** are: .NOT., .AND., .OR., .EQV. and .NEQV..
- **Operands** in a logical expression MUST NOT be derived data types.
- The **operator order of precedence** is as follows: 1. relational operators, 2. .NOT., 3. .AND., 4. .OR., 5. .EQV. and .NEQV..
- Logical expressions may be **assigned** to logical variables.

6.4 Block IF Statement

Syntax for the general form of a block IF statement is shown in Figure 6.5. As explained before, if the logical-expression (the condition) is true, then the statements comprising block-1 will be executed and the statements in block-2 won't (Figure 6.5). If the logical-expression is false, the reverse will happen. Square brackets (i.e., [and]) identify parts of a statement that are optional. Since ELSE . . . block-2 is surrounded by square brackets, it is not required. Therefore, the ELSE part of a block IF is not necessary and it is reasonable to make a block IF such as the one in Figure 6.6 (p. 132). The IF in Figure 6.6 is testing the logical expression:

$$N >= 100 \ .AND. \ Limit > 0.01$$

First N >= 100 is evaluated to either true or false. Next, Limit > 0.01 is evaluated, assuming N >= 100 is true, and finally these two resulting values are "anded" with the .AND. operator according to the truth table in Table 6.4a. The expression N >= 100 .AND. Limit > 0.01 will only be true if both relationships, N >= 100 and

Figure 6.5

```
IF ( logical-expression ) THEN
            block-1
   [ ELSE
            block-2 ]
END IF
```

Figure 6.6

```
IF ( N >= 100 .AND. Limit > 0.01 ) THEN
    PRINT *, "Tolerance not met"
    CALL Error_Mode ( N, Limit )
    Stop_It = .TRUE.
END IF
```

Limit > 0.01, are true. If the result of the logical expression is true, then the three statements inside the IF will be performed. If the result is false, the three statements will be skipped.

6.5 Design Example

Writing a program will help clarify the use of LOGICAL data and the block IF statement. The following program will create three new logical operators that are often needed when working with logic: XOR, NOR and NAND. These logical operators are defined by the truth tables in Table 6.6.

Table 6.6

p	q	p xor q
T	T	F
T	F	T
F	T	T
F	F	F

(a)

p	q	p nor q
T	T	F
T	F	F
F	T	F
F	F	T

(b)

p	q	p nand q
T	T	F
T	F	T
F	T	T
F	F	T

(c)

Program Design

STATE THE PROBLEM:

In this program I want to create three new logical operators: XOR (exclusive or), NOR (not or) and NAND (not and). I will need to get values for LOGICAL variables p and q from the user at the keyboard. The program will present a menu asking the user whether she wants to perform an XOR, NOR or NAND on the values contained in p and q. I want the menu to look as follows:

    1. XOR

    2. NOR

    3. NAND

        Please enter your choice (1, 2 or 3):

Based on the user's choice, the result of the appropriate operation will be printed on the screen . . . I will also print the values in p and q for clarity.

STRUCTURE CHART:

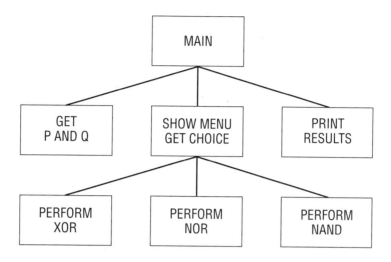

MAIN:

MAIN's purpose is to drive the program by calling the three subroutines under MAIN.

List:

1. Call the subroutine that gets values for p and q.
2. Call the subroutine that presents the menu, which in turn calls subordinate subroutines to perform calculations.
3. Call the subroutine that prints the results.

Data flow to and from the calling program:

None (MAIN isn't called).

English Fortran:

```
PROGRAM Logical_Data
        declarations
        CALL Get_P_And_Q ( p, q )
        CALL Menu ( p, q, Result, Operation )
        CALL Print_Results ( p, q, Result, Operation )
END PROGRAM Logical_Data
```

(Note: this step could have been skipped as the above "English Fortran" is a simple sequential structure and is essentially the same as "List" above.)

GET P AND Q:

Subroutine `Get_P_And_Q ( p, q )` is supplied by the author as a library subroutine:

```
SUBROUTINE Get_P_And_Q ( p, q )
       IMPLICIT   NONE
!
!-------------------------------------------------------------------
!   This subroutine prompts the user to enter values for p and q and stores
!      the typed-in values in p and q.
!
!  p       is a LOGICAL argument of INTENT ( OUT )
!
!  q       is a LOGICAL argument of INTENT ( OUT )
!-------------------------------------------------------------------
!
       LOGICAL,   INTENT ( OUT )       ::   p, q
!
       PRINT *, "Please enter p"
          READ *, p
       PRINT *, "Please enter q"
          READ *, q
       PRINT *, " "                    ! leaves a blank line
       PRINT *, " "                    ! leaves a blank line
!
RETURN
END SUBROUTINE Get_P_And_Q
```

SHOW MENU AND GET CHOICE:

The purpose of this subroutine is to present a user with a menu and get his choice. Depending on the choice, the appropriate subroutine (`XOR`, `NOR`, or `NAND`) will be called.

List:

1. Display the menu with a sequence of `PRINT` statements.
2. Use a `PRINT` statement to prompt the user for a choice.
3. `READ` the choice from the keyboard.
4. Use a selection structure to call the appropriate subroutine.

Data flow to and from the calling program:

p and q come in from the calling program.
`Result` and `Operation` are sent back to calling program.

6.5 Design Example

English Fortran:

```
┌ SUBROUTINE Menu ( p, q, Result, Operation )
│       declarations
│       print "1. XOR"
│       print "2. NOR"
│       print "3. NAND"
│       print " Please enter 1, 2 or 3"
│       read choice
│       ┌ if choice == 1 THEN
│       │    call xor
│       │    Operation = "XOR"
│       └ end if
│       ┌ if choice == 2 THEN
│       │    call nor
│       │    Operation = "NOR"
│       └ end if
│       ┌ if choice == 3 THEN
│       │    call nand
│       │    Operation = "NAND"
│       └ end if
│       return
└ END
```

Notice how nicely the brackets set off the block statements.

PERFORM XOR:

The purpose of this subroutine is to calculate the "exclusive or" of operands p and q.

List:

    1. Store exclusive or of p and q in Result.

Data flow to and from the calling program:

    p and q come in from the calling program.
    Result is sent back to the calling program.

English Fortran:

```
┌ SUBROUTINE XOR ( p, q, Result )
│       declarations
│       Result = ( p .AND. .NOT. q ) .OR. ( .NOT. p .AND. q )
│   return
└ END SUBROUTINE
```

The other two subroutines, NOR and NAND, have the same design. Their respective assignment statements are:

$$\text{Result} = .\text{not.} (p .\text{or.} q)$$

and

$$\text{Result} = .\text{not.} (p .\text{and.} q)$$

PRINT RESULTS:

This final subroutine is also a supplied library subroutine written by the author.

```
SUBROUTINE Print_Results ( p, q, Result, Operation )
          IMPLICIT   NONE
!-----------------------------------------------------------------
!  This subroutine prints the results of the logical operation on the screen.
!
!     p, q          are LOGICAL INTENT ( IN ) arguments
!
!     Result        is a LOGICAL INTENT ( IN ) argument
!                   that contains the result of the operation
!
!     Operation     is a CHARACTER INTENT (IN) argument
!                   that holds the operation
!-----------------------------------------------------------------
!
!     LOGICAL,                    INTENT ( IN )   ::   p, q, Result
!     CHARACTER ( LEN = 4 ),      INTENT ( IN )   ::   Operation
!
          PRINT *, "The Results Are:"
          PRINT *, " "                      ! leaves a blank line
          WRITE ( UNIT = *, FMT = 200 ) p, Operation, q, Result
!
 200      FORMAT ( 15X, L1, 1X, A4, 1X, L1, 2X, "is", 2X, L1 )
!
                !  15X skips the first 15 positions on the screen.
                !  The L1s allow 1 position on the screen for (T)rue
                !  or (F)alse. The A4 provides four positions
                !  for the operation (XOR, NAND etc.). nX skips n
                !  position-putting spaces between operands and
                !  operators. "is" will be printed as itself.
RETURN
END SUBROUTINE   Print_Results
```

The Complete Program

```fortran
        PROGRAM    Logical_Data
                   IMPLICIT NONE
        !
        !-----------------------------------------------------------------
        ! Written by       :    C. Forsythe
        ! Date Written     :    3/25/2004
        !-----------------------------------------------------------------
        ! This program gets values (type LOGICAL) for p and q and evaluates
        !     the XOR, NAND or NOR depending on a menu selection.
        !
        ! p, q             LOGICAL variables to be operands for XOR,
        !                      NAND or NOR
        !
        ! Result           LOGICAL variable to hold the result of the XOR,
        !                      NAND or NOR
        !
        ! Operation        CHARACTER variable to store "XOR",
        !                      "NAND" or "NOR"
        !-----------------------------------------------------------------
        !
             LOGICAL                      ::   p, q, Result
             CHARACTER ( LEN = 4 )        ::   Operation
        !
                CALL Get_P_And_Q ( p, q )
                CALL Menu ( p, q, Result, Operation )
                CALL Print_Results ( p, q, Result, Operation )
        !
        END PROGRAM    Logical_Data
!=====================================================================
        SUBROUTINE    Get_P_And_Q ( p, q )
                      IMPLICIT NONE
        !
        !-----------------------------------------------------------------
        ! This subroutine prompts the user to enter values for p and q and
        !     stores the typed-in values in p and q.
        !
        ! p         is a LOGICAL argument of INTENT ( OUT )
        !
        ! q         is a LOGICAL argument of INTENT ( OUT )
        !-----------------------------------------------------------------
        !
             LOGICAL,    INTENT ( OUT )  ::   p, q
        !
```

continues

```
              PRINT *, "Please enter p"
              READ *, p
              PRINT *, "Please enter q"
              READ *, q
              PRINT *, " "                    ! leaves a blank line
              PRINT *, " "                    ! leaves a blank line
    !
    RETURN
    END SUBROUTINE Get_P_And_Q
!===============================================================================
    SUBROUTINE   Menu ( p, q, Result, Operation )
                 IMPLICIT   NONE
    !
    !-------------------------------------------------------------------------
    !  The purpose of this subroutine is to present a menu to choose XOR,
    !    NOR or NAND and call the appropriate subroutine
    !
    !  p, q           are INTENT ( IN ) arguments of type LOGICAL
    !                     used as operands in XOR, NOR and NAND
    !
    !  Result         is an INTENT ( OUT ) argument of type LOGICAL
    !                     used to hold the results of XOR, NOR or NAND
    !
    !  Operation      is an INTENT ( OUT ) argument of type CHARACTER
    !                     used to hold the operation (XOR, NOR or NAND)
    !
    !  Choice         is a local variable of type INTEGER to hold the
    !                     the user's menu choice
    !-------------------------------------------------------------------------
    !
         LOGICAL,                 INTENT ( IN )  ::   p, q
         LOGICAL,                 INTENT ( OUT ) ::   Result
         CHARACTER ( LEN = 4 ),   INTENT ( OUT ) ::   Operation
    !
         INTEGER                                 ::   Choice
    !
              PRINT *, "1. XOR"
              PRINT *, "2. NOR"
              PRINT *, "3. NAND"
              PRINT *, "     Please enter 1, 2 or 3"
              READ *, Choice
              PRINT *, " "                    ! leaves a blank line
              PRINT *, " "                    ! leaves a blank line
    !
```

```
                IF ( Choice == 1 ) THEN
                    CALL XOR ( p, q, Result )
                    Operation = "XOR"
                END IF
                    IF ( Choice == 2 ) THEN
                    CALL NOR ( p, q, Result )
                    Operation = "NOR"
                END IF
                    IF ( Choice == 3 ) THEN
                    CALL NAND ( p, q, Result )
                    Operation = "NAND"
                END IF
    !
    RETURN
    END SUBROUTINE   Menu
!===============================================================================
    SUBROUTINE    XOR ( p, q, Result )
                  IMPLICIT   NONE
    !
    !-------------------------------------------------------------------------
    !  The purpose of this subroutine is to compute the XOR of p and q.
    !
    !  p, q           are INTENT ( IN ) arguments of type LOGICAL
    !                   used as operands for the XOR computation
    !
    !  Result         is an INTENT ( OUT ) argument of type LOGICAL
    !                   used to hold the results of XOR, NOR, or NAND
    !-------------------------------------------------------------------------
    !
        LOGICAL,                INTENT ( IN  )  ::  p, q
        LOGICAL,                INTENT ( OUT )  ::  Result
    !
            Result = ( p .AND. .NOT. q ) .OR. ( .NOT. p .AND. q )
    !
    RETURN
    END SUBROUTINE   XOR
!===============================================================================
```

continues

```
      SUBROUTINE    NOR ( p, q, Result )
               IMPLICIT   NONE
      !
      !-------------------------------------------------------------------
      !  The purpose of this subroutine is to compute the NOR of p and q.
      !
      !  p, q          are INTENT ( IN ) arguments of type LOGICAL
      !                    used as operands for the NOR computation
      !
      !  Result        is an INTENT ( OUT ) argument of type LOGICAL
      !                    used to hold the results of XOR, NOR or NAND
      !-------------------------------------------------------------------
      !
         LOGICAL,              INTENT ( IN )  ::   p, q
         LOGICAL,              INTENT ( OUT ) ::   Result
      !
            Result = .NOT. ( p .OR. q )
      !
      RETURN
      END SUBROUTINE    NOR
!===================================================================
      SUBROUTINE    NAND ( p, q, Result )
               IMPLICIT   NONE
      !
      !-------------------------------------------------------------------
      !  The purpose of this subroutine is to compute the NAND of p and q.
      !
      !  p, q          are INTENT ( IN ) arguments of type LOGICAL
      !                    used as operands for the NAND computation
      !
      !  Result        is an INTENT ( OUT ) argument of type LOGICAL
      !                    used to hold the results of XOR, NOR or NAND
      !-------------------------------------------------------------------
      !
         LOGICAL,              INTENT ( IN )  ::   p, q
         LOGICAL,              INTENT ( OUT ) ::   Result
      !
            Result = .NOT. ( p .AND. q )
      !
      RETURN
      END SUBROUTINE    NAND
!===================================================================
```

The Complete Program

```fortran
SUBROUTINE    Print_Results ( p, q, Result, Operation )
          IMPLICIT   NONE
!
!-------------------------------------------------------------------
!   This subroutine prints the results of the operation on the screen.
!
!      p, q         are LOGICAL arguments of INTENT ( IN )
!
!      Result       is a LOGICAL argument of INTENT ( IN )
!                   that contains the result of the operation
!
!      Operation    is a CHARACTER argument of
!                   INTENT ( IN ) that holds the operation
!-------------------------------------------------------------------
!
     LOGICAL,                   INTENT ( IN )   ::   p, q, Result
     CHARACTER ( LEN = 4 ),     INTENT ( IN )   ::   Operation
!
        PRINT *, "The Results ARE:"
        PRINT *, " "                   !  leaves a blank line
        WRITE ( UNIT = *, FMT = 200 ) p, Operation, q, Result
!
     200        FORMAT ( 15X, L1, 1X, A4, 1X, L1, 2X, "is", 2X, L1 )
!
                    !  15X skips the first 15 positions on the screen.
                    !  The L1s allow 1 position on the screen for
                    !  (T)RUE or (F)ALSE. The A4 provides 4 positions
                    !  for the operation (XOR, NAND etc.). nX skips
                    !  n position-affording spaces between operands
                    !  and operators. "is" will be printed as itself.
RETURN
END SUBROUTINE    Print_Results
```

continues

```
*************** PROGRAM RESULTS ***************

  Please enter p     <------ prompt from the program
t                    <------ typed by the user at the keyboard
  Please enter q
t


  1. XOR
  2. NOR
  3. NAND
        Please enter 1, 2 or 3
2                    <------ typed by the user at the keyboard


The Results Are:

         T NOR T is F
```

There are several points about this program that should be understood, other than simply how it functions. It contains seven program units, each with a comment block. This makes the program long to type. Why not just lump all the code into the main program and dramatically shorten the program? The reason is that when a programmer creates a program, it should be **easy to understand, easy to modify** and should make use of **reusable generalized subroutines** whenever possible.

Each subroutine in program Logical_Data has one simple purpose and each is **easy to understand**. If the program ever needs to be changed, any of its subroutines can easily be altered because none of them are very complicated and all have good programming style (comments, indentations, blank lines, etc.) This makes the program **easy to modify.**

All the subroutines in program Logical_Data are well documented with comment blocks indicating what arguments need to be supplied and what values are returned in which arguments. If an XOR, NOR or NAND operation is needed in the future, the subroutines are already written and can be used. They work well and include documentation that clearly indicates how to use them. The program has created useful, **reusable generalized subroutines.**

Although the design and implementation of program Logical_Data may seem onerous at the onset, the techniques used will lead to major time savings in the future as useful subroutines from programs are archived. Notice that those two library subroutines Print_Results and Get_P_And_Q didn't have to be designed, coded or debugged. Library routines mean less programming and more engineering and science.

6.6 Multialternative IF

Multialternative IF statements allow an ordered chain of conditions to be tested. **ONLY the first block whose condition is true** is executed. There is no situation that will result in multiple blocks being executed. Syntax for the general form of a multialternative IF is shown in Figure 6.7.

An example of a multialternative IF is illustrated in Figure 6.8. Figure 6.8's multialternative IF works because the conditions are chained together with ELSE IFs. Only ONE block of statements will be performed, the one corresponding to the **first** true condition, and all others will be skipped. Without the multialternative IF, the selection structure in Figure 6.8 would have to be written using the nested-block IF structure seen in Figure 6.9 (p. 144). Figure 6.9's selection structure is functionally exactly the same as the multialternative IF in Figure 6.8. Figure 6.10 (p. 144) shows a complete program that uses a multialternative IF to find the maximum of three integers.

Figure 6.7

```
    IF ( logical-expression-1 ) THEN
        block-1
[ ELSE IF ( logical-expression-2 ) THEN
        block-2 ]
[ ELSE IF ( logical-expression-3 ) THEN
        block-3 ] . . .
[ ELSE
        block-if-all-other-conditions-fail ]
    END IF
```

Figure 6.8

```
!   The following multialternative IF structure creates a series of
!   "gates" to convert a 100-point grade scale to a 4.0 scale.
!   Variable Grade contains an integer value: 0 <= Grade <= 100
!
            IF ( Grade >= 93 ) THEN
                Final Grade = 4.0
            ELSE IF ( Grade >= 88 ) THEN
                Final_Grade = 3.5
            ELSE IF ( Grade >= 83 ) THEN
                Final_Grade = 3.0
            ELSE IF ( Grade >= 78 ) THEN
                Final Grade = 2.5
            ELSE IF ( Grade >= 73 ) THEN
                Final_Grade = 2.0
            ELSE IF ( Grade >= 68 ) THEN
                Final_Grade = 1.5
            ELSE IF ( Grade >= 60 ) THEN
                Final_Grade = 1.0
            ELSE
                Final_Grade = 0.0
            END IF
```

Figure 6.9

```
    IF ( Grade >= 93 ) THEN
        Final_Grade = 4.0
    ELSE
        IF ( Grade >= 88 ) THEN
            Final_Grade = 3.5
        ELSE
            IF ( Grade >= 83 ) THEN
                Final_Grade = 3.0
            ELSE
                IF ( Grade >= 78 ) THEN
                    Final_Grade = 2.5
                ELSE
                    IF ( Grade >= 73 ) THEN
                        Final_Grade = 2.0
                    ELSE
                        IF ( Grade >= 68 ) THEN
                            Final_Grade = 1.5
                        ELSE
                            IF ( Grade >= 60 ) THEN
                                Final_Grade = 1.0
                            ELSE
                                Final_Grade = 0.0
                            END IF
                        END IF
                    END IF
                END IF
            END IF
        END IF
    END IF
```

Figure 6.10

```
    PROGRAM  MultiAlternative_IF
             IMPLICIT   NONE
    !
    !------------------------------------------------------------------
    ! Written by      :    C.  Forsythe
    ! Date Written    :    3/25/2004
    !------------------------------------------------------------------
    ! The purpose of this program is to demonstrate the use of
    !    the multialternative IF statement by writing a program
    !    to find the maximum of 3 numbers.
    !------------------------------------------------------------------
    !
```

Figure 6.10 (continued)

```
        INTEGER                          ::  Num_1, Num_2, Num_3
!
            CALL Get_3_Values ( "Please enter 3 values" , &
                                            Num_1, Num_2, Num_3 )
            CALL Find_Max ( Num_1, Num_2, Num_3 )
!
END PROGRAM    MultiAlternative_IF
!=================================================================
SUBROUTINE   Get_3_Values ( Prompt, Num_1, Num_2, Num_3 )
             IMPLICIT   NONE
!-----------------------------------------------------------------
!   The purpose of this subroutine is to get values for Num_1, Num_2,
!     Num_3, which are INTEGER, INTENT ( OUT ) arguments. Prompt is a
!     CHARACTER argument that contains the prompting message for the
!     READ statement.
!-----------------------------------------------------------------
!
    INTEGER,                 INTENT (OUT )   ::  Num_1, Num_2, Num_3
    CHARACTER ( LEN = * ),   INTENT ( IN )   ::  Prompt
!
       PRINT *, Prompt      ! Generalized prompt
       READ  *, Num_1, Num_2, Num_3
!
END SUBROUTINE    Get_3_Values
!=================================================================
SUBROUTINE   Find_Max ( Num_1, Num_2, Num_3 )
             IMPLICIT   NONE
!-----------------------------------------------------------------
!   The purpose of this subroutine is to print the largest of the 3
!     INTEGER variables, Num_1, Num_2, Num_3... all are INTENT ( IN ).
!-----------------------------------------------------------------
!
    INTEGER,                INTENT ( IN )   ::  Num_1, Num_2, Num_3
!
       PRINT *, "The values are:" , Num_1, Num_2, Num_3
```

continues

Figure 6.10 (continued)

```
                IF ( Num_1 > Num_2 ) THEN
                    IF ( Num_1 > Num_3 ) THEN
                        PRINT *, Num_1, "is largest"
                    ELSE
                        PRINT *, Num_3, "is largest"
                    END IF
                ELSE IF ( Num_2 > Num_3 ) THEN
                    PRINT *, Num_2, "is largest"
                ELSE
                    PRINT *, Num_3, "is largest"
                END IF
        !
        RETURN
        END SUBROUTINE    Find_Max

*************** PROGRAM RESULTS ***************

 Please enter 3 values
-8 5 0
 The values are: -8 5 0
 5 is largest

*************** PROGRAM RESULTS ***************

 Please enter 3 values
27 7 -22
 The values are: 27 7 -22
 27 is largest

*************** PROGRAM RESULTS ***************

 Please enter 3 values
3 66 984
 The values are: 3 66 984
 984 is largest
```

6.7 SELECT CASE Statement

Many situations arise where a variable or an expression needs to be tested to determine its value. That was the situation in subroutine `Menu` in program `Logical_Data` (Figure 6.11). Three `IF` statements from Figure 6.11 are shown on p. 148.

Figure 6.11

```
SUBROUTINE   Menu ( p, q, Result, Operation )
             IMPLICIT   NONE
!
!-------------------------------------------------------------------
!   The purpose of this subroutine is to present a menu
!       to choose XOR, NOR or NAND and call the
!       appropriate subroutine.
!
!   p, q          - are INTENT ( IN ) arguments of type LOGICAL
!                   used as operands in XOR, NOR and NAND
!
!   Result        - is an INTENT ( OUT ) argument of type LOGICAL
!                   used to hold the results of XOR, NOR or NAND
!
!   Operation     - is an INTENT ( OUT ) argument of type CHARACTER
!                   used to hold the operation (XOR, NOR or NAND)
!
!   Choice        - is a local variable of type INTEGER used to hold
!                   the user's menu choice
!-------------------------------------------------------------------
!
       LOGICAL,                    INTENT ( IN )  ::    p, q
       LOGICAL,                    INTENT ( OUT ) ::    Result
       CHARACTER ( LEN = 4 ),      INTENT ( OUT ) ::    Operation
!
       INTEGER                                    ::    Choice
!
            PRINT *, "1. XOR"
            PRINT *, "2. NOR"
            PRINT *, "3. NAND"
            PRINT *, "     Please enter 1, 2 or 3"
              READ *, Choice
            PRINT *, " "                  ! leaves a blank line
            PRINT *, " "                  ! leaves a blank line
!
            IF ( Choice == 1 ) THEN
                CALL XOR ( p, q, Result )
                Operation = "XOR"
            END IF
            IF ( Choice == 2 ) THEN
                CALL NOR ( p, q, Result )
                Operation = "NOR"
            END IF
```

continues

Figure 6.11 (continued)

```
            IF ( Choice == 3 ) THEN
                CALL NAND ( p, q, Result )
                Operation = "NAND"
            END IF
!
RETURN
END SUBROUTINE   Menu
```

```
            IF ( Choice == 1 ) THEN
                CALL XOR ( p, q, Result )
                Operation = "XOR"
            END IF
            IF ( Choice == 2 ) THEN
                CALL NOR ( p, q, Result )
                Operation = "NOR"
            END IF
            IF ( Choice == 3 ) THEN
                CALL NAND ( p, q, Result )
                Operation = "NAND"
            END IF
```

Choice is being tested for three values: 1, 2 and 3. If Choice contains a 1, the block inside the first IF statement is executed and the next two IF statements are skipped. If Choice contains a 2, the first and third IF statements are skipped and the block of statements inside the second IF are executed, and so on. The series of IF statements above tests for different cases of variable Choice.

Fortran supplies a specialized selection structure for testing a series of cases for the same variable or expression against **constants** or **ranges of constants**. It is called the SELECT CASE statement. The general form of a SELECT CASE statement is shown in Figure 6.12. In the SELECT CASE statement in Figure 6.12, the square brackets again surround optional parts of the definition. The **ellipsis** means that a programmer may

Figure 6.12

```
SELECT CASE ( test-expression )
    CASE ( set-expression-1 )
        block-1
    [ CASE ( set-expression-2 )
        block-2 ] . . .
    [ CASE DEFAULT
        block-n ]
END SELECT
```

include as many CASEs as needed. CASE DEFAULT is analogous to the ELSE part of a block IF; if all CASEs fail, the CASE DEFAULT block will be executed, assuming it was included in the SELECT CASE statement.

SELECT CASE works as follows: the test-expression is compared to each set-expression in each CASE in the order they appear from top to bottom. If the value of test-expression is found to be a member of the set defined by set-expression, the corresponding block of statements will be executed. **No other block will be executed**. A warning . . . none of the set-expressions may overlap among all the CASEs in any one SELECT CASE statement.

Test-expression may be a variable or expression of type INTEGER, CHARACTER or LOGICAL **ONLY**. The set-expressions must be the same type as test-expression and may have one of the following forms:

Ranges	1.	(lower bound : upper bound)—is test-expression between the lower and upper bounds inclusively?
	2.	(lower bound :)—is test-expression greater than or equal to the lower bound?
	3.	(: upper bound)—is test-expression less than or equal to the upper bound?
Individual Values	4.	(value-1, value-2, . . .)—is test-expression equal to any of the individual values? Note that the individual values are separated by commas.
A combination of individual values and ranges	5.	E.g., (7, 9, 13 : 17, -9).

All the ranges and individual values **must be constants** or expressions formed of constants. No ranges or values may overlap between all set-expressions in any one SELECT CASE statement.

Take a look at the example SELECT CASE statement in Figure 6.13 (p. 150). There are many problems in Figure 6.13. case-1 overlaps case-2. These set-expressions have -201 and -200 in common; they overlap. case-3, case-4 and case-5 all have 100 in common . . . NO GOOD! Figure 6.14 (p. 150) shows a well-behaved SELECT CASE statement where the set-expressions don't intersect.

Table 6.7 (p. 151) shows various values for How_Does_This_Work (Figure 6.14) and the CASE each satisfies. From Table 6.7 it is clear that only **ONE** CASE's block will be executed for any value of How_Does_This_Work.

Another interesting use of the SELECT CASE statement is comparing CHARACTER data. In Figure 6.15 (p. 151), a user's input is verified, which is called **error checking**, by a SELECT CASE statement. Error checking is, as mentioned before, having a program check for bad data. You never know what users will type as input values or if data has been correctly entered in a data file that is being read. Always verify incoming data whenever possible.

Figure 6.13

```
      . . .
INTEGER       ::       How_Does_This_Work
    PRINT *, "Please enter an integer value"
    READ *, How_Does_This_Work
!
    SELECT CASE ( How_Does_This_Work )
        CASE ( : -200 )            ! case-1
           !block-1
        CASE ( -201 : 0 )          ! case-2
           !block-2
        CASE ( 22 : )              ! case-3
           !block-3
        CASE ( 18, 9, 43, 100 )    ! case-4
           !block-4
        CASE ( 47, 51, 99 : 108, -17, 707 : )
           !block-5            ! case-5 ^^^
        CASE DEFAULT             ! case-6
           !block-6
    END SELECT
      . . .
```

Figure 6.14

```
PROGRAM   Test_Set_Expressions
         IMPLICIT  NONE
!
    INTEGER     ::       How_Does_This_Work
!
       PRINT *, "Please enter an integer value"
       READ *, How_Does_This_Work
!
       SELECT CASE ( How_Does_This_Work )
           CASE ( : -200 )              ! case-1
              PRINT *, "block-1"
           CASE ( -199 : 0 )            ! case-2
              PRINT *, "block-2"
           CASE ( 22 : )                ! case-3
              PRINT *, "block-3"
           CASE ( 18, 9 )               ! case-4
              PRINT *, "block-4"
           CASE ( 1 : 5, 10 : 13 )      ! case-5
              PRINT *, "block-5"
           CASE DEFAULT                 ! case-6
              PRINT *, "block-6"
       END SELECT
!
END PROGRAM    Test_Set_Expressions
```

Table 6.7

How_Does_This_Work	Case # satisfied
-7000	1
-200	1
17	6
-31	2
100	3
12	5
9	4

Figure 6.15

```
   . . .
PRINT *, "Please enter Y or N"
READ  *, Yes_Or_No
!
SELECT CASE ( Yes_Or_No )
    CASE ( "Y", "y" )
        CALL Do_It
    CASE ( "N", "n" )
        CALL Dont_Do_It
    CASE DEFAULT
        PRINT *, "*** INVALID INPUT ***"
END SELECT
```

Important Concepts Review

- A **block** IF conditionally executes blocks of statements.

- ELSE is optional in block IFs and multialternative IFs. If ELSE is included, its block is executed if all other conditions in the selection structure fail.

- **Multialternative** IFs chain conditions together in an "either-or" fashion. Only one block at most will be executed. If more than one condition is true, only the block corresponding to the first true condition will be performed.

- A SELECT CASE statement is a selection structure that checks a test-expression to see if its value is contained in any of the set-expressions.

- A test-expression may be a variable or an expression of type INTEGER, CHARACTER or LOGICAL.

- `Set-expressions` must be the same type as `test-expression` and may be:
    1. A range of constants.
    2. Individual constants separated by commas.
    3. A combination of individual constants and ranges of constants.

- All `set-expressions` **must be mutually exclusive** within any `SELECT CASE` statement.

EXERCISES

6.1 Rewrite subroutine `Menu` from Figure 6.11 using a `SELECT CASE` selection structure.

6.2 Design a program that will read three values: an arithmetic operator, `CHARACTER ( LEN = 2 )`, and two `REAL` operands. Use a `SELECT CASE` to figure out what the operator is and perform the appropriate operation (`**`, `*`, `/`, `+`, `-`). The following subroutine is supplied:

```
SUBROUTINE  Get_Expression ( Operation, Operand_1, Operand_2 )
     IMPLICIT NONE
!
!--------------------------------------------------------------------
!  The purpose of this subroutine is to read in an arithmetic operation
!     and two REAL operands.
!
!  Operation       -a CHARACTER ( LEN = 2 ) variable to hold the
!                     arithmetic operation
!
!  Operand_1, Operand_2 - The operation represented in variable
!                         Operation will be performed on these 2 REAL
!                         variables
!--------------------------------------------------------------------
!
     CHARACTER ( LEN = 2 ),   INTENT ( OUT )  ::  Operation
     REAL,                    INTENT ( OUT )  ::  Operand_1, Operand_2
!
         PRINT *, "Please enter the arithmetic operation &
                                        &(**, *, /, +, -)"
         READ  *, Operation
         PRINT *, "Please enter two REAL values"
         READ  *, Operand_1, Operand_2
!
     RETURN
     END SUBROUTINE   Get_Expression
```

6.3 Modify subroutine `Get_Expression` from Exercise 6.2 so that it verifies the contents of variable `Operation` and ensures that it contains a legitimate arithmetic operator. If the value in `Operation` is not valid, store "NG" in variable `Operation`.

6.4 Modify the main program in Exercise 6.2 so that it doesn't attempt to perform any operation if the value returned in variable `Operation` from subroutine `Get_Expression` is "NG".

6.5 The general form of a quadratic equation is

$$y = ax^2 + bx + c$$

The roots of a quadratic equation are

$$\frac{-b \pm \sqrt{b^2 - 4ac}}{2a}$$

The discriminant is

$$b^2 - 4ac$$

If the discriminant is negative, the roots are imaginary because the parabola does not cross the x axis. If the discriminant is zero, there is one "twin" root. If the discriminant is positive, there are two real roots. Assume that the value of the discriminant for a quadratic equation is stored in variable `Discr`. Write a multialternative `IF` to print out the equation's root type.

6.6 Write the selection structure in Exercise 6.5 as a `SELECT CASE`.

6.7 Design a program to find which two out of three Cartesian points are closest to each other.

6.8 According to Archimedes' principle, a boat floats only if it weighs less than the water it displaces. Design a program that will determine whether a boat made out of a given material will float or not. Assume water weighs 0.9 kg/l and that the volume of water displaced by a boat is described by the following formula:

$$\frac{\pi w^2 l}{4}$$

l is the length of the boat in meters from bow to stern and w is the boat's width in meters. The surface area of the boat is given by the following formula:

$$\frac{\pi w l}{2}$$

The hull is 3 cm thick. Read in l, w, the name of the construction material and its density in g/cm. Print all relevant information and whether the boat will float.

6.9 Write a selection structure(s) that will find the smallest of five `REAL` numbers. *(Hint: maintain a `"Current_Smallest"` and compare each of the real numbers to it.)*

6.10 Read in four binary digits into `CHARACTER ( LEN = 1 )` variables `bin_1`, `bin_2`, `bin_3` and `bin_4`. Assume these four variables comprise a binary number where bin_1 contains the most significant digit and bin_4 the least significant digit. Using the selection structure(s) of your choice, convert the binary number to a base 10 integer.

6.11 Use selection structure(s) to ensure that the CHARACTER data being read in Exercise 6.10 is only "0" or "1". If incorrect data is read, print an error message and don't attempt to convert the errant binary number.

6.12 In the following program segment, what gets printed?

```
              INTEGER,         PARAMETER :: v1 = 34, v2 = 17
              INTEGER                    :: r1
              REAL                       :: r2
              r1 = v2 /v1
              SELECT CASE ( r1 )
                  CASE ( 1 : )
                      r2 = 3.1415927
                  CASE ( : 0 )
                      r2 = 2.7182818
                  CASE DEFAULT
                      r2 = 0.0
              END SELECT
              IF ( r2 <= 1.0 .AND. 4 * v2 / v1 == 3 ) THEN
                  PRINT *, "Results #1"
              ELSE IF ( r2 <= 1.0 .OR. 4 * v2 / v1 == 3 ) THEN
                  PRINT *, "Results #2"
              ELSE IF (v2 < 20 ) THEN
                  PRINT *, "Results #2 and Results #3"
              ELSE
                  PRINT *, "Results #4"
              END IF
```

6.13 What is wrong with the following SELECT CASE statement?

```
              INTEGER          ::      lower, upper, Test_It = 50
          !
              READ *, lower, upper
              SELECT CASE ( Test_It )
                  CASE ( lower : upper )
                      PRINT *, "Test_It is within range"
                  CASE DEFAULT
                      PRINT *, "Test_It is NOT within range"
              END SELECT
```

6.14 Paleontologists study fossils in an attempt to understand the ancestry of today's life forms. One of the methods they use to help see the relationships between species is clatograms. Clatograms are diagrams that show related species based on physical and environmental characteristics. If it has feathers, it's a bird. If it's cold blooded and has scales, it's a reptile; WRONG! It could be a fish. Design a program to classify an unknown animal. The values read into the program are: 1. A CHARACTER variable that contains the environment ("WATER", "NOT WATER", or "DON'T KNOW"), 2. One that contains body covering ("HAIR", "SCALES", "FEATHERS" or "DON'T KNOW") and 3. An INTEGER variable that contains the number of legs (0, 2, 4 or -1 representing "DON'T KNOW"). Design selection structures that determine, based on the values in the above three variables, whether the animal is a land mammal, a sea mammal, a duckbilled platypus, a bird, a reptile, a fish or can't be determined. Assume there are no reptiles that live in water.

6.15 What is the difference between a block and a block statement?

6.16 What is wrong with the following SELECT CASE statement?

```
INTEGER             ::      N, Counter_1 = 0
!
READ *, N
SELECT CASE ( N )
    CASE ( : -1 )
        DO WHILE ( Counter_1 < N )
            N = N + 1
            PRINT *, N
    CASE ( 0 )
        N = N - 1
        PRINT *, N
    CASE ( 1 : )
        N = N
        PRINT *, N
    CASE DEFAULT
            Counter_1 = Counter_1 + 1
        END DO
END SELECT
```

Looping Structures

Introduction

Things are about to explode with possibilities, because this chapter introduces the power of looping structures. They are the last of the four fundamental flow-control structures. *All procedural programs* can be written using some combination of the four basic flow-control structures: CALL, Sequential, Selection and Looping. Each structure is equally important. For instance, whole sets of problems can't be solved if selection structures aren't available. In each of the preceding three chapters at least one of the four structures was unavailable, which limited the programs that could be designed.

So why are looping structures numbered among the notable four fundamental structures? It's because looping allows you to repeat blocks of structures in a controlled way rather than reproducing the blocks multiple times. Consider the following program segment:

```
PRINT *, "Please enter length, width"
READ  *, Length, Width
Area = Length * Width
PRINT *, "Area =", Area
```

The four statements calculate the area of some two-dimensional rectangle. How about doing it three times? We could run the program three times or we could construct the algorithm seen below:

7.1 DO WHILE **loops**

7.2 Error checking

7.3 Reading an unknown number of values

Important Concepts Review

7.4 Series summation

7.5 Design example

7.6 Libraries revisited

Important Concepts Review

Exercises

156

Introduction

```
                PRINT *, "Please enter length, width"
                READ  *, Length, Width
                Area = Length * Width
                PRINT *, "Area =", Area
!
                PRINT *, "Please enter length, width"
                READ  *, Length, Width
                Area = Length * Width
                PRINT *, "Area =", Area
!
                PRINT *, "Please enter length, width"
                READ  *, Length, Width
                Area = Length * Width
                PRINT *, "Area =", Area
```

This modification is not too much trouble with a little editor cut-and-paste, but what if we need to calculate the areas for 1,000 sets of *lengths* and *widths* or worse yet, *n* sets? This would be either cumbersome or impossible to do. You solve this kind of problem with a looping structure, such as the one below:

```
PRINT *, "Please enter the # of data sets"
READ  *, N
!
Counter = 1
!
DO WHILE ( Counter <= N )
     PRINT *, "Please enter length and width"
     READ  *, Length, Width
     Area = Length * Width
     PRINT *, "Area =", Area
!
     Counter = Counter + 1    ! Counter "counts up" to N
END DO
```

Looping structures allow you to execute a block of statements over and over until some condition is met. ● ● ●

7.1 DO WHILE **Loops**

As seen in the introduction, DO WHILE loops enable a block of statements to be repeated as many times as needed. Blocks are any reasonable collection of statement structures and they may follow or be placed completely within other blocks. Fortran's basic looping structure is the `DO WHILE` loop, which is a block statement; it operates on blocks. Syntax for the general form of a `DO WHILE` loop is illustrated in Figure 7.1. Simply put, the loop continues **iterating** (repeating) and doing "`block-of-statements`" as long as "`condition`" is true. This type of loop is called a **conditional loop** because it continues to iterate as long as its condition is satisfied.

Condition has the same meaning in a `DO WHILE` loop as it does in the selection structures explained in Chapter 6; it is a logical expression. Relational operators (`==`, `/=`, `<`, `>`, `>=`, `<=`) and logical operators (`.NOT.`, `.AND.`, etc.) together with operands form **logical expressions**. A loop's logical expression is evaluated, and the loop continues looping or stops depending on the value of its logical expression.

The `DO WHILE` structure is also called a **top-tested** or **pretested** loop. This means that when program flow encounters the loop, the loop's condition is tested to see whether it is true or false **before** flow enters the loop. If the condition is false, the block of statements inside the loop is skipped and flow is continued at the statement immediately following `END DO`. If the condition is true, program flow enters the loop and its block of statements is executed repeatedly until the condition becomes false (Figure 7.2).

`DO WHILE` is the top of the loop and the bottom is `END DO`. `END DO` is the last line of the `DO WHILE` block statement and serves two purposes:

1. It marks the physical end of the loop.
2. It transfers flow back to its corresponding `DO WHILE` so that the condition can be tested again to decide whether the block of statements contained in the loop should be performed another time.

Looping continues until the condition becomes false. What does that imply? It suggests that *something inside the loop must eventually change the value of the condition* so the loop can end. In the loop of Figure 7.3, nothing inside the loop affects the condition.

N is given a value of 10 just before getting to the loop, and the loop's condition is `N > 0`. The condition is true because `10 > 0`, so control passes into the loop where "`I love Fortran`" is printed. `END DO` sends control back to the `DO WHILE` where `N > 0` is tested again, and the condition is still true since N still has a value of 10. Control again passes into the loop.

Nothing inside the loop ever changes the value of N, so `N > 0` will ALWAYS be true and the loop will never end. This kind of loop is aptly called an **infinite loop**. The loop in Figure 7.3 could be corrected as seen in Figure 7.4. Infinite loops can be tricky to identify, especially if there are no `PRINT` statements inside. When a programmer sits in front of his terminal saying, "The system sure is slow," an infinite loop should be suspected.

Figure 7.1

```
DO WHILE ( condition )

        block-of-statements

END DO
```

figure 7.2

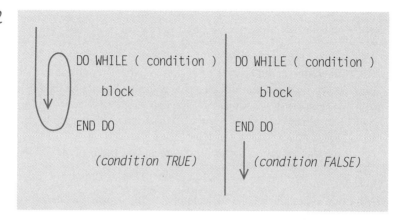

figure 7.3

```
N = 10
!
DO WHILE ( N > 0 )
     PRINT *, "I love Fortran"
END DO
```

To diagnose an infinite loop, a programmer should first **break out of the program**. The way this is done varies from compiler to compiler. Most compilers will print messages indicating approximately where the program was terminated. If the location is inside or near a DO WHILE loop, make sure the condition CAN become false so the loop can end. Put a PRINT statement inside the loop that displays the values involved in the condition and run the program again. It will become clear what went wrong. There is an in-depth discussion of error removal, which is called **debugging**, in Appendix C.

figure 7.4

```
N = 10
!
DO WHILE ( N > 0 )
     PRINT*, "I love Fortran"
     N = N - 1
                    ! N is reduced by 1 each time
                    ! the loop iterates. After
                    ! 10 passes through the loop,
                    ! N will be zero, making the
                    ! condition false and ending
                    ! the loop.
END DO
```

7.2 Error Checking

Error checking is the process of verifying, whenever possible, that any value entering a program from an external source, such as the keyboard or a data file, is correct. Programs that don't use error checking can behave in undesirable ways. In Figure 7.5, an algorithm that does not use error checking is shown. A block IF is used to decide which subroutine to call based on a response typed by a user.

If the user types a "Y", the IF statement will perform the following CALL:

CALL The_Yes_Option

If the user types an "N", the IF statement will perform the other CALL:

CALL The_No_Option

This seems to work properly, but what if the user types "J"? The IF statement will have a false condition because Yes_Or_No is not equal to "Y", and therefore the ELSE block, CALL The_No_Option, will be executed. What did the user mean by typing "J"? Did he mean "Y" or "N"? Obviously it's not clear, but the program is behaving as if the user had typed "N". Thus the algorithm in Figure 7.5 is unacceptable.

Good programming techniques dictate that before any action is taken based on a value entered by a user, that value must be checked for errors if possible. The general form, in English terms, for error checking a value typed by a user is illustrated in Figure 7.6.

This is a straightforward algorithm. A user types in his response to a prompt and, if it's not valid, the loop is entered where an error message is displayed and another response is requested. The loop will keep looping until an acceptable response is entered *making the condition false* and ending the loop.

A loop is used in error checking algorithms because users may misenter values more than once. When control finally passes through the bottom of the error checking loop, the response received IS valid; there is no way to exit the loop other than typing an acceptable response.

Figure 7.5's errant algorithm could be fixed as seen in Figure 7.7. The first CALL followed by the DO WHILE accomplishes the error checking and ensures that the value contained in Y_N will be either "Y", "y", "N" or "n". A SELECT CASE then tests the value

Figure 7.5

```
PRINT *, "Please enter Y or N"
READ *, Yes_Or_No
!
IF ( Yes_Or_No == "Y" ) THEN
    CALL The_Yes_Option
ELSE
    CALL The_No_Option
END IF
```

7.2 Error Checking

Figure 7.6

```
PROMPT FOR A VALUE
READ A RESPONSE FROM THE USER

DO WHILE IT'S A BAD RESPONSE

    PRINT A HELPFUL ERROR MESSAGE
    PROMPT FOR A NEW VALUE
    READ ANOTHER RESPONSE FROM THE USER

END DO
```

Figure 7.7

```
PROGRAM Error_Check
        . . .
    CALL Get_A_Response ( "Please enter Y or N", Y_N )
!
    DO WHILE ( Y_N /= "Y" .AND. Y_N /= "y" .AND. Y_N /= "N" .AND. Y_N /= "n" )
        PRINT *, " *** INCORRECT RESPONSE *** Please enter either Y or N"
            CALL Get_A_Response ( "Please enter Y or N", Y_N )
    END DO
!
!   At this point, the response is a valid one
!
    SELECT CASE ( Y_N )
        CASE ( "Y", "y" )
            CALL The_Yes_Option
        CASE ( "N", "n" )
            CALL The_No_Option
    END SELECT
        . . .
END PROGRAM  Error_Check
!--------------------------------------------
SUBROUTINE  Get_A_Response ( Prompt, Response )
        . . .
    PRINT *, Prompt
    READ  *, Response
        . . .
RETURN
END SUBROUTINE
        . . .
```

in Y_N and calls the appropriate subroutine to accomplish further tasks. There are other situations that require error checking that have different forms. They will be explained as they come up in the text.

7.3 Reading an Unknown Number of Values

In the introduction to this chapter, an example of calculating *area* from *length* and *width* was demonstrated (Figure 7.8). Figure 7.8's algorithm for doing n sets of data is problematical. It requires a user to know exactly how many areas he needs to calculate before doing any calculations. This isn't convenient if there are so many sets of data that they can't be counted at a glance. One solution to this problem is to allow users to start entering data without concern for how much there will be, and to have the program end when either Length or Width is entered as a negative value. These negative values are used to "**flag**" the end of **input** (values entering the program). Using negative values as flags is sensible for the example in Figure 7.8 because valid *lengths* and *widths* are always positive. This "flag" method of stopping input is achieved with a conditional loop. Consider the example in Figure 7.9.

Figure 7.9 shows an elegant use of conditional loops and illustrates some very interesting and useful programming techniques:

1. How to process an **arbitrary** amount of data.
2. How to give users an **"anytime" escape** from an algorithm.
3. The **standard input algorithm**.

Technique 1 was already explained. Technique 2, offering users an escape, is a vital feature of well-designed programs. In Figure 7.8, a user is asked to enter the number of times he wants to perform the *area* calculation. What if he enters 50 and decides that after 20 calculations he wants to stop? (Users are unpredictable.) Figure 7.8's program will

Figure 7.8

```
PRINT *, "Please enter the # of data sets"
READ *, N
!
Counter = 1
!
DO WHILE ( Counter <= N )
     PRINT *, "Please enter length and width"
     READ *, Length, Width
     Area = Length * Width
     PRINT *, "Area =", Area
!
     Counter = Counter + 1 ! Counter "counts up" to N
END DO
```

Figure 7.9

```
PROGRAM   Flag_End_Of_Input
          IMPLICIT   NONE
!
!-------------------------------------------------------------------------
! Written by       :   C. Forsythe
! Date Written     :   3/25/2004
!-------------------------------------------------------------------------
! The purpose of this program is to demonstrate "flagging" the end of
!    "input" by "watching" for a flag value (any negative number
!    in this program) by using a conditional loop.
!
! Length and Width are REAL variables
!
! Area is a REAL variable that holds the value of Length * Width
!-------------------------------------------------------------------------
!
      REAL              ::       Length, Width, Area
!
      PRINT *, "Please enter length and width"
      READ  *, Length and Width
!
      DO WHILE ( Length >= 0.0. .AND. Width >= 0.0 )
          Area = Length * Width
          PRINT*, "Area =", Area
!
          PRINT *, "Please enter length and width"
          READ *, Length, Width
      END DO
!
END PROGRAM   Flag_End_Of_Input

************** PROGRAM RESULTS ***************

 Please enter length and width
4.2 3                <----------- Typed by the user
 Area =  12.5999994
 Please enter length and width
34 34                <----------- Typed by the user
 Area =    1.1560000E+03
 Please enter length and width
.0001 .00006         <----------- Typed by the user
 Area =    5.9999996E-09
 Please enter length and width
-1 -1                <----------- The program stops after
                                  the user typed this line.
```

force users to respond 30 more times—no good. On the other hand, Figure 7.9's program allows a user to enter the flag value any time to stop input. If the flag is entered as the first data set for *length* and *width*, the program will end without doing any calculations.

Technique 3, the **standard input algorithm**, is what facilitates techniques 1 and 2. Examine the standard input algorithm stated in clear English terms in Figure 7.10. Read a value . . . start a conditional loop whose condition is that the just-read value isn't the flag . . . process the value, and just before bouncing back to the top of the loop, read another value. This method of reading data will be seen throughout the text.

Figure 7.10

```
READ A VALUE
DO WHILE ( THE VALUE ISN'T THE "STOP-THIS-LOOP" FLAG )
       PROCESS THE VALUE
           . . .
       READ THE NEXT VALUE
END DO
```

Important Concepts Review

- The `DO WHILE` loop enables a programmer to repeat a block of statements as many times as needed.

- **Conditional loops** continue to iterate as long as their condition is satisfied.

- **Iterate** means to repeat.

- Loop **conditions** are logical expressions constructed the same as `IF` statement logical expressions.

- **Top-tested** or **pretested** loops are conditional loops that have the condition at the top of the loop. This is the only type of conditional loop that Fortran offers.

- `END DO` is the physical bottom of the loop and returns control to the top of the loop.

- An **infinite loop** is when a loop condition never becomes false.

- **Something inside a conditional loop** must have the potential to change the condition of the loop or an infinite loop will result.

- **Error checking** is the act of preventing erroneous data from entering a program.

- When requesting input from users, **keep asking** until an acceptable response is typed—use a conditional loop.

- An **algorithm** is a method of solution that provides a step-by-step solution to a problem.

- **Syntax** is grammar and determines the correct way to construct program statements.

- Conditional loops facilitate something being done an **unknown** number of times, such as reading an unknown amount of data.

- A **flag** is a specialized value that can be "watched" for by a conditional loop.

7.4 Series Summation

Series summation is one of the most useful algorithmic skills that a beginning program designer can develop. Through the next several pages, the idea of writing program statements to preform series summation will be developed from three principles: the mathematical definition of summation, the incremental assignment statement, and looping structures.

A general mathematical expression for summation is shown below:

$$\sum_{i=1}^{n} t_i$$

Sigma (Σ) is the mathematical symbol for summation and is defined as follows:

$$\sum_{i=1}^{n} t_i = t_1 + t_2 + t_3 + \cdots + t_{n-1} + t_n$$

The term t_i is the general ith term of some series. Term t_i may be anything from a simple numeric constant to a very involved arithmetic expression.

To accomplish summation with a computer program, the **incremental assignment statement** is generalized. It is extremely important that the incremental assignment is clearly understood. Figure 7.11 is a reproduction of Figure 4.3 and illustrates how the incremental assignment statement works. `Counter_3 = Counter_3 + 1` is the incremental assignment statement. It increments whatever value is in `Counter_3` by 1. Essentially, the "old" `Counter_3` value is involved in an expression, `Counter_3 + 1` in this example, whose ultimate value is stored into `Counter_3` creating a "new" `Counter_3` value. `Counter_3`'s old value is destroyed. This can be symbolized as:

$$\text{Counter\_3}_{new} = \text{Counter\_3}_{old} + 1$$

(Note: you wouldn't type this into a program—there is no concept of subscripts such as "old" and "new." This is only a conceptualization.)

Figure 7.11

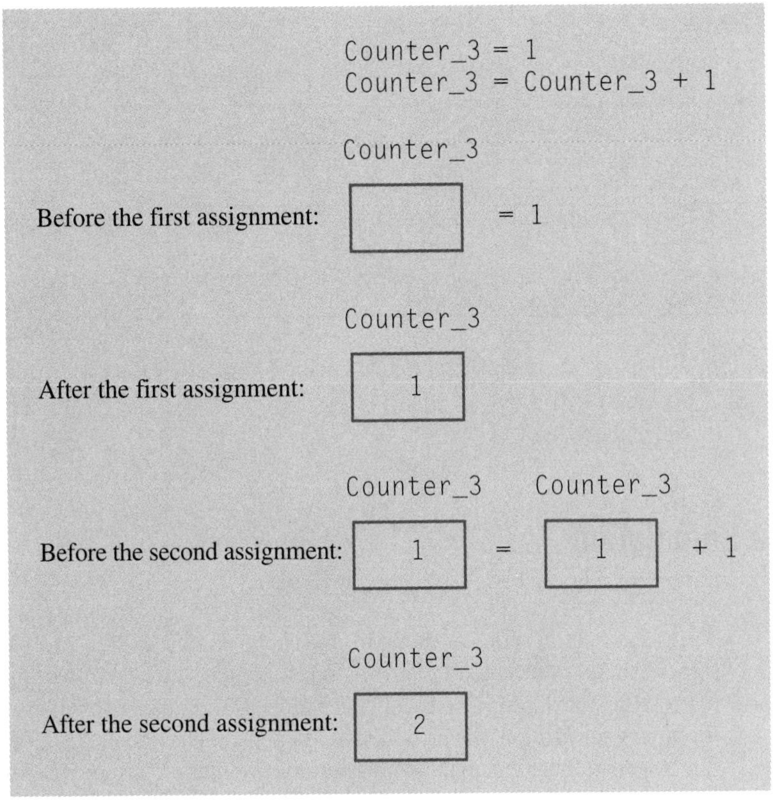

```
Counter_3 = 1
Counter_3 = Counter_3 + 1
```

Before the first assignment: Counter_3 [ ] = 1

After the first assignment: Counter_3 [1]

Before the second assignment: Counter_3 [1] = Counter_3 [1] + 1

After the second assignment: Counter_3 [2]

Counter_3 + 1 becomes the new value for Counter_3 ... the 1 has been summed (accumulated) into Counter_3. Take this idea one step further and consider the following block of assignment statements:

```
Counter_3 = 0
Counter_3 = Counter_3 + 1
Counter_3 = Counter_3 + 1
Counter_3 = Counter_3 + 1
```

The first assignment statement sets Counter_3 to 0. The next assignment statement sums a 1 into Counter_3 making it 1; the third assignment sums another 1 into Counter_3 changing it to 2, and the last statement adds yet another 1 into Counter_3, generating its final value of 3. This works as shown in Figure 7.12.

Each of the underlined sums on the right side of Figure 7.12 represents the current value of Counter_3 before the assignment is done, which are the combined results of all previous accumulative assignments. The block of assignment statements is summing (accumulating) a **series** of ones: 1+1+1. This could be written mathematically as follows:

$$\sum_{i=1}^{3} 1_i = 1_1 + 1_2 + 1_3$$

Figure 7.12

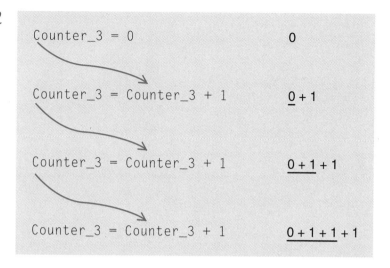

The summation above is a specific case of the general series summation formula seen before where t_i is 1:

$$\sum_{i=1}^{n} t_i = t_1 + t_2 + t_3 + \cdots + t_{n-1} + t_n$$

Figure 7.12's block of assignment statements could also be generalized by using a variable, t, to hold the 1 in each assignment (Figure 7.13). Finally, there are three occurrences of the assignment statement:

```
Counter_3 = Counter_3 + t
```

So why type it three times? Whenever the same statement or block of statements occurs several times in succession, **loops** should occur to the programmer. Create a loop to do the assignment statement n times. In this example, n is only 3, but it could easily be a much larger value (Figure 7.14, p. 168).

Figure 7.14's algorithm is moderately more complicated than Figure 7.13's, but Figure 7.14's is much more generalized. It can be easily modified to sum a different number of 1s by simply changing the value of n. The looping structure in Figure 7.14 can be mathematically represented as follows (think of the general *i*th *t* as 1).

$$\sum_{i=1}^{n} t_i = t_1 + t_2 + t_3 + \cdots + t_{n-1} + t_n$$

Figure 7.13

```
Counter_3 = 0
t = 1
Counter_3 = Counter_3 + t
Counter_3 = Counter_3 + t
Counter_3 = Counter_3 + t
```

Figure 7.14
```
Counter_3 = 0
t = 1
n = 3
LoopCounter = 1
DO WHILE ( LoopCounter <= n )
      Counter_3 = Counter_3 + t
      LoopCounter = LoopCounter + 1
            !
            ! LoopCounter is used to
            ! keep track of how many
            ! times the loop has
            ! repeated. Notice how
            ! it is used in the
            ! loop condition.
END DO
```

The summation at the bottom of p. 167 should look familiar, as it is the one shown at the beginning of the discussion on series summation. The progression of steps in this section, from the basic incremental assignment statement through the looping structure in Figure 7.14, leads to the following splendid conclusion:

$$\sum_{i=1}^{n} t_i \equiv$$
```
      Series_Accumulator = 0
      i = 1
      DO WHILE ( i <= n )
            Series_Accumulator = Series_Accumulator + t
            i = i + 1
      END DO
```

N is the number of terms to sum and t is the general ith term of the series. T **can be an expression of arbitrary complexity**.

7.5 Design Example

In this design example, a program to calculate e^x will be designed. (e is the base of natural logarithms: $\approx 2.718281828459045....$) e^x is defined by the following series:

$$\sum_{i=0}^{n} \frac{x^i}{i!}$$

"$i!$" means "i factorial." $i!$ is mathematically defined as follows:

$$1 \cdot 2 \cdot 3 \cdot \ldots \cdot (i-1) \cdot i$$

0! is defined to be 1.

Program Design

STATE THE PROBLEM:

I want to design a program to accumulate n terms of the series for e^x. The values for x and n come from the keyboard. (I can probably reuse a previously written generalized input subroutine.) A subroutine to calculate factorials has been supplied. When printing the results, I will also print out the value of e^x using the built-in (intrinsic) Fortran function EXP(X) as a comparison. For extra accuracy, I will use DOUBLE PRECISION variables. The main program should have a loop that ends when the value read for n is negative.

STRUCTURE CHART:

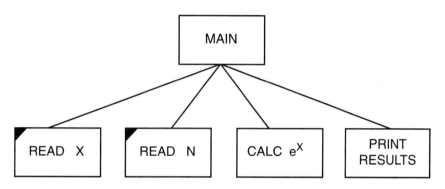

(Note: the shaded corners of the left two structure chart blocks indicate a "shared" program unit—the same program unit is being used in two places with different actual arguments.)

MAIN:

MAIN's purpose is to drive the program by calling the four subroutines under it. When n becomes negative, end the program.

List:

1. Read the first n (*standard input algorithm*).
2. Start a loop that continues as long as n is not negative.
3. If n is $>=$ 0, perform steps 4, 5, 6, 7 and 8.
4. Call a subroutine to get a value for x.
5. Call a subroutine to calculate e^x.
6. Call a subroutine to print the results.
7. Call the subroutine to get the next value for n.
8. Go back to the top of the loop.

Data flow to and from the calling program:

None (MAIN isn't called).

English Fortran:

```
┌─ PROGRAM Series_Summation
│      declarations
│      CALL Get_Generic_INTEGER_Input    n
│  ┌─ DO WHILE ( n >= 0 )  ! end if n < 0
│  │      CALL Get_Generic_INTEGER_Input    x
│  │      CALL Calculate_e2x
│  │      CALL Print_Results
│  │      CALL Get_Generic_INTEGER_Input    n
│  └─ END DO
└─ END PROGRAM Series_Summation
```

GENERIC INTEGER INPUT:

No design necessary; I just stole the subroutine from Figure 7.7:

```
SUBROUTINE   Get_Generic_INTEGER_Input ( Prompt, Response )
             IMPLICIT   NONE
!
!---------------------------------------------------------------
!  This subroutine presents the prompt contained in CHARACTER variable
!     Prompt and then reads a value from the keyboard into INTEGER
!     variable Response.
!---------------------------------------------------------------
!
      CHARACTER ( LEN = * ),     INTENT ( IN )    ::    Prompt
      INTEGER,                   INTENT ( OUT )   ::    Response
!
      PRINT *, Prompt
      READ  *, Response
!
RETURN
END SUBROUTINE   Get_Generic_INTEGER_Input
```

This subroutine will be used to get values for both x and n.

CALCULATE e^x:

The purpose of this subroutine is to sum up the first n terms of the series for *e*, the base of natural logarithms.

List:

1. Set a **series-accumulation** variable to zero.
2. Set a **loop-counter** variable named i to zero.

3. Start a DO WHILE loop whose condition for continuing is that i is less than n (i will go from 0 to n-1 ... a total of n times).
4. Get the factorial of i.
5. Make an expression for the ith term of the series for e^x.
6. Write an assignment statement inside the loop that will accumulate terms of the series.
7. Increment i, the loop counter.

Data flow to and from the calling program:

x and n come in from the calling program.
e2x is sent back to calling program.

English Fortran:

```
SUBROUTINE Calculate_e2x ( x, n, e2x )
   declarations
      Series_Accumulator = 0
      i = 0
      DO WHILE ( i < n )   ! i goes from 0 to n-1
            CALL Factorial ( i, i_Factorial )
            t = x ** i / i_Factorial   ! general ith term for e^x.
            Series_Accumulator = Series_Accumulator + t
            i = i + 1
      END DO
   RETURN
END SUBROUTINE   Calculate_e2x
```

PRINT RESULTS:

The purpose of this subroutine is to print the results of the calculated value for e^x and Fortran's value for e^x together with n and x.

List:

1. Print n.
2. Print x.
3. Print the calculated e^x.
4. Print Fortran's e^x.

Data flow to and from the calling program:

x, n and e2x come from the calling program.

English Fortran:

```
 ┌─ SUBROUTINE  Print_Results ( x, n, e2x )
 │        Print "n = ", n
 │        Print "x = ", x
 │        Print "series e ** x = ", e2x
 │        Print "Fortran's e ** x = ", EXP ( x )
 │  RETURN
 └─ END SUBROUTINE  Print_Results
```

The Complete Program

```
PROGRAM  Series_Summation
         IMPLICIT   NONE
!
!-------------------------------------------------------------------
! Written by    :   C. Forsythe
! Date Written  :   3/25/2004
!-------------------------------------------------------------------
! The purpose of this program is to demonstrate the process of series
!    summation.  The series for e ** x is: x**0/0!+x**1/1!+...+x**n/n!
!
! n            is the number of terms to sum - INTEGER
!
! x            is the power e is raised to - INTEGER
!
! e2x          holds the calculated value for e ** x - DOUBLE PRECISION
!-------------------------------------------------------------------
       ! BLACK BOX:  Library subroutine Print_Blank_Lines ( n ) ...
       !
       ! The purpose of this subroutine is to print n blank lines
       !
       ! n             INTENT ( IN ) INTEGER argument
       !------------------------------------------------------------
!
       DOUBLE PRECISION        ::      e2x
       INTEGER                 ::      x, n
!
```

```
            PRINT *, "     This program calculates e ** x"
            CALL Print_Blank_Lines ( 1 )   ! prints 1 blank line
            CALL Get_Generic_INTEGER_Input &
                 ( "Please enter the number of terms ( < 0 to quit)", n )
                      ! &s allow program statements to be
                      ! continued on subsequent lines.
            DO WHILE ( n >= 0 )
                 CALL Get_Generic_INTEGER_Input &
                      & ( "Please enter an integer value for x ", x )
                 CALL Calculate_e2x ( x, n, e2x )
                 CALL Print_Results ( x, n, e2x )
                 CALL Print_Blank_Lines ( 2 )   ! prints 2 blank lines
                 CALL Get_Generic_INTEGER_Input &
                 &( "Please enter the number of terms ( < 0 to quit)", n )
            END DO
    !
    END PROGRAM  Series_Summation
!===============================================================================
    SUBROUTINE  Get_Generic_INTEGER_Input ( Prompt, Response )
                IMPLICIT    NONE
    !
    !-------------------------------------------------------------------
    !  This subroutine presents the prompt contained in CHARACTER argument
    !     Prompt (in) and then reads a value from the keyboard into INTEGER
    !     argument Response (out).
    !-------------------------------------------------------------------
    !
        CHARACTER ( LEN = * ),    INTENT ( IN )   ::    Prompt
        INTEGER,                  INTENT ( OUT )  ::    Response
    !
            PRINT *, Prompt
            READ  *, Response
    !
    RETURN
    END SUBROUTINE  Get_Generic_INTEGER_Input
!===============================================================================
```

continues

```
SUBROUTINE Calculate_e2x ( x, n, e2x )
          IMPLICIT   NONE
!-------------------------------------------------------------------
!  The purpose of this subroutine is to sum n terms of the series
!     for e ** x.
!
!  n           is the number of terms to sum - INTEGER, INTENT ( IN )
!
!  x           is the power that e is raised to - INTEGER, INTENT ( IN )
!
!  e2x         holds the calculated value for e ** x -
!                  DOUBLE PRECISION, INTENT ( OUT )
!
!  i           is a local variable used to count the number of times
!                  the loop has repeated - INTEGER
!
!  t           is a local variable used to hold the value of the
!                  current (ith) term - DOUBLE PRECISION
!
!  Series_Accumulator  is a local DOUBLE PRECISION variable used to
!                  accumulate the series
!
!  i_Factorial is a local DOUBLE PRECISION variable used to receive
!                  the value of i!
!-------------------------------------------------------------------
     !  BLACK BOX:  Library subroutine Factorial_s ( n, n_Factorial )
     !
     !  The purpose of this subroutine is to calculate n! and store it
     !     into n_Factorial.
     !
     !  n              INTENT ( IN ) INTEGER argument
     !
     !  n_Factorial    INTENT ( OUT ) DOUBLE PRECISION argument
     !-------------------------------------------------------------------
!
     INTEGER,             INTENT ( IN )    ::   x, n
     DOUBLE PRECISION,    INTENT ( OUT )   ::   e2x
     DOUBLE PRECISION                      ::   Series_Accumulator
     DOUBLE PRECISION                      ::   i_Factorial, t
     INTEGER                               ::   i
!
```

continues

```
            Series_Accumulator = 0.0D0
            i = 0
            DO WHILE ( i < n )   ! i goes from 0 to n-1
                CALL Factorial_s ( i, i_Factorial )
                t = x ** i / i_Factorial   ! the ith term of the series
                Series_Accumulator = Series_Accumulator + t
                                            ! accumulate next term
                i = i + 1      ! count terms accumulated
            END DO
            e2x = Series_Accumulator
    !
    RETURN
    END SUBROUTINE   Calculate_e2x
!===============================================================================
    SUBROUTINE   Print_Results ( x, n, e2x )
                 IMPLICIT   NONE
    !
    !-------------------------------------------------------------------
    !  The purpose of this subroutine is to display the results of the
    !     e ** x calculation and a comparison value using Fortran's EXP()
    !     function.
    !
    !  n           is the number of terms summed - INTEGER, INTENT ( IN )
    !
    !  x           is the power that e is raised to - INTEGER, INTENT ( IN )
    !
    !  e2x         holds the calculated value for e ** x -
    !                  DOUBLE PRECISION, INTENT ( IN )
    !
    !  rx          is a REAL variable used to convert x from INTEGER to
    !                  DOUBLE PRECISION because EXP() requires a
    !                  non INTEGER numeric argument
    !-------------------------------------------------------------------
         !  BLACK BOX:  Library subroutine Print_Blank_Lines ( n ) ...
         !
         !  The purpose of this subroutine is to print n blank lines.
         !
         !  n              INTENT ( IN ) INTEGER argument
         !-------------------------------------------------------------------
    !
         INTEGER,                INTENT ( IN )    ::    x, n
         DOUBLE PRECISION,       INTENT ( IN )    ::    e2x
         DOUBLE PRECISION                         ::    rx
    !
```

continues

```
            rx = x    !  EXP() below requires a real argument
            PRINT *, "n = ", n
            PRINT *, "x = ", x
            CALL Print_Blank_Lines ( 1 )    ! prints 1 blank line
            PRINT *, "Series e ** x =    ", e2x
            PRINT *, "Fortran's e ** x = ", EXP ( rx )
   !
   RETURN
   END SUBROUTINE  Print_Results
```

************** PROGRAM RESULTS ***************

```
       This program calculates e ** x

 Please enter the number of terms ( < 0 to quit)
 6
 Please enter an integer value for x
 1
  n =  6
  x =  1

  Series e ** x =       2.7166666666666663
  Fortran's e ** x =    2.7182818284590451
  Please enter the number of terms ( < 0 to quit)
 20
  Please enter an integer value for x
 1
  n =  20
  x =  1

  Series e ** x =       2.7182818284590455
  Fortran's e ** x =    2.7182818284590451
  Please enter the number of terms ( < 0 to quit)
 9
  Please enter an integer value for x
 3
  n =  9
  x =  3

  Series e ** x =       20.0091517857142840
  Fortran's e ** x =    20.0855369231876680
  Please enter the number of terms ( < 0 to quit)
 -1
```

In the program above there are two subroutines that are used and that don't have program statements. So where are they? The subroutines in question are listed below:

1. SUBROUTINE Print_Blank_Lines (n)
2. SUBROUTINE Factorial_s (n, n_Factorial)

These routines' **source code** (program statements) are not shown because they exist in a library of subroutines and don't need to be designed, typed, debugged, or understood in any way other than what their arguments represent. Time to review libraries . . . .

7.6 Libraries Revisited

Subroutines are mini Fortran programs each of which should be designed to accomplish one well-defined task. One subroutine might read in values from a disk file and another might perform calculations on those values. In different programs, the same kinds of jobs often need to be done. When the same work needs to be accomplished in a different program, why rewrite the subroutine? Reuse existing subroutines.

When a useful subroutine is created, save it in a collection of subroutines called a **library**. Programmers can create their own libraries or use existing libraries. Existing libraries are either inherited or purchased.

It isn't necessary to understand *how* library routines work. Understand the description (documentation) of the subroutine's purpose and its arguments, and use it; this is the black box concept. In the complete program of the previous section, library subroutines were documented as seen in Figure 7.15. Figure 7.15 describes a subroutine called `Print_Blank_Lines` that has an `INTEGER INTENT ( IN )` argument, n, which tells the subroutine how many blank lines to print. **Who cares how the subroutine does its job?** It's sufficient to understand how to use it.

The kind of programming that uses existing subroutines is the way to go. Create large programs where a few hundred lines of program statements invoke thousands of lines of existing, working subroutines. Never write a subroutine that has been written before; reuse subroutines whenever possible.

To create a Fortran library, type the subroutines one after another like a Fortran program without a **main program**. As time progresses and useful subroutines are created, they can be added to the library. This collection of subroutines is called the **source library**

Figure 7.15

```
!---------------------------------------------------------------
!  The purpose of this program is to demonstrate the process of series
!     summation.  The series for e ** x is: x**0/0!+x**1/1!+...+x**n/n!
!
!  n          is the number of terms to sum - INTEGER
!
!  x          is the power that e is raised to - INTEGER
!
!  e2x        holds the calculated value for e ** x - DOUBLE PRECISION
!---------------------------------------------------------------
       !  BLACK BOX:   Library subroutine Print_Blank_Lines ( n ) . . .
       !
       !  The purpose of this subroutine is to print n blank lines.
       !
       !  n             INTENT (IN) INTEGER argument
       !---------------------------------------------------------
```

because it contains the actual Fortran program statements for the subroutines—the **source code**. The source library is then compiled to create an **object library**, which is the machine language version of the source library (Table 7.1). An object library can be made available to any program requiring one or more of the library subroutines. Only the required subroutines will be added to a program—not the whole library.

Table 7.1

SOURCE LIBRARY	COMPILER ==>	OBJECT LIBRARY
Subroutine Traj (...) ... End Subroutine Traj		1001001010011101110101 00000 1010011101000101110110 01110
**********************		0100010010011101000000 11010
Subroutine I_O (...) ... End I_O		0000110100100010010101 10100 0010100111011001101010 10010
**********************		0010011010010000110101 10111
Subroutine Random ... End Random		1111101010010100110101 010
**********************		etc.
etc.		

When a Fortran program is compiled, a machine language version of the program is created but the program still isn't quite ready to run. A second process called **linking** incorporates any required library routines into the program to create an executable version. Figure 7.16 shows the whole process of getting a Fortran program into an executable form, including the linking stage where required subroutines are added to the program from libraries.

Figure 7.16

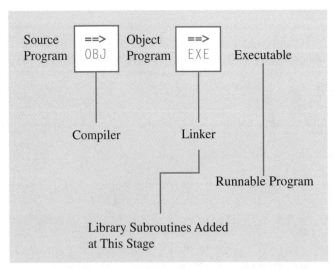

The source code for library routines `Factorial_s` and `Print_Blank_Lines` are presented here to help demystify them:

```
        SUBROUTINE   Factorial_s ( n, n_Factorial )
                     IMPLICIT    NONE
!
!-----------------------------------------------------------------------
!  The purpose of this subroutine is to calculate n! and store it into
!     n_Factorial.
!
!  n              INTENT ( IN ) INTEGER argument
!
!  n_Factorial    INTENT ( OUT ) DOUBLE PRECISION argument
!
!  Local_n        INTEGER local variable to protect n's original value
!                    for the calling program
!-----------------------------------------------------------------------
!
        INTEGER,              INTENT ( IN )    ::    n
        DOUBLE PRECISION,     INTENT ( OUT )   ::    n_Factorial
        INTEGER                                ::    Local_n
!
            IF ( n >= 0 ) THEN    ! error checking - n MUST be >= 0
                n_Factorial = 1.0D+0
                Local_n = n
                DO WHILE ( Local_n > 1 )
                    n_Factorial = n_Factorial * Local_n
                    Local_n = Local_n - 1
                END DO
            ELSE
                PRINT *, "ERROR IN FACTORIAL SUBROUTINE - n &
                                      &MUST be >= 0 for n!"
            END IF
!
RETURN
END SUBROUTINE  Factorial_s
!=======================================================================
        SUBROUTINE   Print_Blank_Lines ( n )
                     IMPLICIT    NONE
!
!-----------------------------------------------------------------------
!  The purpose of this subroutine is to print n blank lines.
!
!  n              INTENT ( IN ) INTEGER argument
!
!  i              an INTEGER local variable used as a loop counter
!-----------------------------------------------------------------------
```

continues

```fortran
!
        INTEGER,            INTENT ( IN )      ::      n
        INTEGER                                ::      i
!
        IF ( n >= 0 ) THEN   ! error checking ... n MUST be positive
            i = 1
            DO WHILE ( i <= n )
                PRINT *, " "
                i = i + 1
            END DO
        ELSE
            PRINT *, "CAN'T PRINT A NEGATIVE NUMBER OF &
                    &LINES - Print_Blank_Lines"
        END IF
!
RETURN
END SUBROUTINE  Print_Blank_Lines
```
!===

Take the time to type these subroutines into the computer just as they appear. Get a personal subroutine library started. Whenever a library program unit is used, just **cut and paste its documentation** from the source library into the program being designed and use the subroutine accordingly. Another useful subroutine that came out of the e^x program is shown below:

```fortran
SUBROUTINE   Get_Generic_INTEGER_Input ( Prompt, Response )
        IMPLICIT   NONE
!
!----------------------------------------------------------------------
!   This subroutine presents the prompt contained in CHARACTER
!     argument Prompt, INTENT ( IN ), and then reads a value from the
!     keyboard into INTEGER argument Response INTENT ( OUT ).
!----------------------------------------------------------------------
!
        CHARACTER ( LEN = * ),   INTENT ( IN )    ::    Prompt
        INTEGER,                 INTENT ( OUT )   ::    Response
!
        PRINT *, Prompt
        READ  *, Response
!
RETURN
END SUBROUTINE  Get_Generic_INTEGER_Input
```
Add this subroutine to the library as well—why create it again?

Important Concepts Review

- **Shading the upper left corner** of structure chart blocks indicates that a subroutine is being called from more than one place in a program.

- The **standard input algorithm** provides the ability to read an unknown amount of data with an "anytime" escape from the input algorithm.

- **Series summation** is based on the incremental assignment statement.

- $n!$ is defined as: $1 \cdot 2 \cdot 3 \cdot \ldots \cdot (n-1) \cdot n$.

- **Libraries** are collections of useful subroutines.

- **Source libraries** contain Fortran statements.

- **Object libraries** are the machine language versions of source libraries.

- **Compiling** is the process where source code—Fortran program statements—is converted to machine language.

- **Linking** is the process where libraries are searched to supply any missing routines for a program and an executable version of the program is created.

- **Document library routines**—black boxes—below the program unit's comment block as seen in the design example.

EXERCISES

7.1 Design a program to calculate the average of n positive integers. Terminate input when the incoming value for n is -1, which is the flag value.

7.2 Modify Exercise 7.1 so that if a negative value other than -1 is entered, the program will print an error message and not include the negative value in the calculation. The program should still end when a -1 is entered and shouldn't present an error message for the -1.

7.3 Design a complete Fortran program to sum the series for sin(θ). Sin(θ) is defined by the series:

$$\sin(\theta) = \sum_{i=0}^{\infty} \frac{(-1)^i \, \theta^{2i+1}}{(2i+1)!}$$

7.4 Write a `DO WHILE` loop that will print a table of powers of 2 from `2**0` to `2**16`. Use `INTEGER` variables.

7.5 Design a program to *calculate* and display the first eight rows of Pascal's triangle. Don't just print the rows. Any number in the triangle, other than the 1s, is calculated by adding the number above to the left and the number above to the right. The first six rows are shown below:

```
row
0                     1
1                   1   1
2                 1   2   1
3               1   3  (3)+(1)
4             1   4   6  (4)  1
5           1   5  10  10   5   1
```

It is noteworthy that a given row of Pascal's triangle represents the coefficients of polynomial expansion:

$$(x+y)^5 = \underline{1}x^5 + \underline{5}x^4y + \underline{10}x^3y^2 + \underline{10}x^2y^3 + \underline{5}xy^4 + \underline{1}y^5$$

The coefficients come from the 5th row of Pascal's triangle. It is acceptable to display the triangle as follows:

```
1
1  1
1  2   1
1  3   3   1
1  4   6   4   1
1  5  10  10   5   1
```

7.6 Design a subroutine that calculates an approximation of π using Gregory's series for π/4. The series is defined as follows:

$$\pi/4 = 1 - \tfrac{1}{3} + \tfrac{1}{5} - \tfrac{1}{7} + \cdots + (-1)^n/(2n+1)$$

In this subroutine, *prompt* the user for a tolerance *t*, type `REAL`, and calculate an approximation of π, not π/4, to within tolerance *t*. (*Hint: stop the loop when* `ABS(your-current-value-for-π - 3.1415927)` *is less than tolerance t.*) Error check *t* to ensure that it isn't smaller than `1E-6`.

7.7 *Prompt* the user to enter a positive integer *N*. "Echo" the input (display it on the screen). Using this value as a "seed," print out the Hailstones series, which is defined as follows:

N_{i+1} gets the value: $N_i/2$, if N_i is even, but gets $3N_i + 1$, if N_i is odd. The series ends when N_{i+1} becomes 1. Example: if N is read in as 11, the series is

11 34 17 52 26 13 40 20 10 5 16 8 4 2 1

Use the *Hint* in Exercise 7.12 to figure out whether N_i is even or odd.

7.8 Generate a table of pendulum periods for lengths from 1 to 100 cm for the earth and moon. Use 9.8 m/sec² for earth's gravity and 1/6th of earth's gravity for the moon's gravity (think PARAMETERs). The period P of a pendulum of length L, REAL, in a gravitational field G, REAL, is given by the formula:

$$P = 2\pi \sqrt{\frac{L}{G}}$$

7.9 Euclid's method for finding the GCD, greatest common divisor, of two integers N_1 and N_2 is calculated by successively subtracting the smaller of the two numbers from the larger until they become equal. Read in values for N_1 and N_2. Echo the inputs. Write out the GCD. Use a conditional loop to determine when the GCD is found. Do a few examples by hand to see how Euclid's algorithm works.

```
        24   18
       -18          18 < 24
         6          6 < 18
           -6
           12       6 < 12
           -6
      GCD   6       6 = 6
```

7.10 An approximation for $\sqrt{n}$ is defined as follows:

$$G_{new} = \frac{G_{old} + \left(\dfrac{n}{G_{old}}\right)}{2}$$

G is some positive initial guess. Design a program to calculate $\sqrt{n}$ using the expression above. Use a conditional loop to repeat the calculation until G is within $1E-10$ of the $\sqrt{n}$. Use DOUBLE PRECISION variables.

7.11 Write a looping structure that will print a table of squares and square roots of the numbers 1.0 through 25.0.

7.12 Design a program to determine whether a read-in value, n, is prime. n is prime if it is divisible only by 1 and itself. (*Hint 1:* compare n/num to rn/rnum where n and num are INTEGER variables and rn and rnum are the REAL counterparts of n and num. If they compare equal, n is divisible by num. Hint 2: use Fortran's MOD function to determine whether two numbers are evenly divisible or not.)

7.13 Design a *main* program that will repeatedly present the following menu until the quit option is selected.

```
1. Calculate density given mass and volume.
2. Calculate mass given density and volume.
3. Calculate volume given density and mass.
4. Quit.
```

Include error checking so that if an incorrect menu option is selected, an error message will be presented and the menu redisplayed.

7.14 Design appropriate subroutines to make the program menu options in Exercise 7.13 work.

7.15 Define error checking.

7.16 Add error checking to the subroutines of Exercise 7.14.

7.17 Design a program that will read x, type `DOUBLE PRECISION`, and print the value of $\ln(x)$ (the natural logarithm of x). Use Fortran's intrinsic function `LOG(x)` to calculate $\ln(x)$. If $x <= 0$, print an error message and end the program. After $\ln(x)$ is printed, ask the user whether he would like to input another x. Repeat the process if the response is yes; end the program if the response is no. Use error checking whenever possible.

7.18 Write a subroutine to find the prime factors of n. An example: the prime factors of 5040 are 2·2·2·2·3·3·5·7.

7.19 Include the prime number subroutine from Exercise 7.12 in your subroutine library.

Spreadsheets: Special Section 1

- Elements of a Spreadsheet
- Navigating a Spreadsheet
- Cells and What They Can Hold
- The Active Cell
- Making an Entry in a Cell
- Editing a Cell Entry
- Cell Attributes
- Important Concepts Review
- Working with Groups of Cells
- Moving Groups of Cells
- Copying Cells
- Creating a Series
- Important Concepts Review
- Exercises

Introduction

Spreadsheets are fascinating tools that simplify the analysis and graphing of tabular data. Figure S1.1 shows a generic spreadsheet—The Forsythe Spreadsheet. There are many spreadsheet programs on the market and they all do the same basic things. The Forsythe Spreadsheet, which is only a facsimile of a real spreadsheet, is shown as a **GUI** style spreadsheet that is used in a **windowing** environment. GUI means *G*raphical *U*ser *I*nterface ("GUI" is pronounced "gooey"). These environments are controlled with a pointing device called a **mouse**. Various features can be activated by pointing at graphic items on the screen and clicking a mouse button.

Figure S1.1

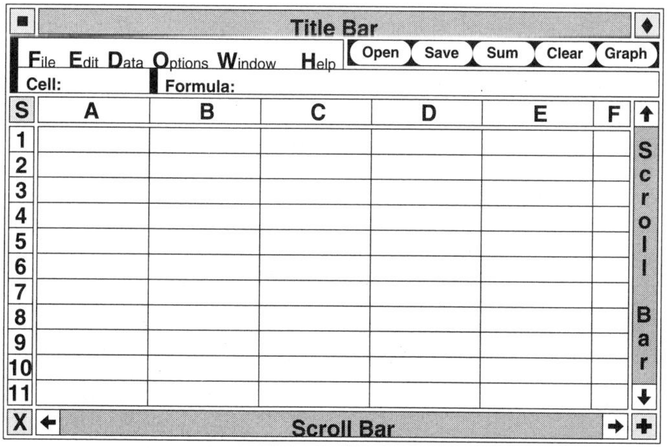

Mouse operations can usually be accomplished with keystrokes on the keyboard, but it is a clumsy method. In this chapter's explanation of spreadsheets, only mouse operations will be explained.

The first spreadsheet was created in the mid to late 1970s as a software package for small businesses. Spreadsheets were originally designed to maintain financial records, such as ledgers. VisiCalc®, the earliest commercially successful spreadsheet, was shipped with Apple® II computers in 1977 under the wise guidance of Mike Markkula, the CEO of Apple® Computers, Inc.

Markkula realized that his soon-to-be partners, Steven Jobs and Steve Wozniak, who invented Apple® computers, only saw their invention as something computer enthusiasts might buy as an electronic kit. Markkula knew that Jobs' and Wozniak's invention had much more potential than that. They had created a computer that could be used by businesses if it was coupled with the right **software**.

Business people fell in love with the idea of desktop computers that could maintain financial records, do payroll, help forecast business trends and do a host of other things to make them more productive. It was businesses that brought desktop micro computers into public use, and spreadsheets offered the most enticing features that attracted the business world.

A typical business operation for a spreadsheet is to total ledger values contained in rows and columns. An example is shown in Figure S1.2. Businesses need to track their sales by maintaining tabular records, such as the ones for the "Shoe Business" in Figure S1.2. This information can also be conveniently graphed by a spreadsheet to view sales trends.

One fantastic thing about spreadsheets is that values in a spreadsheet can be calculated based on other values in the sheet. For example, the "Total" line in Figure S1.2 is not figured "by hand." Each total value for a given column is calculated by entering a **formula** in the corresponding "total" **cell** whose purpose is to sum the values in the column above. A **cell** is any one block in the body of a spreadsheet. Cell C10 (column C, row 10 showing $10,353) in Figure S1.2 might have a formula that looks like this:

$$SUM(C6:C8)$$

Figure S1.2

Cell:	Formula:				
	A	B	C	D	E	F
1			Shoe Business			
2			SALES			
3						
4		Jan	Feb	Mar	Apr	
5						
6	Shoes	$9,234	$8,127	$11,213	$18,901	
7	Socks	$403	$375	$485	$879	
8	Boots	$1,667	$1,851	$1,581	$987	
9		--------	--------	--------	--------	
10	Total	$11,304	$10,353	$13,279	$20,767	
11						

This formula results in the value $10,353 being displayed (C6+C7+C8). The technique of using formulas that refer to other cells eliminates calculation errors that might be introduced by clumsy fingers on a calculator.

Why are you aspiring engineers and scientists being offered this knowledge? It is because spreadsheets don't have to do just financial calculations. Spreadsheets provide fast and easy data analysis through powerful mathematical and statistical functions and wonderfully flexible graphing tools . . . **without having to design and code a Fortran program** to do the work.

Always use the simplest and most effective tools to get a job done when you do your science and engineering. Why take five to ten hours creating a customized statistical analysis program to interpret your data? Why design and code a primitive graphing program that displays your data in one graph style? With spreadsheets, you can import data quickly and analyze it in many different ways. Then click the mouse or press a few keys, and view your data in different graph styles and choose the one that best shows the implications of your data.

Spreadsheets are very visual, dynamic computing tools. As a result, this chapter is full of "how to" figures to help you get functional with spreadsheets as quickly as possible. It is a good idea, but not necessary, to sit at a computer with the spreadsheet of your choice running as you read this chapter. You can try the examples in the figures as you read.

The generic Forsythe Spreadsheet will look different from your spreadsheet, but the actions described should work as explained for most spreadsheets. If you find a difference between your spreadsheet and the Forsythe Spreadsheet that you can't figure out by trial and error, use your spreadsheet's help facility to figure out the specific steps you need to follow. If all else fails, read the spreadsheet's documentation.

Skills presented in this chapter are the very basic basics. Spreadsheets can do much more than what is shown here. It would require an entire text to explain all that spreadsheets can do. After studying this chapter, however, you will be able to use spreadsheets effectively at a fundamental level. Later you can learn more from your software's help facility and documentation as your needs evolve.

Spreadsheets can't replace designing programs, but when the situation fits, spreadsheet data analysis outspeeds creating programs 100 to 1. If you can get answers faster with a spreadsheet, do so; it'll save time and you can balance your checkbook, too.

• • •

Elements of a Spreadsheet

A spreadsheet is a two-dimensional grid of blocks. Each block in the spreadsheet is called a **cell**. Cells can hold items that will be defined soon. Every cell is identified by the intersection of a column and row. The highlighted cell, Item 6 in Figure S1.3, is cell B3.

Other than cells, spreadsheets have features that are used to control the behavior and appearance of the sheet. The elements of the generic Forsythe Spreadsheet are shown in Figure S1.3 (p. 188) and described as follows:

1. **Informational region**—supplies information such as the coordinates of the active cell.

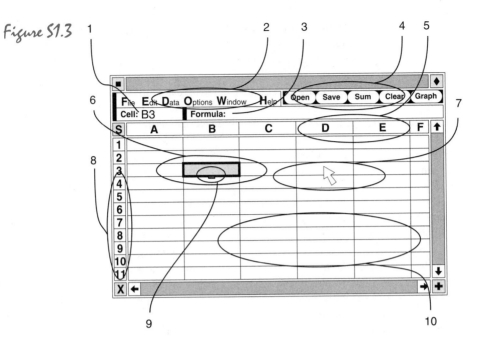

Figure S1.3

2. **Menubar**—offers various options for manipulating the sheet.
3. **Formula area**—allows formulas or values to be created/edited.
4. **Toolbar**—provides menu shortcut buttons.
5. **Columns**—designate the vertical stacks of adjacent cells.
6. **Active cell**—indicates where the next entry will be stored.
7. **Pointer**—points at an item to be activated. It is controlled by a mouse or trackball.
8. **Rows**—designate the horizontal line of adjacent cells.
9. **Anchor point**—allows you to create or modify spreadsheet series.
10. **Inactive cells**—indicate cells other than the active cell.

More details on the spreadsheet features shown in Figure S1.3 are presented in the following list:

1. The **informational region** reflects which cell receives the typed-in information or can be edited. Other information about the active cell or the current operation may be shown in this area as well.
2. A **menubar** is a common feature of GUI environments. In a spreadsheet, the menubar offers access to features such as the ability to save a sheet on disk, cut and paste cells, get help, print the sheet, etc. (Figure S1.4). In Figure S1.4, the "Edit" menubar option has been clicked with a **pointing device**, which displays a drop-down menu that offers various selectable options.
3. The **formula area** shows formulas or values that can be entered or edited in this region. The results go into the active cell.

Figure S1.4

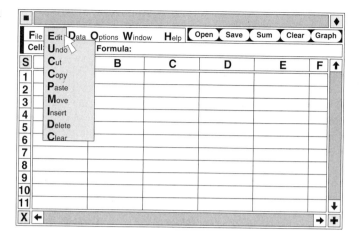

4. A **toolbar** is a set of buttons that can be clicked with a mouse. Each button is set up to accomplish a complete menubar task without having to go through all the point-and-click steps required by menubar drop-down menus.
5. The **columns** each have a letter or two at the top so they can be referenced. A column represents a vertical stack of adjacent cells. Columns can be sized to any width to fit various widths of data.
6. The **active cell** is highlighted by a contrasting rectangle. Anything that is typed in the formula area will enter the active cell. An active cell's contents may also be edited in the formula area.
7. The **pointer arrow** is controlled by a mouse or trackball and is used to point at various graphic features such as buttons, menubar options, cells, etc. A feature is activated by pointing at it and clicking the mouse.
8. The **rows** are numbered, as opposed to columns being identified by letters, and represent a horizontal line of adjacent cells. Rows may be made taller or shorter (sized) to fit different heights of data.
9. The **anchor point** is used to create, extend or contract spreadsheet series.
10. The **inactive cells** are not necessarily empty. They may contain entries but they are not currently being edited or having their contents replaced.

Navigating a Spreadsheet

Spreadsheets are vast. Commonly, they have 250+ columns and 15,000+ rows. It is impossible to see a whole spreadsheet at once on a computer screen. A computer screen is like a window that allows a spreadsheet user to view a small group of the spreadsheet's cells (Figure S1.5).

Figure S1.5

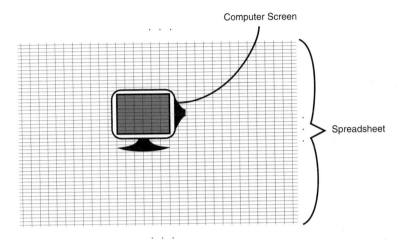

Figure S1.6

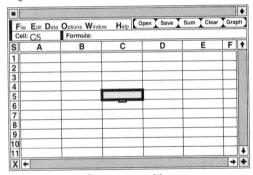
Current position

Right arrow

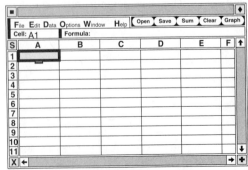
Home key

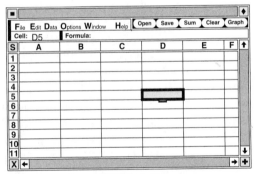
Go to command (cell E10)

Since the spreadsheet is so large, there need to be ways to position the sheet so that the desired set of cells is visible on the computer screen. The ways to move around in the Forsythe Spreadsheet, and most real spreadsheets, are described below:

1. Use **arrow** keys to move the active cell to a new location. Holding down an arrow key will cause it to repeat and quickly scroll the active cell in the arrow's direction.
2. Use **scroll bars** to shift the cells up, down, left or right. The active cell stays at the same coordinates and may scroll off the screen out of view when a sheet is positioned with scroll bars. To make a cell in the newly visible range of cells the active cell, click the desired cell with the mouse.
3. Press the **"Home"** key to make cell A1 the active cell. This is a key that can usually be configured in spreadsheets; the "Home" key may be defined to do something other than making A1 the active cell.
4. Use the **"Go To"** command, which is often the F5 function key, to specify a cell location (cell F90, for example). The specified cell becomes the active cell.
5. **"PgUp"** and **"PgDn"** keys move the active cell up or down approximately one screen.

Some navigational methods described above are illustrated in Figure S1.6.

Cells and What They Can Hold

In the chapter on variables, a variable was described as a container that can hold a value of the variable's declared type. For example, a Fortran `INTEGER` variable named `Time_It` can hold an integer value such as `2300`. Spreadsheet cells are similar in that they are containers, but cells are much more clever than simple Fortran variables—cells can hold a variety of entries:

1. **Number**—a numeric quantity (constant) that is limited to approximately the same precision as Fortran's `DOUBLE PRECISION` type.
2. **Label**—a string of characters like a Fortran `CHARACTER` constant.
3. **Formula**—an expression composed of **operands** and **operators**.

   Spreadsheet **operands** may be any of the following:
   - **Constants.**
   - **Cell references.**
   - **Functions.**

   Spreadsheet **operators** are described below:

   ^   exponentiation, H81^X56 <==> $H81^{X56}$

   *   multiplication

   /   division

   +   addition

   −   subtraction

All the rules of **arithmetic order of precedence** for expression evaluation in Fortran apply to formulas created in a spreadsheet, except for exponentiation. Exponentiation is evaluated from left to right in the Forsythe spreadsheet. Parentheses may be used to override order of precedence.

The Active Cell

The **active cell** is always highlighted as in Figure S1.6. When a cell is the active cell, entries go into that cell. In addition, when a cell is made the active cell and it already contains an entry, that entry will appear in the formula area (Item 3 of Figure S1.3) and can be edited. If the active cell contains an entry and a new one is entered, the old entry is lost just as with assignment to Fortran variables.

Another useful feature of the active cell is the ability to undo undesirable changes. This can only be done if an entry hasn't been registered (see the next section to understand registering an entry). If a cell is being filled or edited and a user wants to return the active cell to its original contents, pressing the **"Esc"** key (escape key) will do the job. Two reasons one might want to abort changes, for example, are if a severe editing error is made or changes are accidentally being made to the wrong cell.

Making an Entry in a Cell

To make an entry (which can be a number, label or formula) in a cell it is necessary to do the following things:

1. **Make the cell that is to receive the entry the active cell**. Move the active cell highlight to the appropriate cell using the positional methods described before.
2. **Type the entry**. It will simultaneously appear in the formula area and in the active cell.
3. **Register the entry**. Registering an entry means telling the spreadsheet that the entry in the formula area is correct, complete and should be put in the active cell now. There are two main ways to do this:
    - **Press the "Enter" key** on the keyboard—the active cell stays the same.
    - **Make a different cell the active cell**—click another cell, use the arrow keys, etc.

Once an entry is registered, all formulas in the spreadsheet are recalculated and their new values shown.

When working on a large spreadsheet that takes several seconds or minutes to recalculate, it is possible to configure the sheet so it will not recalculate after each modification. Manual recalculation is usually accomplished with the F9 function key or appropriate menu selections.

Editing a Cell Entry

Cells often contain formulas that calculate values based on the contents of other cells. Sometimes formulas are created incorrectly and need to be changed. The formula region makes changes easy.

Assume the spreadsheet in Figure S1.7 is supposed to show powers of 2. Labels in cells B1 and C1 describe what kind of information is in columns B and C. Obviously, the powers of 2 are incorrect ... wrong formulas in column C should be suspected.

To debug this problem, make cell C2 the active cell and examine the formula in the formula editing area (Figure S1.8). C2's formula is seen in the formula editing area: =B2^2 (Forsythe Spreadsheet formulas start with an equal sign: "="). If cell C3 is made

Figure S1.7

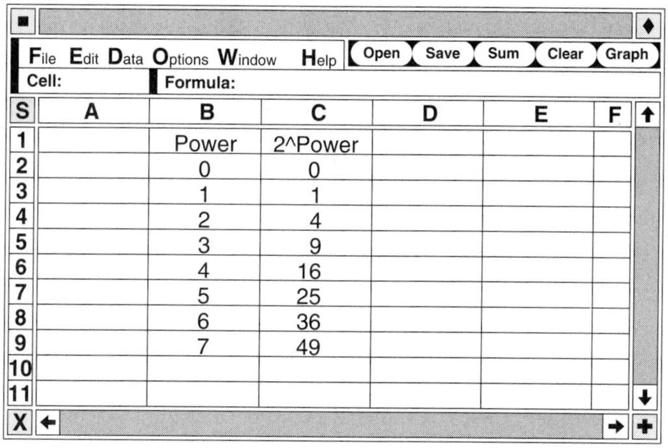

	A	B	C	D	E	F
1		Power	2^Power			
2		0	0			
3		1	1			
4		2	4			
5		3	9			
6		4	16			
7		5	25			
8		6	36			
9		7	49			
10						
11						

Figure S1.8

Cell: C2   Formula: = B2^2

	A	B	C	D	E	F
1		Power	2^Power			
2		0	=B2^2			
3		1	1			
4		2	4			
5		3	9			
6		4	16			
7		5	25			
8		6	36			
9		7	49			
10						
11						

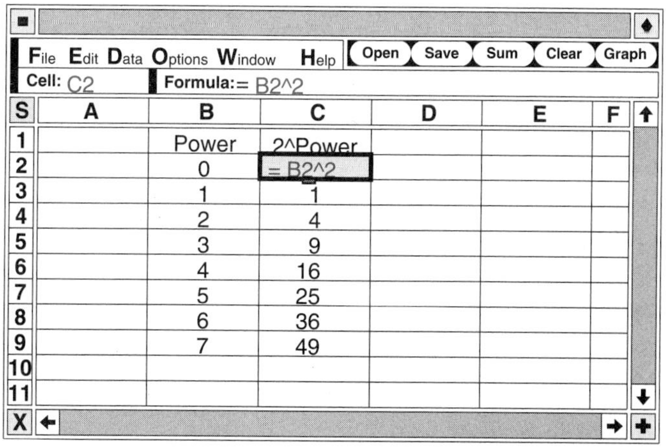

the active cell, the formula area would show C3's formula: =B3^2. After seeing these formulas, it is clear that the error here is that all the formulas in column C are reversed; the B column cells are being squared instead of raising 2 to the power of their values.

One way to remedy this situation is to edit the formulas in column C. To do this, move the pointer into the formula editing area where the pointer changes from an arrow to an **I-beam**. Click in the formula and change it so it is correct (Figure S1.9). After the corrections are made to cell B2, the corrected formula can easily be copied to the other column C locations, C3 through C9, creating a spreadsheet with correct powers of 2 (Figure S1.10).

Figure S1.9

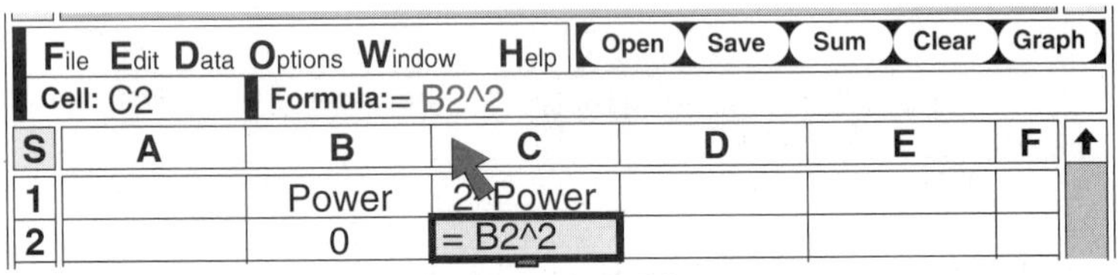

Move the pointer towards formula editing area.

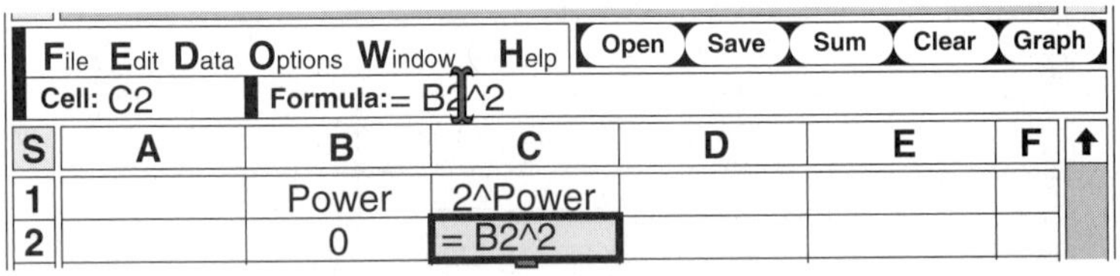

As the pointer enters the formula area, it turns into an I-beam indicating editing can be performed.

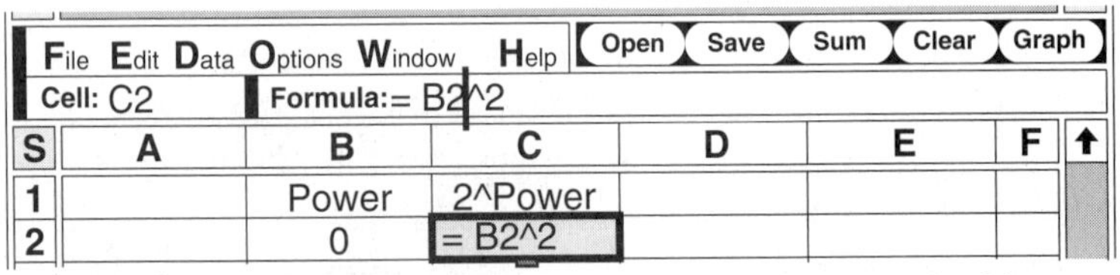

Click the mouse and a verticle cursor appears where the I-beam was (between B2 and ^).

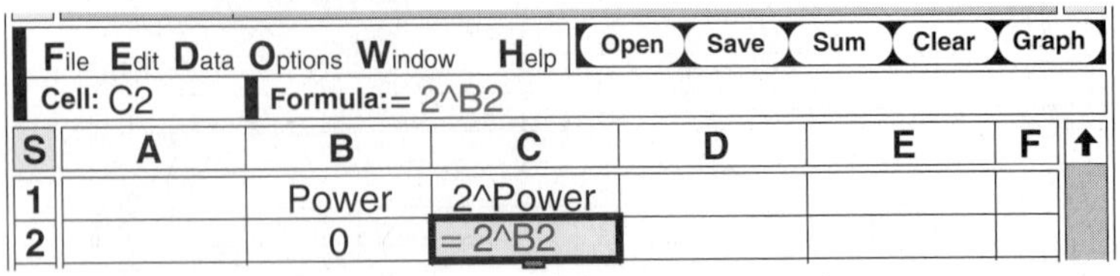

Using Backspace, Delete and the left and right arrow keys, correct the formula.

Figure S1.10

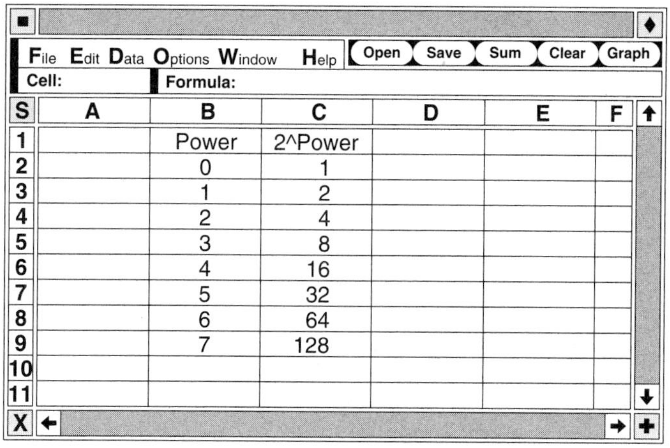

```
┌─────────────────────────────────────────────────────────────┐
│ ■                                                         ◆ │
├─────────────────────────────────────────────────────────────┤
│ File  Edit  Data  Options  Window  Help │ Open │ Save │ Sum │ Clear │ Graph │
│ Cell:              │ Formula:                               │
├───┬───────┬───────┬───────┬───────┬───────┬──┤
│ S │   A   │   B   │   C   │   D   │   E   │F │↑
├───┼───────┼───────┼───────┼───────┼───────┼──┤
│ 1 │       │ Power │2^Power│       │       │  │
│ 2 │       │   0   │   1   │       │       │  │
│ 3 │       │   1   │   2   │       │       │  │
│ 4 │       │   2   │   4   │       │       │  │
│ 5 │       │   3   │   8   │       │       │  │
│ 6 │       │   4   │   16  │       │       │  │
│ 7 │       │   5   │   32  │       │       │  │
│ 8 │       │   6   │   64  │       │       │  │
│ 9 │       │   7   │  128  │       │       │  │
│10 │       │       │       │       │       │  │
│11 │       │       │       │       │       │  │↓
│ X │ ←                                            → │ +
└───┴─────────────────────────────────────────────────┘
```

Cell Attributes

In Figure S1.10, cells B2 through B9 contain the **numbers** 0 through 7, cells B1 and C1 contain **labels** (strings of characters), and cells C2 through C9 contain the **formulas**: 2^B2, 2^B3 ... 2^B9. These different entries represent the complete range of entry types for spreadsheet cells: numbers, labels and formulas.

Besides holding entries, cells may also be given **attributes** to enhance their appearance. Table S1.1 summarizes commonly used cell attributes.

Table S1.1

Cell Attributes	
Attribute	Purpose
Font	Fonts define how the characters in cells look. The appearance of a cell's characters can also be changed to different styles such as italics, bold, underline, etc.
Size	The size of cell contents can be made larger or smaller.
Alignment	Entries can be left justified, centered, or right justified.
Format	Formats are most often used to specify the way numbers look . . . how many digits to the right of the decimal? Display numbers as currency? How should negative numbers be indicated? Show commas for numbers > than 1,000?
Column Width, Row Height	Column widths and row heights can be adjusted to accommodate larger or smaller cell entries.

Cell attributes can be applied to a single cell or a group of cells. To see how the cell attributes from Table S1.1 work, examine the spreadsheet in Figure S1.11. Cells F3 through F5 have several attributes: italics, bold, increased size and currency format. Cells B2 through B4 illustrate alignment attributes. The value stored in cell D10 was too large to fit. When a number is too large to be displayed in a column, "#"s are displayed in the cell. Widening column D solves the problem.

Figure S1.11

	A	B	C	D	E	F
1						
2		Right			Italic	
3		Left				$1.00
4		Center				$2,401.00
5				BIG		$22.00
6						
7	BOLD					
8			small			
9						
10				###		
11						

Important Concepts Review

- A **cell** is a container that can hold a number, label or formula. Cells are identified by the intersection of a column (designated by a letter) and a row (designated by a number).
- **Rows** are horizontal lines of adjacent cells.
- **Columns** are vertical stacks of adjacent cells.
- Basic elements of a spreadsheet include:
    1. **Informational region**—supplies information such as the coordinates of the active cell.
    2. **Menubar**—offers various options for manipulating the sheet.
    3. **Formula area**—allows formulas or values to be created/edited.
    4. **Toolbar**—provides menu shortcut buttons.
    5. **Columns**—designate the vertical stacks of adjacent cells.
    6. **Active cell**—indicates where the next entry will be stored.

7. **Pointer**—points at an item to be activated. It is controlled by a mouse or trackball.
8. **Rows**—designate the horizontal line of adjacent cells.
9. **Anchor point**—allows you to create or modify spreadsheet series.
10. **Inactive cells**—indicate cells other than the active cell.

Navigating a spreadsheet:
1. Use **arrow** keys to move the active cell to a new location.
2. Use **scroll bars** to shift the cells up, down, left or right. The active cell's position doesn't change.
3. Press the "**Home**" key to make cell A1 the active cell.
4. Use the "**Go To**" command and specify a cell to make active.
5. "**PgUp**" and "**PgDn**" keys move the active cell up or down approximately one screen.

- **What can a cell hold?**
    1. **Numbers**—numbers in spreadsheets are limited to approximately the same precision as Fortran's DOUBLE PRECISION type.
    2. **Labels**—strings of characters.
    3. **Formulas**—expressions composed of **operands** and **operators**.

- **Formulas** are expressions composed of **operands** and **operators**.

- **Operands** may be any of the following:
    1. Constants.
    2. Cell references.
    3. Functions.

- **Arithmetic operators** are as follows:
    - ^ exponentiation
    - * multiplication
    - / division
    - + addition
    - - subtraction

- To **make an entry** in a cell, do the following:
    1. Make the cell that is to receive the entry the active cell.
    2. **Type the entry.**
    3. **Register the entry.**

- To **register an entry**, use one of these methods:
    1. **Press the "Enter" key** on the keyboard—the active cell's position remains the same.
    2. **Make a different cell the active cell**—click another cell, use the arrow keys, etc.

- **Cell attributes** are used to enhance the appearance of cell contents.

Working with Groups of Cells

It is frequently necessary to do operations such as applying a numeric format or font to a **group** of cells. A **selected** group of cells can also be moved, copied, erased, etc. To operate on groups of cells, it is necessary to understand how to **select a range of cells**. A range of cells is usually a rectangular group of adjacent cells. There are four basic methods of selecting ranges of cells. The techniques are outlined in Table S1.2. Figure S1.12 shows how each of these methods works.

Once the cells are selected, the next operation performed applies to all cells in the range. If, for example, a spreadsheet user chooses a numeric format that specifies two digits to the right of the decimal, all cells in the selected range inherit that format.

Another point that should be understood about applying attributes to a range of cells is that the attributes are applied to all cells in the range even if some cells are empty. If a

Figure S1.12

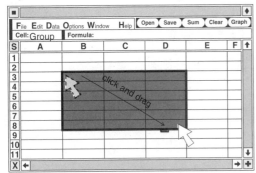
Click B3 and drag to D8

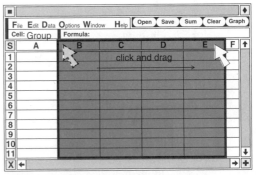
Click the B column header and drag to the E column header

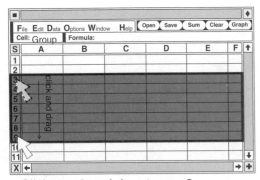
Click row 3 and drag to row 9

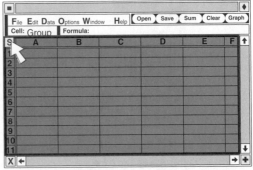
Click the "Select Whole Sheet" button

Table S1.2

	Selecting a Range of Cells		
Method	Effect		How To
Click-and-Drag from Cell to Cell	Selects a rectangle of cells from the first clicked cell diagonally to the cell where the mouse button is released.		Click inside a cell and while holding the mouse button down, drag the pointer to another cell to highlight the appropriate rectangle of cells.
Click-and-Drag Across Column Header Letters	Selects all cells in a set of adjacent columns.		Click on a column header letter and drag to another column. All intervening columns will be selected including the starting and ending columns.
Click-and-Drag Down Row Numbers	Selects all cells in a set of adjacent rows.		Click on a row number and drag to another row. All intervening rows will be selected as well as the starting and ending rows.
Click to Select the Whole Sheet	Selects all cells in the spreadsheet.		Click the "Select Whole Sheet" button.

value is subsequently put into one of those empty cells, the value's appearance will take on previously applied cell attributes such as alignment, font, etc.

Moving Groups of Cells

When spreadsheets are created, they often need to be reorganized as additional information is added. Groups of cells may need to be moved to make way for new data. Cells may also need to be moved simply to make a sheet more readable.

To **move a group of cells** from one position to another, use the following steps:

1. **Select the cells**.
2. **Place the pointer on the border** of the selected cells. The pointer will change appearance when it is exactly on the border.
3. **Click-and-drag** the selected group of cells to a new location (Figure S1.13).

Figure S1.13

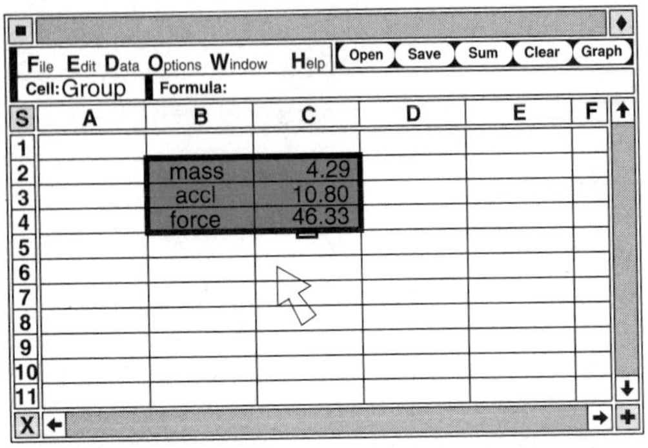

Move the pointer to the boarder of the selected cells

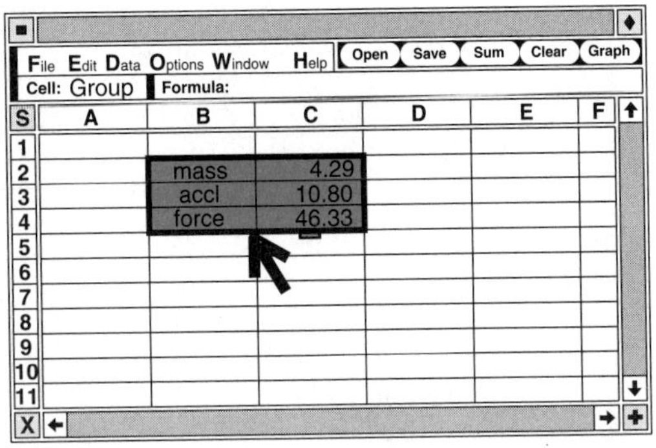

The pointer changes appearance when the boarder is touched

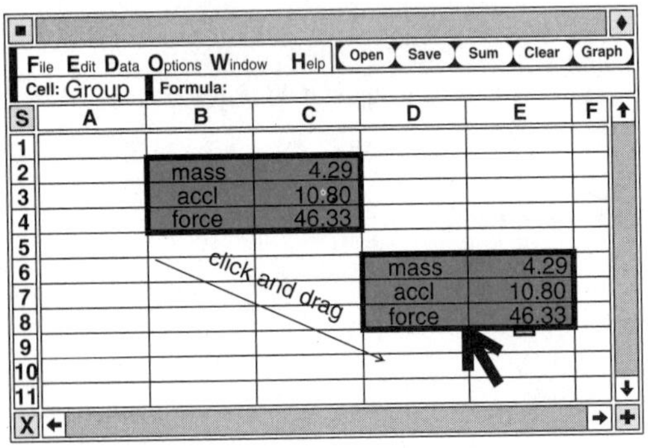

Click and drag the group of cells to the new location

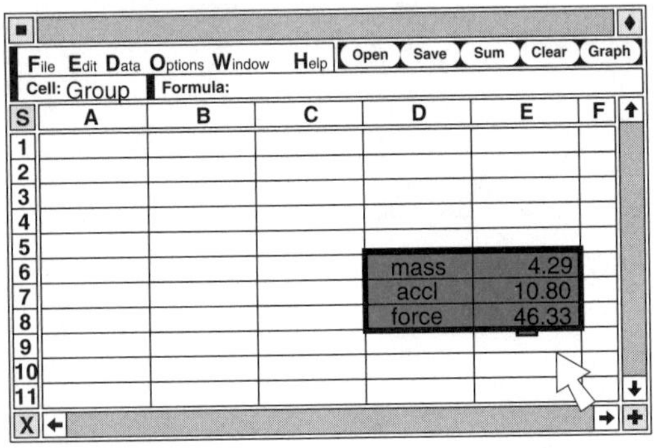

Release the mouse button to complete The move operation

Copying Cells

An ability to copy cells quickly and easily from one place to another makes spreadsheets very fast as data analysis tools. The value of copying cells with ease will become more apparent when you need to replicate formulas, which is explained later in this chapter. There are several ways to copy cells from one place to another. The two most common and useful methods are explained below.

Copying a cell or group of cells into a selected range of cells is the style of copying used most often. To copy a cell into a selected range, use the following steps:

1. **Select the cell**(s) to be copied (source cells).
2. **Choose** "Copy" from the "Edit" drop-down menu on the menubar (or use a toolbar button for "copy" if one exists).
3. **Select the destination range** that will receive the copies of the source cell(s).

Figure S1.14

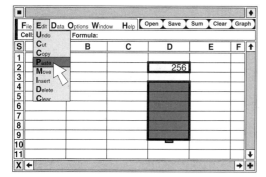

Select the cell to be copied and choose "Copy" from the Edit menu

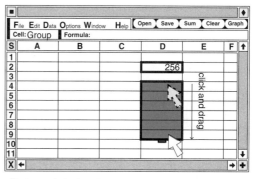

Select the destination range using the click-and-drag technique

Choose "Paste" from the Edit menu

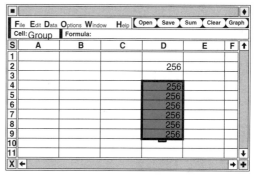

Cell D2 is copied into all the cells in the selected destination range

4. **Choose** "Paste" from the "Edit" drop-down menu on the menubar (or use a toolbar button for "paste").

Examine Figure S1.14 to see how copying a cell(s) into a selected range of cells works.

The second method of copying cells makes a duplicate copy of *selected* cells. To copy a group of cells from one location to another, follow these steps:

1. **Select the cells** to be copied.
2. **Choose** "Copy" from the "Edit" drop-down menu on the menubar.
3. **Click a cell that is in the upper left corner** of the destination range for the source cells.
4. **Choose** "Paste" from the "Edit" drop-down menu on the menubar.

See Figure S1.15 (p. 202) for an example of this copying method.

Figure S1.15

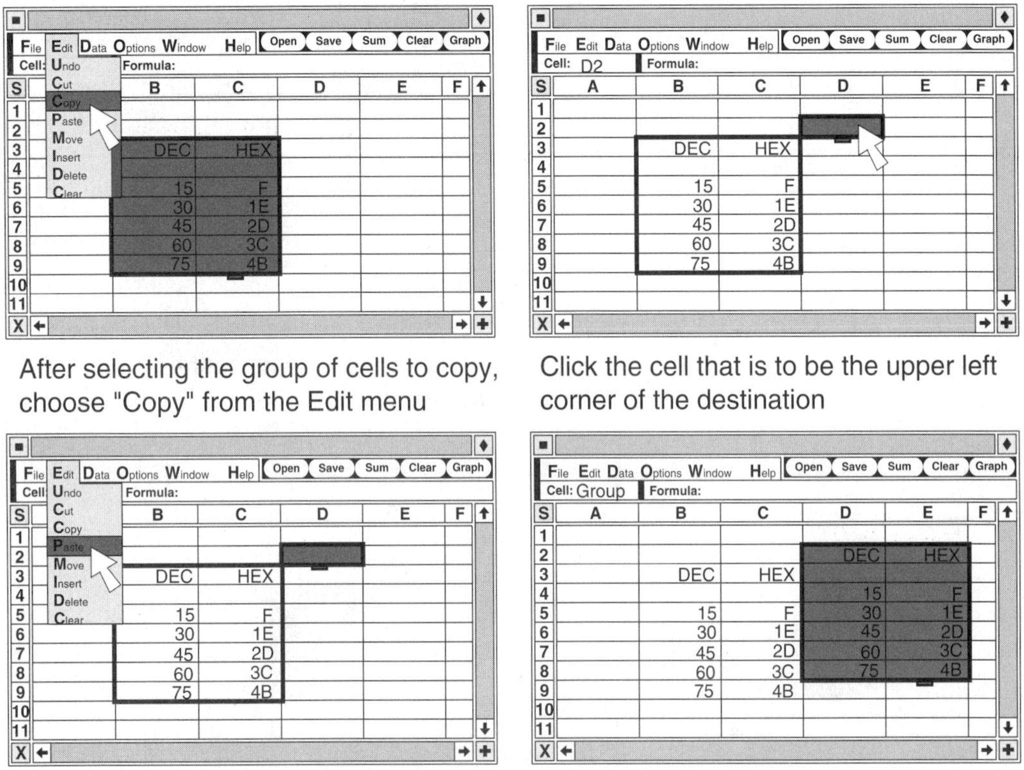

After selecting the group of cells to copy, choose "Copy" from the Edit menu

Click the cell that is to be the upper left corner of the destination

Choose "Paste" from the Edit menu

The cells are copied

Creating a Series

Many basic spreadsheet skills were explained in the previous sections. As always, much must be understood to use powerful computing tools such as spreadsheets even at a primitive level. The skill presented in this section is the most powerful so far; it is the method of **creating a series**.

A spreadsheet series is generated by putting two or more values in either a row or column and having the spreadsheet produce a linear extrapolation of those terms. To create a series, follow these four steps:

1. **Enter at least the first two terms**.
2. **Select the terms** as a range.
3. **Move the pointer to the anchor point of the selected range** where the pointer changes appearance. Click-and-drag the anchor point to extend the selected range.
4. **Release the mouse button** and the selected range fills with the series.

Creating a Series 203

Figure S1.16 illustrates how series are created. If a series is accidentally made too long, grab the anchor point with the pointer and click-and-drag it shorter. As spreadsheets were originally designed for financial endeavors involving fiscal years, monthly inventories, quarterly reports, etc., spreadsheets can also create time-related series. One can create a series of days of the week, months of the year, etc. (Figure S1.17, p. 204).

Figure S1.16

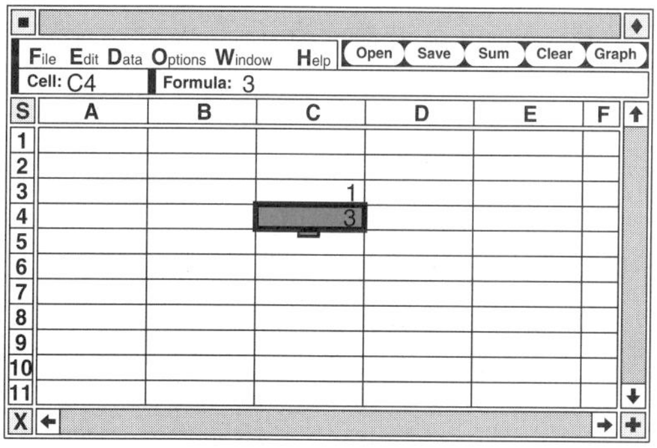

Enter the first two terms of a linear series

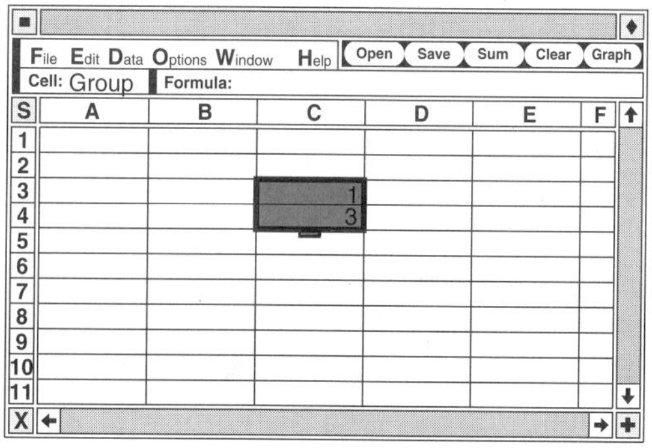

Select the first two terms as a range

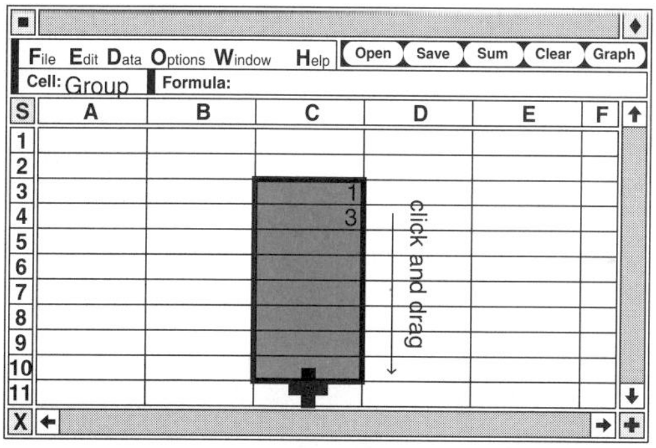

Move the pointer to the anchor point where it changes shape; click-and-drag

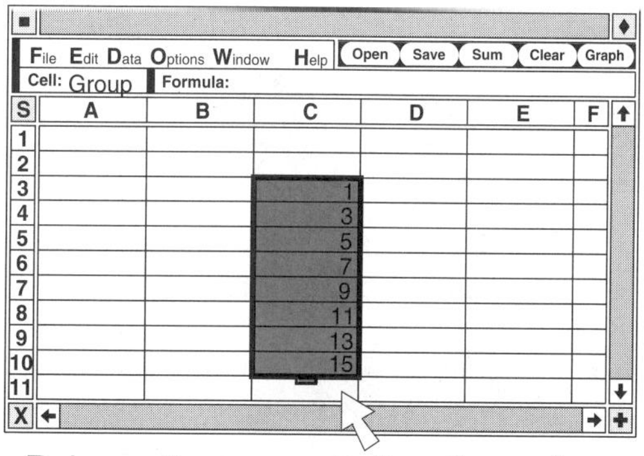

Release the mouse button; the series is extended to fill the selected range

Figure S1.17

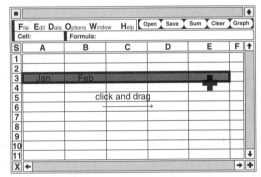

Enter the first two terms of a linear series

Select the first two terms as a range

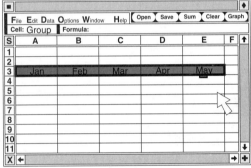

Move the pointer to the anchor point and click-and-drag

Release the mouse button

Important Concepts Review

- **Select a range of cells**—specify a rectangular group of cells by highlighting them. Operations can be performed on the selected cells.

- **Moving a group of cells**
  1. **Select the cells.**
  2. **Place the pointer on the border** of the selected cells. The pointer changes appearance, indicating that the pointer is exactly on the border.
  3. **Click-and-drag** the selected group of cells to a new location.

- **Copying a cell into a selected range**
    1. **Select the cell**(s) to be copied (source cells).
    2. **Choose "Copy"** from the "Edit" drop-down menu on the menu bar (or use a toolbar button for "copy" if one exists).
    3. **Select the destination range** that will receive copies of the source cell(s).
    4. **Choose "Paste"** from the "Edit" drop-down menu on the menu bar (or use a toolbar button for "paste" if one exists).

- **Creating a series**
    1. **Enter at least two terms.**
    2. **Select the terms** as a range.
    3. **Move the pointer to the anchor point** where the pointer changes appearance. Click-and-drag the anchor point to extend the selected range.
    4. **Release the mouse button** and the selected range fills with the series.
    5. The series can be lengthened or shortened by grabbing the anchor point and dragging it.

EXERCISES

S1.1 Spreadsheet navigation: Do the following steps in order.
  a) Enter π in cell HG8976
  b) With cell HG8976 as the active cell, make cell A1 the active cell in one keystroke.
  c) Move the active cell from A1 down three screens without using the keyboard.
  d) Enter a formula in the active cell to calculate the sine of cell HG8976; the active cell location will vary depending on how step c) worked.
  e) Change HG8976's value to $\pi/2$ and return to the cell in step d).
  f) Enter a formula for cosine of HG8976 four cells to the right of and one cell down from the cell that has the formula for sine of HG8976.

S1.2 What three kinds of entries can be stored in a cell? Give an example of each.

S1.3 Create a table of powers of 16 that shows 16^0 to 16^7. Italicize the powers and bold the integers used for the exponents. Enter labels to clarify the spreadsheet. Change the alignment of the labels and/or the numbers so the table of values is readable.

S1.4 Size the powers column from Exercise S1.3 if necessary so all the values show. Format the numbers in an integer format so that groups of three digits are separated by commas.

S1.5 Apply a standard currency format to ALL cells in a spreadsheet.

S1.6 Create a spreadsheet that calculates the average of a column of numbers. (*Hint:* use the "COUNT" function ... it counts the number of occupied cells in a range.)

Select the column of numbers, not including the average, and move it three columns to the right and two rows down. Make the cell containing the average the active cell. Look at how the formula has changed.

S1.7 Select the whole sheet and erase all cells.

S1.8 Make the series $-1, -9, -17, -25 \ldots$ starting with -1 and -9. Extend it to 25 rows. Contract it to 10 rows.

S1.9 Generate a series for Mon, Wed, Fri that repeats in consecutive cells for 10 weeks: Mon, Wed, Fri, Mon, Wed, Fri, Mon, Wed, Fri. (*Hint:* use more than two terms as the basis for the series.)

S1.10 Create a series of radians that goes from 0 to 2π in steps of $\pi/20$.

S1.11 Put a formula for secant (1/cosine) in the cell to the left of 0 in the spreadsheet from Exercise S1.10. Move the column of radians if necessary. Copy the secant formula down to 2π.

Spreadsheets: Special Section 2

- Working with Formulas
- Functions in Formulas
- Copying Formulas
- Copying Relative Cell References
- Copying Absolute Cell References
- Important Concepts Review

- Graphing
- Importing Data into a Spreadsheet
- Importing Free Format Data
- Importing Tabular Data
- Important Concepts Review
- Exercises

Introduction

In Special Section 1, you were introduced to spreadsheet basics such as entering values into a cell, moving and copying groups of cells, creating series, etc. These fundamentals are essential for manipulating spreadsheets, but they are not glamorous. In this Special Section, the more exciting features are explicated. Through powerful formula creation and replication, and amazing graphing tools, you will be on your way to creating graphs such as the one seen in Figure S2.1 below. To quickly produce a graph such as the one in Figure S2.1 is an incredibly powerful facility. Think how difficult it would be to write a Fortran program to do the same job. There is even more power here than meets the eye.

Figure S2.1

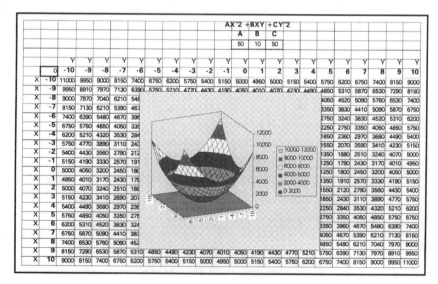

The paraboloid in Figure S2.1 is static as you see it above. When you actually work with such a graph, you can size it, change the equation coefficients and see an instantaneous change in the curve or, by using a pointing device, you can grab a corner of the graph and rotate it to see the curve from any perspective.

Before graphing is explained, formula, formula replication and functions will be explored so that you will be able to create the data that you want to graph. Formulas allow you to create expressions that refer to other cells, constants and/or functions. Functions generate some useful value based on an argument. An example of a trigonometric function is COS(cell-reference); this function will generate the cosine of whatever numeric value is contained in (cell-reference).

Working with Formulas

Spreadsheets would have been abandoned years ago without the ability to create and store formulas in cells. Formulas allow calculations to be made based on the contents of other cells.

In the example in Figure S2.2, only column A contains numeric constants. Columns B and C contain formulas that calculate the displayed numbers. Formula calculations in Figure S2.2 are based on the numbers in column A. Figure S2.3 shows a behind-the-scenes view that reveals the underlying formulas in Figure S2.2.

Figure S2.3 shows actual cell contents. Cells B5 through B8 and C5 through C8 contain formulas that display the calculated numbers seen in Figure S2.2. There is no difference between the displayed values and actual cell contents in cells B3, C2, C3, and A5 through A8. These cells contain either numeric constants or strings of characters.

In Figure S2.3, the cells that contain formulas work as follows: after the equal sign, which starts all formulas, an arithmetic expression involving a cell from column A defines

Figure S2.2

	A	B	C	D	E	F
1						
2			Square			
3		Square	Root			
4						
5	1	1	1.0000			
6	2	4	1.4142			
7	3	9	1.7321			
8	4	16	2.0000			
9						
10						
11						

Figure S2.3

	A	B	C	D	E	F
1						
2			Square			
3		Square	Root			
4						
5	1	= A5 * A5	= A5 ^ 0.5			
6	2	= A6 * A6	= A6 ^ 0.5			
7	3	= A7 * A7	= A7 ^ 0.5			
8	4	= A8 * A8	= A8 ^ 0.5			
9						
10						
11						

how to calculate the values displayed in columns B and C. For example, cell C6 holds the following formula:

$$=A6\wedge0.5$$

The value for A6 in the formula above is supplied by the contents of cell A6. After that value is substituted into the formula, the arithmetic expression becomes the following:

$$2\wedge0.5$$

This expression evaluates to 1.4142 ($\approx$ the square root of 2) and is displayed in cell C6 in Figure S2.2. Notice that the "types" of numeric constants in expressions are unimportant in spreadsheets. Unlike Fortran, spreadsheets do ALL their arithmetic operations in the equivalent of Fortran's `DOUBLE PRECISION` type. Then cell formats can display cell values in various ways, such as reals with four decimal places, currency, integers, etc.

A cell reference in a spreadsheet formula works just like a Fortran variable. Cell references have names such as D10 that represent values, and similarly, Fortran variables have names and can hold values.

As mentioned before, formulas are expressions composed of **operands** and **operators**. **Operands** may be any of the following:

1. Constants.
2. Cell references.
3. Functions.

Operators are:

    ^   exponentiation, H81^X56 $\equiv$ H81^{X56}
    *   multiplication
    /   division
    +   addition
    –   subtraction

The definitions above for operands and operators are not that different from Fortran's. Differences include spreadsheets' exponentiation operator, which is "^" instead of "**", and functions as operands. Fortran functions will be introduced as operands in Fortran expressions in Chapter 8 "Functions."

All the rules of **arithmetic order of precedence** in Fortran apply to expressions in spreadsheet formulas except for the evaluation of exponentiation. Exponentiation still has the highest precedence in spreadsheets, but it is evaluated from **left to right** instead of Fortran's method of **right to left**. This definitely makes a difference, as seen in the following example:

$$2\wedge 3\wedge 2 = 64 \quad \text{in a spreadsheet—left to right}$$
$$2**3**2 = 512 \quad \text{in Fortran—right to left}$$

Parentheses may be used to override order of precedence, just as in Fortran.

Functions in Formulas

A **function** is a **specialized variable** whose value is derived from the function's algorithm operating on any accompanying argument(s). For example, LOG10(x) is a spreadsheet function (and a Fortran function) that calculates $\log_{10}(x)$. LOG10, the function name, is like a Fortran variable and is set to the value of $\log_{10}(x)$. LOG10 then has a defined value that is used in whatever expression the LOG10 function occurred.

This process happens dynamically as expressions are evaluated. Some spreadsheet functions such as PI (which produces π: $\approx 3.1415927\ldots$) and RAND (which generates an apparently random decimal value between zero and one) don't require arguments.

Look at the following expression for hyperbolic cosine:

$$\cosh(x) = \frac{e^x + e^{-x}}{2}$$

The hyperbolic cosine of x is defined by the expression on the right-hand side of the equal sign. Spreadsheets have a function, EXP(X), that can calculate e^x (as does Fortran). Therefore, $\cosh(x)$ can be expressed as the following formula:

$$\frac{(\text{EXP}(\text{cell-reference}) + \text{EXP}(-\text{cell-reference}))}{2}$$

In the above expression, x is stored in cell "cell-reference." So how does this work? How will the spreadsheet evaluate the expression? The answer is that functions are evaluated first *before* any arithmetic operations are done. "cell-reference" is given to the EXP function as an argument and the function does whatever it does to calculate $e^{\text{cell-reference}}$ and puts the results in the expression. After all functions are evaluated, the arithmetic operations are performed according to the rules of arithmetic order of precedence (Figure S2.4). First, B3 and $-$B3 are supplied to the respective EXP() functions in the formula. Then $e^{3.1}$ and $e^{-3.1}$ are calculated. The resulting numbers are put into the expression in place of the EXPs. Then the expression is resolved by completing the addition in the parentheses and dividing by 2.

Figure S2.4

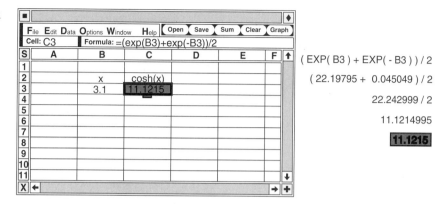

This leads to the unavoidable conclusion for expression evaluation: there is a **new** arithmetic order of precedence (Table S2.1).

Table S2.1

	Revised Arithmetic Order of Precedence for Spreadsheets	
Order	Operation	Direction
first	function evaluation	from left to right
second	^	from left to right
third	*, /	from left to right
fourth	+, –	from left to right

Copying Formulas

Copying cells that contain formulas does not work as one might think. Remember how copying was done in Special Section 1? A cell is made the active cell, "Copy" is chosen from the "Edit" menu, a destination range is selected, and finally, "Paste" is chosen from the "Edit" menu. The figure that illustrated this in Special Section 1 is reproduced here as Figure S2.5 (p. 212).

It all works in a very straightforward way. Copying a numeric constant such as 256 IS straightforward. Copying labels is also easy to understand because labels are constants too . . . character constants. But when formulas containing cell references are copied, the results are surprising, and pleasantly so. The way cell references are defined in a formula determines how those references are copied.

There are four ways to define cell references in formulas.

    1.   **Relative cell references.**   —D450
    2.   **Absolute cell references.**   —$D$450
    3 & 4.   **Combination cell references.**   —$D450, D$450

Figure S2.5

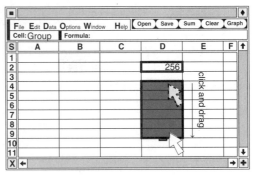

Select the cell to be copied and choose "Copy" from the Edit menu

Select the destination range using the click-and-drag technique

Choose "Paste" from the Edit menu

Cell D2 is copied into all the cells in the selected destination range

The example cell references, D450, $D$450, $D450 and D$450 refer to the same cell, but each of the four cell designations will be copied very differently. The two basic cell references, **relative** and **absolute**, are explained in the next two sections.

Copying Relative Cell References

Relative cell references are most common. When a cell reference is typed into a formula, it is a relative reference by default. (**Default** means a value or response that is automatically supplied if none is provided.)

To show how copying relative cell references works, a spreadsheet for "work" will be created. The work done, w, by applying force, f, at angle, θ, to the direction of the motion of an object over horizontal distance, d, is given by the formula

$$w = f \cos(\theta) d$$

A spreadsheet that shows various values for work is shown in Figure S2.6. Isn't it interesting that a variety of values are displayed in cells D4 through D9? Why are the values different? The formula in D3, which refers to cells A3, B3 and C3, was copied into cells D4 through D9; shouldn't all these cells contain the same formula and show the same values?

The answer lies in relative cell references. First, look at the formula originally entered into cell D3:

$$=B3*COS(A3*PI/180)*C3$$

B3's force value is multiplied by the cosine of θ, which is in A3. θ is converted from degrees to radians by multiplying it by PI/180 where PI is a spreadsheet function that always produces π: ≈3.14159265358979. Finally, the expression is multiplied by the distance value in cell C3, giving the result: 32.400.

Other formulas in cells D4 through D9 are shown in a behind-the-scenes view of the spreadsheet in Figure S2.7 (p. 214). When relative cell references in a formula such as the formula in cell D3 are copied into successive rows, the row parts of the cell references are incremented by one in each successive cell in the destination range. Cells B3, A3, and C3 in cell D3's formula become B4, A4, and C4 in cell D4, etc. This is called **relative cell replication**.

Figure S2.6

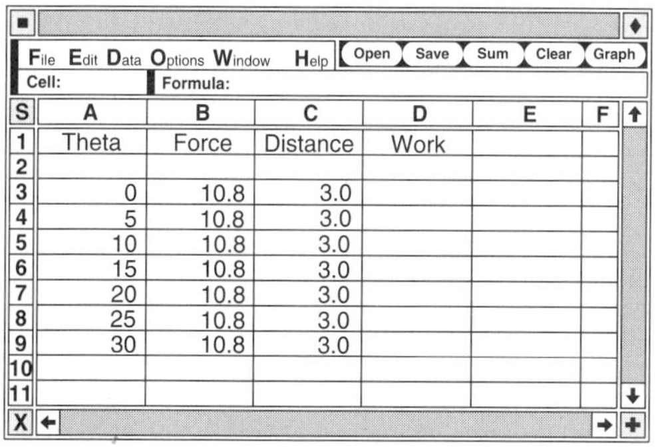

Values are entered for theta, force and distance

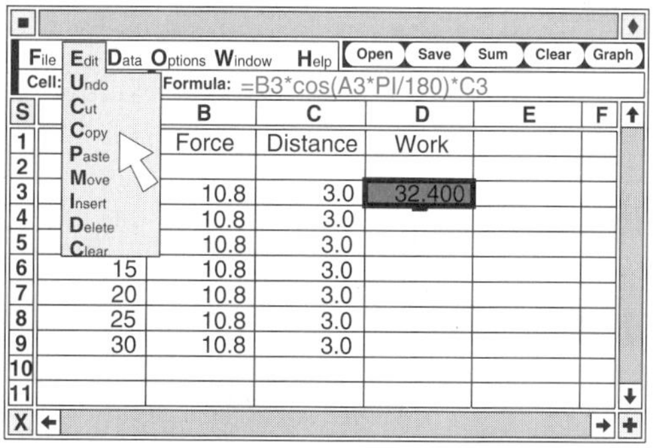

The formula for work is entered in D3 and Copy is chosen from the Edit menu

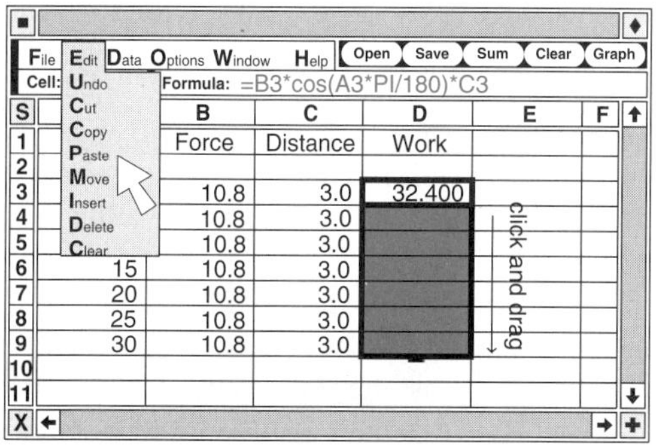

Click-and-drag the destination range and choose Paste from the Edit menu

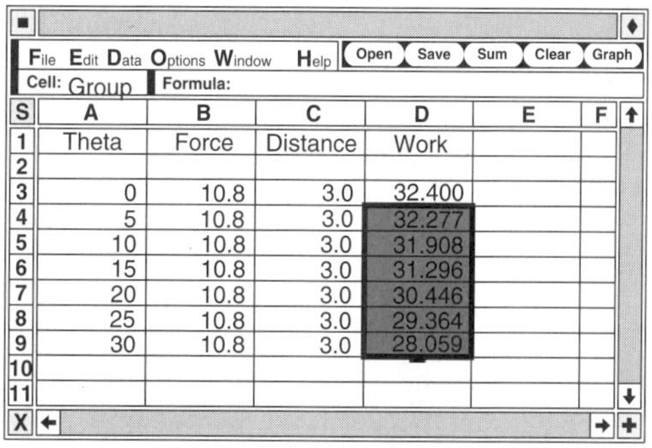

The formula in D3 is copied into the destination range and the corresponding values for "work" are displayed

Figure S2.7

Formula:				
B	C	D	E	F
		=B3*cos(A3*PI/180)*C3		
		=B4*cos(A4*PI/180)*C4		
		=B5*cos(A5*PI/180)*C5		
		=B6*cos(A6*PI/180)*C6		
		=B7*cos(A7*PI/180)*C7		
		=B8*cos(A8*PI/180)*C8		
		=B9*cos(A9*PI/180)*C9		

Figure S2.8

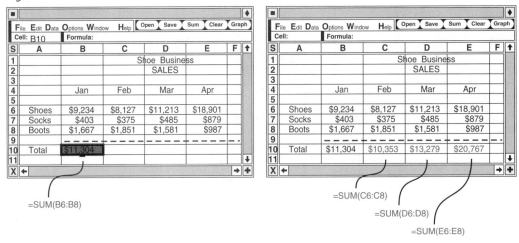

Relative cell replication works virtually the same way when copying into adjacent columns (Figure S2.8). In Figure S2.8, the formula in cell B10 contains a **SUM** function that totals up the values contained in cells B6, B7, and B8 (=SUM(B6:B8)). Rather than re-creating individual formulas in each of the total cells (C10, D10, E10, . . .) to calculate the various totals, cell B10 is simply copied into cells C10, D10, E10, ... The formula (=SUM(B6:B8)) contains relative cell references that will be copied in a relative way. As the formula in B10 is being copied across columns, the row numbers remain the same but the column letters change in a "relative" way, producing appropriate formulas for each total:

    Cell C10   Cell D10   Cell E10 . . .

    =SUM(C6:C8)   =SUM(D6:D8)   =SUM(E6:E8) . . .

Copying Absolute Cell References

When cell references are typed into a formula, they are relative references by default and will change when they are copied. Sometimes, however, it is necessary to copy formulas without all the cell references changing.

An example of this is calculating percentages. Assume a column of numbers is totaled and that value is shown in a cell at the bottom of the column. What percentage of the total is each number in the column? This example is shown in Figure S2.9.

Formulas in cells C3 through C7 contain the following:

C3	=B3/$B$9*100
C4	=B4/$B$9*100
C5	=B5/$B$9*100
C6	=B6/$B$9*100
C7	=B7/$B$9*100

Cell references that have their column or row indicators preceded by a dollar sign, such as $B$9, DON'T CHANGE when they're copied. This style of cell reference is called **absolute cell reference**, and copying absolute cell references is called **absolute cell replication**.

The combination of relative and absolute cell referencing makes creating the percentage column in Figure S2.9 very easy. First, enter the following formula in cell C3:

$$=B3/\$B\$9*100$$

Then use standard copying techniques to replicate the formula in cells C4 through C7. The formula being copied has a relative cell reference, B3, which will change according to the rules of relative cell replication. On the other hand, $B$9 is an absolute cell reference and will NOT change. The formula just needs to be entered once in cell C3 and then copied into C4 through C7. Relative and absolute cell referencing takes care of everything else in the copying process.

Figure S2.9

S	A	B	C	D	E	F
1						
2			Percent			
3		34.132	8.7	%		
4		76.980	19.7	%		
5		97.009	24.8	%		
6		93.451	23.9	%		
7		89.771	22.9	%		
8						
9	Total =	391.343				
10	Average =	78.269				
11						

Important Concepts Review

- **Formulas** start with an equal sign followed by an expression made of **operands** and **operators**.

- **Spreadsheet formula operands** may be:
    1. Constants.
    2. Cell references.
    3. Functions.

- **Spreadsheet formula arithmetic operators** are:

    ^   exponentiation, H81^X56 <==> H81^{X56}
    *   multiplication
    /   division
    +   addition
    −   subtraction

- A **function** is a specialized variable whose value is derived from the function's algorithm operating on any accompanying argument(s). For example, LOG10(x) is a spreadsheet function that calculates $\log_{10}(x)$. LOG10, the function name, is like a Fortran variable and is set to the value of $\log_{10}(x)$.

- **Some functions** such as PI (which produces π: ≈3.1415927) and RAND (which generates an apparently random decimal value between zero and one) **don't require arguments**.

- **Arithmetic order of precedence is revised!** All function evaluation is completed before any arithmetic operations are performed.

- **Relative cell replication** occurs when copying formulas and means that cell references change relative to the location of the cells being copied.

- **Absolute cell replication** means that cell references that have a dollar sign preceding the row or column designator, such as $J$67, don't change when copied—they refer to an "absolute" location on the spreadsheet.

Graphing

Graphing is an extremely vital part of data analysis. The implications of a set of data may not be immediately obvious by looking at the values, but the data can be interpreted quickly when it is represented graphically. There are few tools that can outperform a spreadsheet when it comes to creating graphs.

Basic steps for graphing spreadsheet data are listed below. These steps vary from spreadsheet to spreadsheet. It may be necessary to use the spreadsheet's help facility or documentation to figure out the *exact* steps to follow to create a graph in a particular spreadsheet.

1. **Select the columns or rows** that contain the data to be graphed.
2. **Click the graph tool** on the spreadsheet's toolbar or make the appropriate menu selections. Three things appear: a rectangle for the graph, a dialogue box to configure the graph and a graph-style toolbar.
3. **Fill in the dialogue box**. Answer the question: **is the first data set to be graphed or used as the values for the x axis scale?** Supply axis labels as needed.
4. **Click the graph style** that best reveals the important features of the data.
5. **Click different graph-style toolbar buttons** to change the appearance of the graph.

As an example of spreadsheet graphing, a graph for a simple quadratic equation is created.

$$y = x^2$$

First, generate the data for the graph by using spreadsheet series and relative cell replication (Figure S2.10, p. 218). Next, actually create a graph of the spreadsheet data (Figure S2.11, p. 219). Once the basic graph has been created, the data can be viewed in different graph styles by simply clicking the different buttons on the graph-style toolbar (Figure S2.12).

Think about what's going on here . . . how easy it is to create and change graphs in spreadsheets. Spreadsheets make creating the *data* for a graph effortless with **series** and **formula replication**. Creating the *graph* is also simple. Select the data to be graphed, click the graph-style toolbar button, make a few choices to configure the graph, and a scaled graph appears on the screen. Spreadsheet graphing is amazingly effortless and flexible. Imagine what a daunting task it would be to create a Fortran program that could graph like a spreadsheet!

The next interesting feature of spreadsheet graphing is that multiple data sets can be graphed on the same graph. To accomplish this, put the y values for each curve in adjacent columns. If those columns are all selected before the graph-style toolbar button is clicked, all the curves are plotted together.

One example of data that would be appropriate to graph as multiple curves on the same chart is the "Shoe Business" spreadsheet presented in the introduction to Special Section 1. There are three items being sold in the Shoe Business: shoes, socks and boots. It might be meaningful from a business point of view to graph each of the three products on the same chart to see how their sales relate to each other over time (Figure S2.13, p. 220). Figure S2.13's graph clearly shows that as the weather warms up, shoe sales increase while boot sales drop off. These sales trends are probably not particularly interesting to nascent engineers and scientists, but plotting multiple data sets on the same graph is also used regularly in technological applications.

Figure S2.10

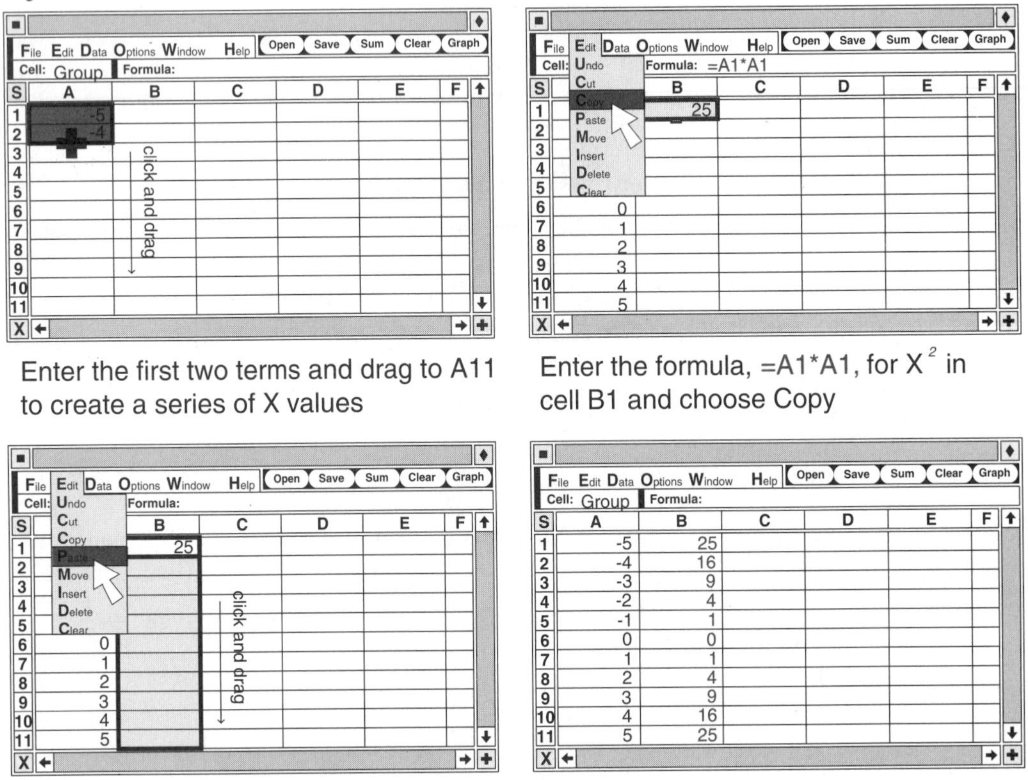

[Spreadsheet panel 1] Enter the first two terms and drag to A11 to create a series of X values

[Spreadsheet panel 2] Enter the formula, =A1*A1, for X^2 in cell B1 and choose Copy

[Spreadsheet panel 3] Click and drag the destination range and choose Paste

[Spreadsheet panel 4] Relative cell replication causes the various values for X^2 to be displayed

Taking spreadsheet graphing a step further to demonstrate its flexibility, the graph of the simple parabola in Figure S2.12, $y = x^2$, will be generalized to plot any quadratic equation. The general form of a quadratic equation is shown below:

$$y = ax^2 + bx + c$$

A graph will be created by designating three cells in the spreadsheet as the coefficients a, b and c. When these coefficients are changed, the impact on the curve will be observed. Because the formulas for this spreadsheet use both relative and absolute cell replication, the formulas' creation is shown in detail in the next two figures.

$ax^2 + bx + c$ is represented as a formula in cell B1 (Figure S2.14, p. 221). This formula is composed of both relative and absolute cell references. Cell B1 is copied into cells B2 through B11. After the copying process is finished, the formulas in column B appear as seen in the behind-the-scenes view of the spreadsheet in Figure S2.15 (p. 221). Notice how the x references (A1, A2, . . . , A11) change according to the rules of relative cell replication, while the coefficients ($C$2, $D$2 and $E$2 . . . a, b, and c, respectively) don't change because they are absolute cell references. Once the formulas are copied as seen in Figure S2.15, the data is graphed (Figure S2.16, p. 222). When different graph-style tool-

Figure S2.11

Click-and-drag to select the values to be graphed

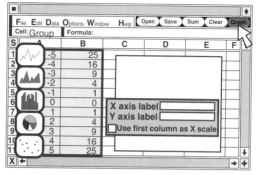
Click the "Graph" toolbar button; a graph-style toolbar, rectangle for the graph and dialogue box appear

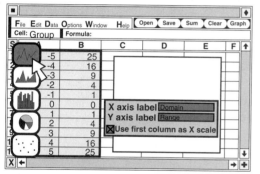
Fill in the dialogue box and click a graph-style button

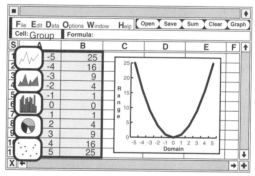
Presto! The data is graphed

bar buttons were clicked in Figure S2.12, the graph style changed immediately. There is also an immediate response when a new value, which affects the curve's data, is entered in the sheet. The sheet is recalculated and all graphs are redrawn to reflect the changes (unless the sheet is configured to not recalculate automatically).

Figure S2.17 (p. 222) shows how various values for a, b and c affect the parabola in Figure S2.16. When changes are made to the coefficients, the curve is automatically adjusted. ANY change in the graph's data will cause the same kind of immediate change in the curve. If, for instance, the domain in column A is changed to a different set of values, the range values change immediately and so does the curve (Figure S2.18, p. 223).

Figure S2.18 shows two new sets of domain values. After each change in the domain, the graph is immediately redrawn and scaled to reflect the new domain and range values.

The next graphing example illustrates some additional powerful mathematical and logical spreadsheet functions. A graph will be created that shows sine, cosine and tangent of θ all on the same chart. θ's values will range from 0 to 8π to give four complete periods of the sine/cosine curves. Data for the graph is created the same way as before: series creation

Figure S2.12

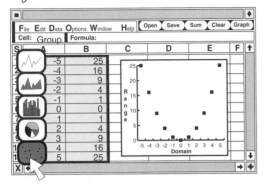
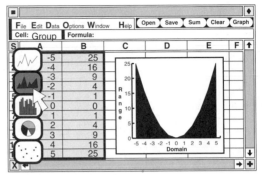
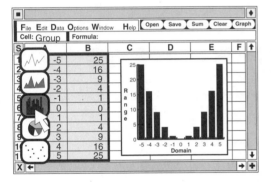
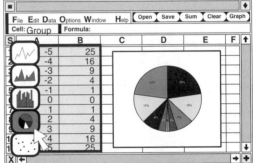

Figure S2.13

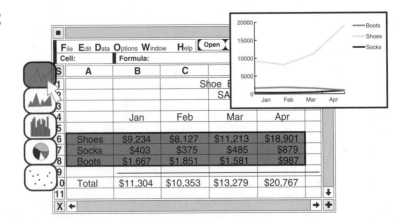

Figure S2.14

```
┌─────────────────────────────────────────────────────────────┐
│ ■                                                         ◆ │
│ File Edit Data Options Window Help  Open Save Sum Clear Graph│
│ Cell: B1        Formula: =$C$2*A1^2+$D$2*A1+$E$2            │
│  S    A        B         C         D         E      F   ↑  │
│  1   -5      146         A         B         C             │
│  2   -4                  5        -2        11             │
│  3   -3                                                     │
│  4   -2                                                     │
│  5   -1                                                     │
│  6    0                                                     │
│  7    1                                                     │
│  8    2                                                     │
│  9    3                                                     │
│ 10    4                                                     │
│ 11    5                                                  ↓  │
│ X ←                                                    → + │
└─────────────────────────────────────────────────────────────┘
```

$$=\$C\$2*A1\char94 2+\$D\$2*A1+\$E\$2$$
$$A * X^2 + B * X + C$$

Figure S2.15

```
┌─────────────────────────────────────────────┐
│ Formula:                                    │
│    B              C         D       E   F ↑ │
│ =$C$2*A1^2+$D$2*A1+$E$2                     │
│ =$C$2*A2^2+$D$2*A2+$E$2                     │
│ =$C$2*A3^2+$D$2*A3+$E$2                     │
│ =$C$2*A4^2+$D$2*A4+$E$2                     │
│ =$C$2*A5^2+$D$2*A5+$E$2                     │
│ =$C$2*A6^2+$D$2*A6+$E$2                     │
│ =$C$2*A7^2+$D$2*A7+$E$2                     │
│ =$C$2*A8^2+$D$2*A8+$E$2                     │
│ =$C$2*A9^2+$D$2*A9+$E$2                     │
│ =$C$2*A10^2+$D$2*A10+$E$2                   │
│ =$C$2*A11^2+$D$2*A11+$E$2                ↓ │
│                                         → + │
└─────────────────────────────────────────────┘
```

and formula replication. Values for θ fill many more cells than are visible on the sheet because the increment for θ is 0.1 radians.

Figure S2.19 (p. 223) shows the spreadsheet after the basic data is entered. (The reason for column D is thoroughly explained after Figure S2.20 on p. 224.) Column E contains the actual values for tan(θ), which is a formula that calculates the tangent of the A column entries. However, there is a problem with these values. As θ approaches any multiple of π/2, tan(θ)'s magnitude gets very large (Figure S2.20). This causes the characteristic peaks of a tan(θ) graph. The values of θ will not be close enough to a multiple of π/2 to cause a division-by-zero error in the spreadsheet, but the tan(θ) peaks WILL cause the graph to look wrong because the spreadsheet automatically scales the graph to the highest and lowest points.

Figure S2.16

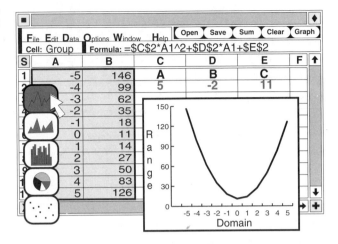

Figure S2.17

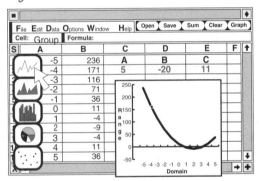

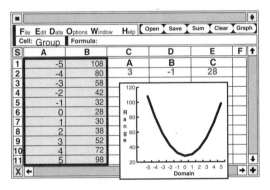

Notice how tan(θ)'s magnitude peaks on row 111 of the spreadsheet ($-225.9508 =$ tan($\approx 7 * \pi/2$)). If the E column values for tan(θ) were used for the graph, the three curves—sine, cosine and tangent—would be scaled to -225.9508, the largest magnitude of all three curves. Because the other two curves—sin(θ) and cos(θ)—have a maximum peak (amplitude) of 1, they would be lost in a graph where tan(θ)'s peaks are scaled correctly.

Figure 52.18

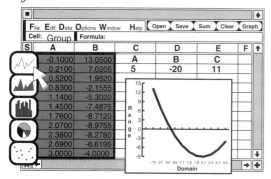
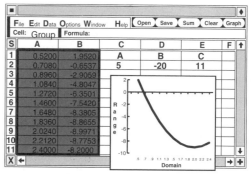

Figure 52.19

S	A	B	C	D	E	F
1	Theta	sin	cos	clipped tan	tan	
2						
3	0.0000	0.0000	1.0000	0.0000	0.0000	
4	0.1000	0.0998	0.9950	0.1003	0.1003	
5	0.2000	0.1987	0.9801	0.2027	0.2027	
6	0.3000	0.2955	0.9553	0.3093	0.3093	
7	0.4000	0.3894	0.9211	0.4228	0.4228	
8	0.5000	0.4794	0.8776	0.5463	0.5463	
9	0.6000	0.5646	0.8253	0.6841	0.6841	
10	0.7000	0.6442	0.7648	0.8423	0.8423	
11	0.8000	0.7174	0.6967	1.0296	1.0296	

=sin(A3) =cos(A3) =IF(E3>3,3,IF(E3<-3,-3,E3)) =tan(A3)

To solve this problem, a spreadsheet function that can perform selection operations is used to clip off the tangent peaks so that their maximum amplitude is 3. The Forsythe Spreadsheet selection function works as follows:

IF (logical-expression, value-if-true, value-if-false)

This function is equivalent to the following Fortran block-if selection structure:

```
IF ( logical-expression ) THEN
        display value-if-true in the cell
ELSE
        display value-if-false in the cell
END IF
```

Figure S2.20

S	A	B	C	D	E	F
1	Theta	sin	cos	clipped tan	tan	
106	10.5000				1.8499	
107	10.6000				2.3947	
108	10.7000				3.2841	
109	10.8000				5.0478	
110	10.9000				10.4312	
111	11.0000				-225.9508	
112	11.1000				-9.5414	
113	11.2000				-4.8234	
114	11.3000				-3.1828	
115	11.4000				-2.3363	

To graph the specific data in Figure S2.20, a more complicated selection structure is needed—a nested "IF" function is used to create the column D values:

$$IF(E3 > 3, 3, IF(E3 < -3, -3, E3))$$

To understand how this function usage works, examine the Fortran equivalent:

```
IF ( E3 > 3 ) THEN
        display 3 in cell D3
ELSE IF ( E3 < -3 ) THEN
        display -3 in cell D3
ELSE
        display cell E3's contents in cell D3
END IF
```

The IF formulas in column D change the values of the tangent curve so that if it exceeds 3 in either a positive or negative direction, it is "clipped" to 3 or −3, as seen in Figure S2.21.

Figure S2.21

S	A	B	C	D	E	F
1	Theta	sin	cos	clipped tan	tan	
106	10.5000			1.8499	1.8499	
107	10.6000			2.3947	2.3947	
108	10.7000			3.0000	3.2841	
109	10.8000			3.0000	5.0478	
110	10.9000			3.0000	10.4312	
111	11.0000			-3.0000	-225.9508	
112	11.1000			-3.0000	-9.5414	
113	11.2000			-3.0000	-4.8234	
114	11.3000			-3.0000	-3.1828	
115	11.4000			-2.3363	-2.3363	

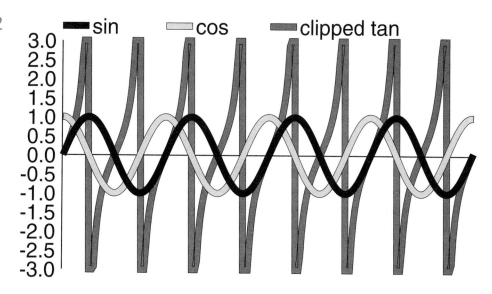

Figure S2.22

Column D now contains reasonable values for the tan(θ) graph that will not dwarf the other curves. The final plot of the functions is the splendid diagram in Figure S2.22.

There is no real limit to the graphs that can be created in a spreadsheet, and the process of creating a graph such as the one in Figure S2.22 is relatively simple once spreadsheet series and formula replication are mastered.

Importing Data into a Spreadsheet

In the real world, scientists and engineers often have large sets of data that need to be analyzed. When the data need the kind of analysis that spreadsheets offer, the information can be **imported** into a spreadsheet and processed. The imported data is divided into sequential columns and is then at the will of the spreadsheet user to analyze and graph as needed.

Importing data into the Forsythe Spreadsheet depends on how the data file looks. There are two types of file formats that are commonly imported into a spreadsheet:

1. **Free format** data. This style of data file usually has variable length lines with values separated by some special character, such as a comma.
2. **Tabular** data. Each line of the data file is the same length and all lines' values are spaced the same way.

Importing Free Format Data

As mentioned above, free format files use a special character to separate values from each other within lines of a data file. This special character is called a **delimiter** (*delimit* means to establish boundaries) and is most often a **space**, **comma** or **TAB**. Words in this sentence are delimited by spaces. Sentences are delimited by a period and a space, etc.

Figure S2.23's data file uses commas to delimit individual datums. It doesn't matter whether there are extra spaces before or after the commas. Sequential commas as in the last line of file `final.dat` (7,,,6) cause columns in the spreadsheet to be skipped when the data is imported.

Steps for importing delimited, free format data files are as follows:

1. Select "**Import**" from the "**File**" menubar.
2. Check the "**Free Format**" check box.
3. Check the appropriate **delimiter** check box for the data file.
4. Select the file name.
5. Click the "**Import**" button.

As explained previously, other "real" spreadsheets will probably have slightly different steps to follow. Consult the documentation and help facility if the steps above don't work exactly right.

Figure S2.24 illustrates how the importing process works. The data file from Figure S2.23 is imported into the Forsythe Spreadsheet. At this point, the data from the data file

Figure S2.23

Data file: `final.dat`				
10.789,	678.0,	23.	,72	
10.89,	678.0,	23.	,	72
10.9,	678.0,	23.,		72
10.,	678.0,	23. ,		72
33., 7.6, 5.211, 5.9				
3 , 3		, 3 ,	7	
4 ,4 ,4 ,4				
7,,,6				

Figure S2.24

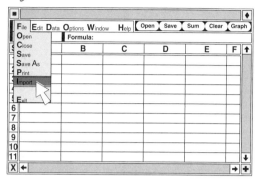

Choose "Import" from the "File" menu

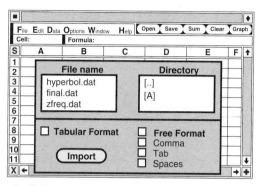

A dialogue box appears

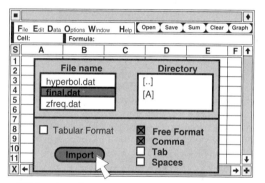

Fill in the check boxes, choose the file and click the Import button

The data is imported into contiguous columns

is in the spreadsheet and can be formatted, analyzed, graphed or manipulated in any way that spreadsheets can process data.

Importing Tabular Data

The second style of data file that is commonly imported into spreadsheets is the tabular data file. A tabular data file is one where the data is evenly spaced or, at least, all of the file lines look the same with respect to spacing. Such a file is seen in Figure S2.25 (p. 228).

Importing tabular data files is the same as importing free format data files except that tabular files only require the "Tabular Data" check box to be checked; a delimiter doesn't need to be specified (Figure S2.26).

Spreadsheets: Special Section 2

Figure S2.25

Data File: regress.dat			
1.3092	.7674	2.1541	.9734
1.2308	.6670	1.9483	.9602
1.1524	.5454	1.7253	.9385
1.0741	.3826	1.4661	.8988
4.9076	2.2734	9.7122	1.0000
1.9901	1.3112	3.7107	.9988
2.6501	1.6301	5.1043	.9999
3.4471	1.9089	6.7460	1.0000

Figure S2.26

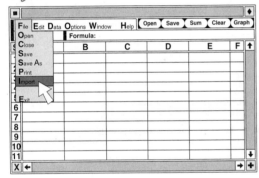

Choose "Import" from the "File" menu

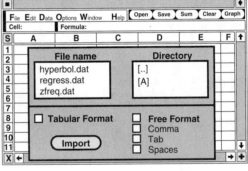

A dialogue box appears

Check the "Tabular Format" check box, choose the file and click the Import button

The data is imported into contiguous columns

Important Concepts Review

- **To create a graph:**
    1. **Select the columns or rows** that contain the data to be graphed.
    2. **Click the graph tool** on the spreadsheet's toolbar. Three things appear: a rectangle for the graph, a dialogue box to configure the graph and a graph-style toolbar.
    3. **Fill in the dialogue box.** Answer the question: **is the first data set to be graphed or used as the values for the *x* axis scale?** Supply axis labels as needed.
    4. **Click the graph style** that best reveals the features of the data.
    5. **Click different graph-style toolbar buttons** to change the appearance of the graph.

- **Graphing general equations** can be accomplished by putting the coefficients in cells that are referred to with absolute cell references.

- When **new values are entered**, the spreadsheet recalculates the values of all formulas and redraws all graphs to reflect the new values (unless the sheet is configured to not recalculate automatically).

- **IF (logical-expression, value-if-true, value-if-false)** is a spreadsheet function that displays different values in a cell depending on a logical expression.

- **Importing data** means bringing an external data set into a spreadsheet for analysis.

- A **free format** data file usually has variable length lines with values separated by some special character.

- A **tabular** data file has lines in the data file that are the same length and its values are evenly spaced.

- A **delimiter** is a special character used to separate values on a line of a data file (*delimit* means to establish boundaries). The most common delimiters are **spaces**, **commas**, and **TABS**.

- **To import data:**
    1. Select "**Import**" from the "**File**" menu bar.
    2. Check the "**Free Format**" or "**Tabular Format**" check box.
    3. Check the appropriate **delimiter** check box for free format data files.
    4. Select the file name.
    5. Click the "**Import**" button.

EXERCISES

S2.1 Make a spreadsheet that has a hundred Fibonacci numbers. Fibonacci numbers are defined such that each number is the sum of the previous two: 1, 1, 2, 3, 5, 8, 13, 21, ...

S2.2 Create a spreadsheet that contains data for graphing a general cubic equation:

$$y = ax^3 + bx^2 + cx + d$$

Put the domain values in row 5 and the range values in row 6. Graph it. Try different graph styles (pie chart, too). Experiment with different coefficients (including zeros).

S2.3 Enter a series of integers from 0 to 25. In a nonadjacent column, generate each integer's factorial. A factorial is defined as follows:

$$n! = 1 \cdot 2 \cdot 3 \cdot \ldots \cdot (n-1) \cdot n$$

Note: 0! is defined to be 1.

S2.4 Using the spreadsheet from Exercise S2.14, create a column of terms of the infinite series for e^x (put x in some cell):

$$e^x = \sum_{i=0}^{n} \frac{x^i}{i!}$$

Sum the column of terms into a cell. In the cell below that sum, enter a formula that calculates e^x using the spreadsheet function EXP(). The two cells should have approximately the same values. Change the value of x a few times. As x gets larger, the series calculation gets less accurate. Add more terms to the series.

S2.5 Create a graph of $y = e^x$ where x goes from 0.05 to 3.0 in increments of 0.05.

S2.6 Explain how function evaluation fits in with arithmetic order of precedence. Show an example.

S2.7 Make a table of electrical power (watts), which is defined as

$$p = ie$$

where i is current and e is voltage. Create i and e using spreadsheet series. Graph p, i and e all on the same chart. Change the values for i and observe the way the curves change. Change the values for e.

S2.8 Enter ten students' names in a column. In the four adjacent columns to the right, enter grades in a 1-to-100 point scale. In the fifth column, calculate each student's average. At the bottom, calculate the class average on each of the four graded assignments.

S2.9 Using the sheet from Exercise S2.8, calculate the total class average.

S2.10 Again, use the spreadsheet from Exercise S2.8 and create a sixth column that will use the spreadsheet IF() function to convert individual student averages to a 4.0 point scale. The 4.0 point scale is defined as follows:

average >= 93,	4.0
88 <= average < 93,	3.5
83 <= average < 88,	3.0
78 <= average < 83,	2.5
73 <= average < 78,	2.0
68 <= average < 73,	1.5
60 <= average < 68,	1.0
average < 60	0.0

S2.11 Make the spreadsheet from Exercise S2.10 "pretty." Add labels, formats, fonts, alignment, etc. Move the cells around as necessary . . . the correct cell referencing will be maintained by the spreadsheet.

S2.12 Fill a column of 25 cells with a formula that will generate random numbers between 0 and 14 to represent pH values. The RAND spreadsheet function will create a random decimal between 0 and 1 noninclusively, i.e., 0.0 < RAND < 1.0. (*Hint: the RAND value must be operated on to give values between 0 and 14 rather than between 0 and 1.*)

Surround this random number *expression* with the INT() function to change the expression to an integer (INT() truncates any decimal part).

In an adjacent column, use an IF function to display "Acid" if the pH is less than 7.0, "Alkaline" if the pH is greater than 7.0, or "Neutral" if the pH is exactly 7.0. Each time the sheet is recalculated (this can be done manually with the F9 key in most spreadsheets), the RAND expressions' values will change.

Functions

Introduction

Somewhere along the educational path that led you to reading this text, you studied mathematical functions such as:

$$y = f(x)$$

The idea behind a mathematical function is that if you give it some x_i, the function will crank out a y_i based on how $f(x)$ is defined. For example, if $f(x)$ is $2x - 6$, a set of y values can be generated for any set of x values:

x_i	y_i
7	8
8	10
9	12
...	

For each x, there is exactly one y. The American Heritage® dictionary defines a mathematical function as:

> **func•tion** (fungk'-shen) *n*. . . . **5.** *Abbr.* **f** *Mathematics.* **a.** A variable so related to another that for each value assumed by one there is a value determined for the other. **b.** A rule of correspondence between two sets such that there is a unique element in the second set assigned to each element in the first set.*

Fortran functions have the same kinds of features as mathematical functions, " . . . give it some x_i, the function will crank out a y_i . . . <u>A variable</u> so related to another that for each value assumed by one, there is a value determined for the other . . . A rule of correspondence. . . . " When you put values into a Fortran

8.1 What is a Fortran function?

8.2 Styles of functions

   Important Concepts
   Review

8.3 Program design example

   Important Concepts
   Review

   Exercises

---

*Copyright © 1996 by Houghton Mifflin Company. Reproduced by permission from *The American Heritage Dictionary of the English Language,* Third Edition.

function, the values are processed through a "rule of correspondence" (the function's algorithm) to produce the function result (the "cranked out y_i"). An example of a simple Fortran function is the absolute value function:

$$ABS (X)$$

The purpose of the ABS function is to produce the absolute value of argument X. This ABS function could, for instance, be used in an assignment statement to make X's contents positive:

$$X = ABS (X)$$

Fortran functions can also be used in more complicated expressions such as:

$$(-B + SQRT (B * B - 4 * A * C)) / 2 / A$$

This expression calculates one root of a general quadratic equation using Fortran's square root function.

Fortran functions add a powerful new dimension to expressions in programs. In this chapter, you will learn how to use Fortran's built-in functions and how to create your own customized functions.

● ● ●

8.1 What Is a Fortran Function?

In the introduction, Fortran functions were likened to mathematical functions. That is a reasonable metaphor, but in Fortran, rather than working with an abstract math function, actual data is being processed. Fortran functions operate on the practical level of algorithms and data structures.

A **function** is a specialized **variable** whose value is dynamically *derived* from the function's algorithm operating on any accompanying **argument**(s). For example, LOG10(X) is a Fortran function that calculates $\log_{10}(x)$. LOG10, the function name, is a variable and is set to the calculated value of $\log_{10}(x)$ within the function.

To understand exactly how this explanation of Fortran functions works, examine Figure 8.1 (p. 234). In Figure 8.1, the value of variable theta is handed off to the COS function's algorithm, which does whatever processing is necessary to calculate the cosine of theta. This "resulting value" is then stored into the function's name, COS, which is a variable. Function COS( ) can be used in any **numeric expression**. Figure 8.2 shows how to use the COS( ) function in an expression to calculate the secant of theta.

In the assignment statement in Figure 8.2, the expression on the right-hand side of the equal sign is evaluated and the result is stored into variable secant. COS (theta) is dynamically evaluated first in the expression. The result of the function's algorithm, which is cosine (theta), is put into the function's name: COS. *Then* the division 1/COS is performed, and that final value is stored into variable secant.

In an expression, all functions are evaluated before any other operations are performed. This makes sense because all variables in an expression must be "defined" (have been given values) before they can be used, and function variables get their values from their algorithms *as* an expression is being evaluated.

This leads to the conclusion that operator order of precedence is different than originally explained (Table 8.1). The reason Table 8.1 has generic entries is that there is an order of precedence for arithmetic, relational and logical operators. Functions are evaluated first in any type of expression.

Since function names are variables, they must have a type and structure as do all Fortran variables. A function variable, which is the function's name, may be any legal Fortran type including INTEGER, REAL, DOUBLE PRECISION, COMPLEX, LOGICAL, CHARACTER or TYPE ().

As explained before, functions derive their results from value(s) that are put in the function's parentheses. These values are called **actual arguments** and may be constants, variables or expressions. Arguments in functions should **always** be INTENT (IN).

Figure 8.1
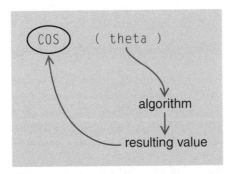

Figure 8.2
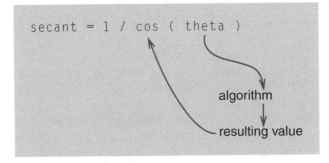

Table 8.1

Revised Order of Precedence	
Order	Operation
1	function evaluation
2	other operators
...	...

In the following program fragment, `String` is an actual argument for Fortran's `LEN_TRIM` function.

```
   . . .
   CHARACTER ( LEN = 10 )    ::    String = "Fortran"
   PRINT *, "The Length of String is:", LEN_TRIM ( String )
   . . .
```

`LEN_TRIM` is an `INTEGER` function that requires a `CHARACTER` type actual argument and calculates how many characters are in it, excluding all trailing blanks. Since `Fortran` has seven characters in it, the `PRINT` statement in the example above displays the following:

```
              The Length of String is: 7
```

Although function names are variables, they can't be used in all the situations that a variable can. A standard variable must be given a value before it is used. This is usually accomplished by a `READ` statement, an assignment statement, a `DATA` statement, etc. A function variable, however, derives its value from the function's algorithm and must NOT be the object of any statement that puts a value into a variable. The following `READ` statement illustrates a **wrong** use of functions.

```
                    READ *, SQRT ( 9 )
```

Fortran's `SQRT` function derives its value from its square root algorithm by processing the actual argument (9 in the example above) and must not be in a `READ` statement. Table 8.2 lists situations where functions **cannot** be used.

Table 8.2

Don't Use a Function . . .
as an input variable in a `READ` statement
in a `DATA` statement
in a declaration statement that assigns an initial value
in a declaration with the `PARAMETER` attribute
in a `PARAMETER` statement
on the left side of the equal sign in an assignment statement
as an `INTENT ( OUT )` or `INTENT ( INOUT )` argument

8.2 Styles of Functions

Fortran functions come in two basic styles:

1. **Intrinsic Functions**—built in.
2. **Program Unit Functions**—custom algorithm.

Intrinsic Functions

The first style of function is the **intrinsic function**. Intrinsic means "built in" or "part of." Intrinsic functions are those that come with a compiler as its library. Intrinsic Fortran functions can be used in any Fortran program unit. In addition, intrinsic functions may be used without declaring or describing them in any way; they are completely "known" to the compiler. The machine language code for intrinsic functions is added to programs in the **linking** process, which produces the executable version of the program.

Program designers should think of intrinsic functions as **black boxes**. To use any black box from a given library, only the black box's **interface** needs to be understood. An interface consists of a name for the black box and the order and type of any required argument(s). To use Fortran's tangent function, for example, the function's name and argument must be understood (Figure 8.3).

Function variable, TAN, is a REAL variable that is set to the value of tan(x) where x is in radians and is a REAL constant, variable, or expression. With this understanding of the TAN function's interface, the function can be **used** without understanding **how** the function calculates tan(x). There are many high-powered mathematical, LOGICAL, and CHARACTER functions in the Fortran intrinsic library. See Appendix B.

Program Unit Functions

Although Fortran's intrinsic library is extensive, situations arise where customized functions must be designed. **Program unit functions** solve this problem and are the second style of functions mentioned in the previous section. An example of a program unit function is illustrated in Figure 8.4. Figure 8.4's function is designed to calculate the distance between any two points on a Cartesian plane. Notice that the function's name is **declared**

Figure 8.3

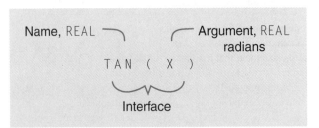

Figure 8.4

```
FUNCTION   Distance_Between_Points (x1, y1, x2, y2)
           IMPLICIT    NONE
!
!------------------------------------------------------------------
!
!   This function calculates the distance between two Cartesian points
!     (x1, y1) and (x2, y2).
!
!   Distance_Between_Points - is a REAL variable (the function's name)
!                             and will be set to the value of the
!                             calculated distance
!
!   x1, y1, x2, y2          - are INTENT ( IN ) REAL arguments that
!                             represent the points (x1, y1) and
!                             (x2, y2)
!------------------------------------------------------------------
!
           REAL,      INTENT (IN)     ::    x1, y1, x2, y2
           REAL                       ::    Distance_Between_Points
!
    Distance_Between_Points = SQRT(( x2 - x1 ) ** 2 + ( y2 - y1 ) ** 2 )
!
END FUNCTION   Distance_Between_Points
```

inside the function as a variable of type REAL. *Function names are variables* and must be assigned a value somewhere within the body of the function so they *must be declared*. All formal arguments have the INTENT (IN) attribute, and the function name is **given a value** in the assignment statement shown below:

```
Distance_Between_Points = SQRT(( x2 - x1 ) ** 2 + ( y2 - y1 ) ** 2 )
```

The function name *must* be given a value in the body of the function. That is how the function derives its value. In function Distance_Between_Points, the Fortran intrinsic function SQRT( ) is used in the expression that calculates the distance. It is perfectly reasonable to use functions within other functions. Program unit functions and intrinsic functions can be used in any program unit.

Intrinsic functions don't need to be declared because they are "known" to the compiler; they are in its library. This is not so for *program unit* functions. Their names, which are variables, must be declared in any program unit that uses them. Figure 8.5 (p. 238) shows an example of how to *use* a program unit function. Distance_Between_Points from Figure 8.4 is implemented in a program designed to find the distance between the closest of four points on a Cartesian plane (Figure 8.5).

Figure 8.5

```
PROGRAM  Closest_Cartesian_Points
              IMPLICIT   NONE
    !
    !------------------------------------------------------------------
    ! Written by      :    C. Forsythe
    ! Date Written    :    3/25/2004
    !------------------------------------------------------------------
    ! This program will find the distance between the two closest points
    !    of four ordered pairs in a Cartesian plane.
    !
    ! Distance_Between_Points - A function that calculates the distance
    !                              between two points
    !
    ! Dist1, Dist2, Dist3, Dist4, Dist5, Dist6  - REAL variables that hold
    !                the various combinations of distances
    !                between points
    !
    ! x1, y1, x2, y2, x3, y3, x4, y4 - REAL  variables that represent the
    !                                    ordered pairs
    !
    ! Shortest_Distance  - REAL variable to hold shortest of all distances
    !------------------------------------------------------------------
        !  BLACK BOX:   Library subroutine ...
        !                    Distance_Between_Points ( x1, y1, x2, y2 )
        !
        !  This function calculates the distance between two Cartesian
        !     points (x1,y1) and (x2,y2).
        !
        !  Distance_Between_Points - is a REAL variable (the function's
        !                             name) and will be set to the value
        !                             of the calculated distance
        !
        !  x1, y1, x2, y2           - are INTENT ( IN ) REAL arguments that
        !                             represent the points (x1, y1) and
        !                             (x2, y2)
        !------------------------------------------------------------------
    !
         REAL     ::    Dist1, Dist2, Dist3, Dist4, Dist5, Dist6
         REAL     ::    x1, y1, x2, y2, x3, y3, x4, y4
         REAL     ::    Shortest_Distance
         REAL     ::    Distance_Between_Points  ! The function's name
                                                 ! is declared here.
    !
```

8.2 Styles of Functions

Figure 8.5 (continued)

```
              CALL  Read_Points ( x1, y1, x2, y2, x3, y3, x4, y4 )
!
! Calculate distance between all combination of points
!
              Dist1 = Distance_Between_Points ( x1, y1, x2, y2 )
              Dist2 = Distance_Between_Points ( x1, y1, x3, y3 )
              Dist3 = Distance_Between_Points ( x1, y1, x4, y4 )
              Dist4 = Distance_Between_Points ( x2, y2, x3, y3 )
              Dist5 = Distance_Between_Points ( x2, y2, x4, y4 )
              Dist6 = Distance_Between_Points ( x3, y3, x4, y4 )
!
! Use Fortran's intrinsic MIN function to find the
!    smallest of the 6 distances.
!
      Shortest_Distance = MIN ( Dist1, Dist2, Dist3, Dist4, Dist5, Dist6 )
!
      PRINT *, "The distance between the closest points:", &
                                           & Shortest_Distance
!
      END PROGRAM  Closest_Cartesian_Points
!================================================================
SUBROUTINE  Read_Points ( x1, y1, x2, y2, x3, y3, x4, y4 )
            IMPLICIT   NONE
!
!----------------------------------------------------------------
! The purpose of this subroutine is to read values for four ordered
!    pairs.
!
! x1, y1, x2, y2, x3, y3, x4, y4 - REAL, INTENT ( OUT ) variables that
!                                   represent the ordered pairs
!
! Counter   - a local INTEGER variable used as a loop counter
!----------------------------------------------------------------
!
      REAL, INTENT ( OUT )    ::     x1, y1, x2, y2, x3, y3, x4, y4
      INTEGER                 ::     Counter
```

continues

In Figure 8.5, assume function Distance_Between_Points (Figure 8.4) was added to the user library of useful functions developed in this text. Distance_Between_Points is now a **black box**; it can be used without re-creating it. When the program in Figure 8.5 is linked, the machine language version of Distance_

Figure 8.5 (continued)

```
                Counter = 1
                DO WHILE ( Counter <= 4 )    ! read four ordered pairs
                    PRINT *, "Please enter ordered pair #", &
                                                    & Counter, "(x,y)"
                    SELECT CASE (Counter)
                        CASE (1)
                            READ *, x1, y1
                        CASE (2)
                            READ *, x2, y2
                        CASE (3)
                            READ *, x3, y3
                        CASE (4)
                            READ *, x4, y4
                    END SELECT
                    Counter = Counter + 1
                END DO
    RETURN
    END SUBROUTINE  Read_Points
```

Between_Points will be added to the program to make the final, executable program. The differences between intrinsic functions and program unit functions are summarized in Table 8.3.

Table 8.3

Comparison of Intrinsic and Program Unit Functions

The Programmer Must . . .	Intrinsic	Program Unit
declare the function's name as a variable in any program unit where the function is used.		X
write the Fortran statements to accomplish the function's purpose.		X
assign a value to the function name within the body of the function.		X
supply the correct number, type(s) and order of any actual argument(s).	X	X
give formal arguments the INTENT (IN) attribute inside the function.		X

Important Concepts Review

- Function names are **variables**.

- A **function** is a specialized variable whose value is derived from the function's algorithm operating on any accompanying **argument**(s).

- In an expression, **function evaluation** is done before any other operations.

- Functions usually have **arguments**. Function arguments should always be INTENT (IN).

- Function variables derive their values from their algorithms and **MUST NOT be the object of any statement that puts a value into a variable**.

- **Intrinsic** functions are built into the compiler and are available to all program units without declaring or describing them in any way.

- **Program unit** functions are created when a custom function is needed because there is no intrinsic function to do a given job.
   Key features for **creating** a function are itemized below:
   - Create a function header (its *interface*—see below).
   - Declare the function's name as a variable within the function.
   - Assign a value to the function's name within the body of the function. This is how the function gets its "derived value."

   Key feature for **using** a program unit function:
   - declare the function as a variable within any program unit that uses it.

- A **black box** is a library program unit whose purpose and interface are all that needs to be understood to use the program unit. **How** it works is unimportant for the programmer to understand.

- An **interface** is a black box's name and the order, purpose and type of required arguments.

8.3 Program Design Example

In this program design example, a program is created to calculate permutations and combinations. **Permutations and combinations** define how many different ways *n* items can be grouped into subgroups of k items where $0 \leq k \leq n$. Permutations and combinations are useful when working with probabilities.

The formula for calculating the combinations of n things taken k at a time is:

$$_nC_k = \frac{n!}{(n-k)!\,k!}$$

$n!$ (n factorial) is defined as follows:

$$n! = 1 \cdot 2 \cdot 3 \cdot \ldots \cdot (n-1) \cdot n$$

and

$$0! = 1$$

There is an implicit example of **combinations** in the program of Figure 8.5, where six distances are calculated to represent all combinations of distances between four points taken two at a time:

```
Dist1 = Distance_Between_Points ( x1, y1, x2, y2 )
Dist2 = Distance_Between_Points ( x1, y1, x3, y3 )
Dist3 = Distance_Between_Points ( x1, y1, x4, y4 )
Dist4 = Distance_Between_Points ( x2, y2, x3, y3 )
Dist5 = Distance_Between_Points ( x2, y2, x4, y4 )
Dist6 = Distance_Between_Points ( x3, y3, x4, y4 )
```

Since four points were combined in the distance calculations using two points at a time, the *number of distances* between all combinations of points could have been figured out using the formula for combinations:

$$_nC_k = \frac{n!}{(n-k)!\,k!} = \frac{4!}{(4-2)!\,2!} = \frac{24}{2 \cdot 2} = \frac{24}{4} = 6$$

There are *six* unique combinations of four points taken two at a time.

Permutations of n things taken k at a time are represented by a formula similar to the one for combinations:

$$_nP_k = \frac{n!}{(n-k)!}$$

The order in which items are combined is important when calculating permutations. For example, examine the comparison of permutations and combinations of three integers taken two at a time and three at a time in Table 8.4.

It is noteworthy that, for example, the combination formula doesn't view -1, 6 and 6, -1 as different groupings but the permutation formula does; **the order** makes a difference in permutations.

Just think, the chances of winning a lottery of 52 numbers requiring 6 correct numbers to win are the *combinations* of 52 numbers taken 6 at a time. *Combinations* are used because the order of the 6 numbers is unimportant.

$$_nC_k = \frac{n!}{(n-k)!k!} = \frac{52!}{(52-6)!6!} = \frac{8.0658175170944 \cdot 10^{67}}{5.5026221598121 \cdot 10^{57} \cdot 720} = 20{,}358{,}520$$

Table 8.4

Permutations and Combinations of: −1, 6, 4			
Permutations of the 3 integers taken 2 at a time	Combinations of the 3 integers taken 2 at a time	Permutations of the 3 integers taken 3 at a time	Combinations of the 3 integers taken 3 at a time
−1, 6	−1, 6	−1, 6, 4	−1, 6, 4
6, −1	6, 4	−1, 4, 6	
−1, 4	−1, 4	6, 4, −1	
4, −1		6, −1, 4	
6, 4		4, 6, −1	
4, 6		4, −1, 6	

One chance out of 20,358,520 . . . great odds! Buying two tickets a week, a person would win approximately once every 195,755 years on the average.

Now that permutations and combinations have been explained, it is time to move on to the program design. The program will be most useful if a menu is created to offer users the choice of calculating permutations or combinations.

Program Design

STATE THE PROBLEM:

In this program, I want to present a menu that has three options:

```
1. Calculate Permutations.
2. Calculate Combinations.
3. Exit.
```

I want to redisplay the menu after each calculation until users choose menu option 3 to exit the program. I also don't want to allow users to be able to enter incorrect menu selections; the program must do error checking. Probably, I will need two subroutines: one for combinations and one for permutations. Both subroutines should prompt a user for values (and error check them) and calculate the appropriate answer. My program will require a factorial function, a combinations function and a permutations function. All three of these functions must be custom program unit functions because Fortran doesn't have any such functions in its intrinsic library. Once the program is done, I'll add the combinations, permutations and factorial functions to my library for future use.

STRUCTURE CHART:

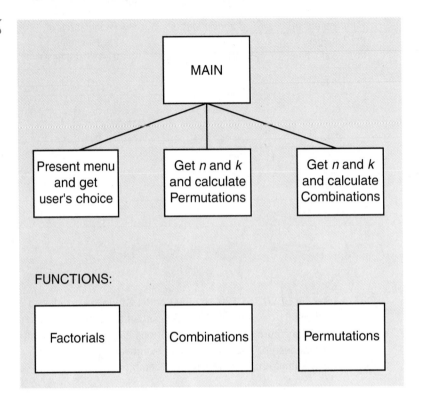

Figure 8.6

Notice that the blocks in Figure 8.6 that represent the three functions are not connected to the structure chart. The reason for this is that functions are variables; variables don't belong on structure charts as attached blocks. On the other hand, program unit functions DO have algorithms that must be designed, so the functions need to be shown to complete the structure chart diagram (Figure 8.6). A function's existence as a block on the structure diagram implies two things:

1. The function is not a part of Fortran's intrinsic library.
2. A detailed design must be created for the function.

MAIN PROGRAM:

Purpose:

The purpose of the main program is to present the menu repeatedly until the "Exit" option (menu option 3) is chosen and to call the subroutines that correspond to the user's choice.

List to do:

1. Present the menu.
2. Make sure the user enters a valid menu choice.

3. If the choice is 3, end the program.

4. If the choice is either 1 or 2, call the appropriate subroutine.

5. Repeat the steps above.

Data to and from calling program unit:

None. The main program isn't called.

English Fortran:

```
┌─ PROGRAM Permutations_And_Combinations
│  ┌─ DO WHILE ! User's Choice Isn't 3
│  │     CALL Present_Menu ( Choice )
│  │  ┌─ SELECT CASE ! ON Choice
│  │  ├─   ! Choice Is 1:  call permutations
│  │  ├─   ! Choice Is 2:  call combinations
│  │  ├─   ! Choice Is 3:  do nothing ... the loop will
│  │  │                        end on the next iteration
│  │  ├─   ! Case if not 1, 2 or 3:
│  │  │              PRINT "Invalid Menu Choice"
│  │  └─ END SELECT
│  └─ END DO
└─ END MAIN
```

PRESENT MENU AND GET CHOICE:

Purpose:

This subroutine is supplied as a black box and doesn't need to be designed.

This is not a subroutine of any great general use, which leads one to ask, why is it a library routine? When programs are coded, there are often several programmers working on the project. Each programmer puts her completed subroutines/functions in a project library so the program units are available to other programmers who need them. Therefore, the menu program unit is of general use within this specific project because each programmer will need it.

```
SUBROUTINE   Present_Menu ( Choice )
             IMPLICIT   NONE
!
!---------------------------------------------------------------
! This subroutine displays the menu, prompts for a selection and reads
!     the user's response into INTEGER variable Choice.
!
!  Choice  -   an INTEGER INTENT ( OUT ) argument to hold the user's
!                menu selection
!---------------------------------------------------------------
```

```
!     BLACK BOX:   SUBROUTINE   Print_Blank_Lines ( n )
!
!   The purpose of this subroutine is to print n blank lines.
!
!   n                INTENT ( IN ) INTEGER argument
!
!   i                an INTEGER local variable used as a loop counter
!------------------------------------------------------------------
!   BLACK BOX:
!            SUBROUTINE Get_Generic_INTEGER_Input ( Prompt, Response )
!
!   This subroutine presents the prompt contained in CHARACTER
!       argument Prompt, INTENT ( IN ), and then reads a value from
!       the keyboard into INTEGER argument Response INTENT ( OUT ).
!------------------------------------------------------------------
    INTEGER, INTENT ( OUT )    ::    Choice

    CALL Print_Blank_Lines ( 3 )   ! ... Library subroutine that prints
                                   !     blank lines to effect vertical spacing
                                   !     ... first seen in the Chapter 7
                                   !     on looping structures, now being used
                                   !     as a black box.  REUSE PROGRAM UNITS!
    PRINT *, "1. Calculate Permutations."
    PRINT *, "2. Calculate Combinations."
    PRINT *, "3. Exit."
    CALL Print_Blank_Lines ( 1 )
    CALL Get_Generic_INTEGER_Input &
         & ( "            Please enter your selection:", Choice )
         !   The above subroutine, Get_Generic_INTEGER_Input, is now a
         !   library subroutine ... a black box.  It was also
         !   developed in the looping chapter.
RETURN
END SUBROUTINE   Present_Menu
```

It may seem like a lot of typing to include the documentation for the black boxes *but it really isn't*. Documentation for subroutines and functions is created when the routines are designed and coded ... their documentation can simply be "cut and pasted" into the comment block of any program unit that uses the black box.

PERMUTATIONS SUBROUTINE:

Purpose:

The purpose of this subroutine is to prompt the user for *n* and *k* and calculate the permutations of *n* things taken *k* at a time. Error checking must ensure that *n* and *k* are positive and that $k \leq n \leq 75$. (If you find that your compiler can't handle 75! ($\approx 2.4809 \cdot 10^{109}$), the limit will need to be made smaller.)

List to do:

1. Prompt for n.
2. Read n
3. Prompt for k.
4. Read k.
5. Check that $k \leq n$.
6. Check that $n \leq 75$.
7. Check that both n and k are positive.
8. Use the permutations function to calculate the permutations.
9. Print the results.

Data to and from calling program unit:

None.

English Fortran:

```
SUBROUTINE Permutations
    n = 0   ! to enter the n error checking loop 1st time
    ! error checking loop for n, condition: n ≤ 0 or n > 75
        CALL Get_Input ('please enter n:', n)   ! Black box
        ! Read n
        If n > 0 .AND. n ≤ 75 THEN
            k = 0   ! enter the k error checking loop
            ! error checking loop for k:   k ≤ 0 or k > n
                CALL Get_Input ('please enter k:', k)
                IF k > 0 .AND. k ≤ n THEN
                    calculate permutations of n k
                ELSE
                    PRINT "k's value is invalid"
                ENDIF
            END DO for the k error checking loop
        ELSE    ! n's value is wrong
            PRINT "n's value is invalid"
        ENDIF
    END DO for the n error checking loop
END SUBROUTINE
```

COMBINATIONS SUBROUTINE:

The combinations subroutine is the same as the permutations subroutine except for the prompt and that this subroutine uses the combinations function.

FACTORIAL FUNCTION:

This is a supplied black box and doesn't need to be designed.

```
FUNCTION   Factorial ( m )
              IMPLICIT   NONE
   !
   !--------------------------------------------------------------
   !  The purpose of this function is to calculate m! . . . if the argument
   !     is negative or > 75, a -1 is returned as the function's value to
   !     aid error checking in any program unit that uses this function.
   !
   !  m             -  an INTEGER INTENT ( IN ) argument
   !
   !  Loop_Counter  -  a local INTEGER variable used to control a DO loop
   !
   !  Factorial     -  the function name: a DOUBLE PRECISION variable
   !
   !--------------------------------------------------------------
   !
        INTEGER,     INTENT ( IN )     ::   m
        INTEGER                        ::   Loop_Counter
        DOUBLE PRECISION               ::   Factorial   ! the function name
   !
        IF ( m >= 0 .AND. m <= 75 ) THEN   ! error checking
            Factorial = 1       ! start the function variable to 1
            Loop_Counter = 2
            DO WHILE ( Loop_Counter <= m )
                Factorial = Factorial * Loop_Counter
                Loop_Counter = Loop_Counter + 1
            END DO
        ELSE
            Factorial = -1   ! an incorrect argument was sent
        END IF
   !
END FUNCTION   Factorial
```

PERMUTATIONS FUNCTION:

Purpose:

This function will calculate the permutations of *n* things taken *k* at a time. This function uses the "Factorial" function above.

List to do:

1. Create an expression for permutations.
2. Assign it to the function name.

Data from the program using the function:

n and k, INTENT (IN) arguments.

English Fortran:

> function permutations
>   permutations = factorial (n) / factorial ($n - k$)
> end function

COMBINATIONS FUNCTION:

This function also uses the factorial function. The combinations function is the same as the permutations function except that the formula for combinations would be:

combinations = factorial (n) / (factorial ($n - k$) * factorial (k))

The Complete Program

```
PROGRAM  Permutations_And_Combinations
            IMPLICIT   NONE
   !
   !----------------------------------------------------------------
   ! Written by Chester Forsythe
   ! DATE:  03/25/2010
   !----------------------------------------------------------------
   ! The purpose of this program is to provide a convenient, menu-driven
   !    tool for calculating permutations and combinations.
   !
   ! Choice      -  an INTEGER variable that holds the user's menu
   !                    selection
   !
   ! Exit_Menu   -  an INTEGER PARAMETER that holds the "EXIT"
   !                    menu choice
   !----------------------------------------------------------------
         !   BLACK BOX:    SUBROUTINE  Present_Menu ( Choice )
         !
         ! This subroutine displays the menu, prompts for a selection
         !    and reads the user's response into INTEGER variable
         !    Choice.
         !
         ! Choice   -  an INTEGER INTENT ( OUT ) argument that holds the
         !                 user's menu selection
         !----------------------------------------------------------
   !
```

```fortran
        INTEGER                 ::      Choice = 1
        INTEGER, PARAMETER      ::      Exit_Menu = 3
!
        DO WHILE ( Choice /= Exit_Menu )
            CALL   Present_Menu ( Choice )
            SELECT CASE ( Choice )
                CASE ( 1 )
                    CALL   Calculate_Permutations
                CASE ( 2 )
                    CALL   Calculate_Combinations
                CASE ( 3 )
                    !!!!! DO NOTHING . . . the menu loop will end now
                CASE DEFAULT
                    PRINT *, "     Your Menu choice was invalid"
                    PRINT *, "     Please try again"
            END SELECT
        END DO
END PROGRAM   Permutations_And_Combinations
!================================================================================
SUBROUTINE   Calculate_Permutations
            IMPLICIT    NONE
!
!-------------------------------------------------------------------------
! The purpose of this subroutine is to prompt the user for n and k and
!   calculate the permutations of n things taken k at a time.
!
! n      - an INTEGER variable that represents the total number of
!           items to permute
!
! k      - an INTEGER variable that represents the number of items in
!           each permutation
!
! Permutations  - function: a DOUBLE PRECISION variable
!
! perms         - a DOUBLE PRECISION local variable that receives the
!                   permutations value
!-------------------------------------------------------------------------
        !   BLACK BOX:   SUBROUTINE   Print_Blank_Lines ( n )
        !
        !   The purpose of this subroutine is to print n blank lines.
        !
        !   n            INTENT ( IN ) INTEGER argument that defines how
        !                   many blank lines to print
        !-------------------------------------------------------------------
```

```
      !     BLACK BOX:
      !        SUBROUTINE Get_Generic_INTEGER_Input ( Prompt, Response )
      !
      !     This subroutine presents the prompt contained in CHARACTER
      !        argument Prompt, INTENT ( IN ), and then reads a value from
      !        the keyboard into INTEGER argument Response, INTENT ( OUT ).
      !-----------------------------------------------------------------------
      !     BLACK BOX:   FUNCTION  Factorial ( m )
      !
      !     The purpose of this function is to calculate m! . . . if the
      !        argument is negative or > 75, a -1 is returned as the
      !        function's value to aid in error checking in any program
      !        unit that uses this function.
      !
      !     m  -  an INTEGER INTENT ( IN ) argument
      !
      !     Factorial  -  the function name: a DOUBLE PRECISION variable
      !-----------------------------------------------------------------------
!
!     AGAIN, THE BLACK BOX DOCUMENTATION ABOVE IS JUST CUT AND PASTED FROM
!        THE SUBROUTINES'/FUNCTIONS' ORIGINAL DOCUMENTATION.
!
      INTEGER                       ::    n, k
      DOUBLE PRECISION              ::    Permutations ! a function
      DOUBLE PRECISION              ::    perms
                  !
      n = 0       ! This will let flow enter the "n" error checking loop
                  !    on the first iteration.
                  !
      DO WHILE ( n <= 0 .OR. n > 75 )          ! while n is invalid
          CALL  Print_Blank_Lines ( 2 )
          CALL  Get_Generic_INTEGER_Input ( "Please enter n:", n )
                      !
                      ! The two calls above use black boxes developed in the
                      !    looping chapter.
                      !
          IF ( n > 0 .AND. n <= 75 ) THEN
                          !
              k = 0       ! This will let flow enter the "k" error
                          !    checking loop on the first iteration.
                          !
```

```
                        DO WHILE ( k <= 0 .OR. k > n )  ! while k is invalid
                           CALL Print_Blank_Lines ( 1 )
                           CALL Get_Generic_INTEGER_Input("Please enter k:",k)
                           IF ( k > 0 .AND. k <= n ) THEN
                              perms = Permutations ( n, k )
                              CALL Print_Blank_Lines ( 1 )
                              PRINT *, "      **** Permutations =", perms
                           ELSE
                              CALL  Print_Blank_Lines ( 1 )
                              PRINT *, "k's value is invalid . . . 1 <= k <= n"
                              PRINT *, "     Please try again."
                           END IF
                        END DO   ! k error checking loop
                  ELSE  ! n's value is invalid
                     CALL  Print_Blank_Lines ( 1 )
                     PRINT *, "n's value is invalid . . . 1 <= n <= 75"
                     PRINT *, "     Please try again."
                  END IF
            END DO   ! n error checking loop
      RETURN
      END SUBROUTINE  Calculate_Permutations
!=================================================================================
      SUBROUTINE  Calculate_Combinations
                  IMPLICIT    NONE
!
!-------------------------------------------------------------------------
!   The purpose of this subroutine is to prompt the user for n and k and
!      calculate the combinations of n things taken k at a time.
!
!  n       - an INTEGER variable that represents the total number of
!               things to combine
!
!  k       - an INTEGER variable that represents the number of items in
!               each combination
!
!  Combinations  - function: a DOUBLE PRECISION variable
!
!  combs        - a DOUBLE PRECISION local variable that receives the
!                     combinations value
!-------------------------------------------------------------------------
            !   BLACK BOX:  SUBROUTINE  Print_Blank_Lines ( n )
            !
            !  The purpose of this subroutine is to print n blank lines.
            !
            !  n            INTENT ( IN ) INTEGER argument that defines how
            !                  many blank lines to print
            !-----------------------------------------------------------
```

```
!     BLACK BOX:
!             SUBROUTINE  Get_Generic_INTEGER_Input ( Prompt, Response )
!
!  This subroutine presents the prompt contained in CHARACTER
!     argument Prompt, INTENT ( IN ), and then reads a value from
!     the keyboard into INTEGER argument Response, INTENT ( OUT ).
!----------------------------------------------------------------------
!     BLACK BOX:   FUNCTION  Factorial ( m )
!
!  The purpose of this function is to calculate m! . . . if the
!     argument is negative or > 75, a -1 is returned as the
!     function's value to aid in error checking in any program
!     unit that uses this function.
!
!  m    -   an INTEGER INTENT ( IN ) argument
!
!  Factorial  -  the function name: a DOUBLE PRECISION variable
!----------------------------------------------------------------------
!
        INTEGER                         ::    n, k
        DOUBLE PRECISION                ::    Combinations ! a function
        DOUBLE PRECISION                ::    combs
                   !
        n = 0      !   This will let flow enter the "n" error checking loop
                   !       on the first iteration.
                   !
        DO WHILE ( n <= 0 .OR. n > 75 )          ! while n is invalid
            CALL  Print_Blank_Lines ( 2 )
            CALL  Get_Generic_INTEGER_Input ( "Please enter n:", n )
                       !
                       !  The two calls above use black boxes developed in the
                       !       looping chapter.
                       !
            IF ( n > 0 .AND. n <= 75 ) THEN
                           !
                k = 0      !  This will let flow enter the "k" error
                           !  checking loop on the first iteration.
                           !
```

```
                        DO WHILE ( k <= 0 .OR. k > n )  ! while k is invalid
                            CALL Print_Blank_Lines ( 1 )
                            CALL Get_Generic_INTEGER_Input("Please enter k:",k)
                            IF ( k > 0 .AND. k <= n ) THEN
                                combs = Combinations ( n, k )
                                CALL Print_Blank_Lines ( 1 )
                                PRINT *, "     **** Combinations =", combs
                            ELSE
                                CALL  Print_Blank_Lines ( 1 )
                                PRINT *, "k's value is invalid . . . 1 <= k <= n"
                                PRINT *, "     Please try again."
                            END IF
                        END DO  !  k error checking loop
                    ELSE  ! n's value is invalid
                        CALL  Print_Blank_Lines ( 1 )
                        PRINT *, "n's value is invalid . . . 1 <= n <= 75"
                        PRINT *, "     Please try again."
                    END IF
                END DO  ! n error checking loop
        RETURN
        END SUBROUTINE  Calculate_Combinations
!===============================================================================
        FUNCTION  Permutations ( n, k )
                  IMPLICIT    NONE
        !
        !-------------------------------------------------------------------------
        !  The purpose of this function is to calculate the permutations of n
        !    things taken k at a time.
        !
        ! n, k           -  INTEGER, INTENT ( IN ) arguments
        !
        ! Permutations   -  the function name: a DOUBLE PRECISION variable
        !
        ! Factorial      -  a DOUBLE PRECISION function variable
        !-------------------------------------------------------------------------
              !  BLACK BOX:  FUNCTION  Factorial ( m )
              !
              !  The purpose of this function is to calculate m! . . . if the
              !     argument is negative or > 75, a -1 is returned as the
              !     function's value to aid error checking in any program
              !     unit that uses this function.
              !
              !  m  -  an INTEGER INTENT ( IN ) argument
              !
              !  Factorial  -  the function name: a DOUBLE PRECISION variable
              !-------------------------------------------------------------------
        !
```

```
      INTEGER, INTENT ( IN )         ::     n, k
      DOUBLE PRECISION               ::     Permutations, Factorial
!
      Permutations = Factorial ( n ) / Factorial ( n - k )
!
END FUNCTION  Permutations
!================================================================================
FUNCTION  Combinations ( n, k )
            IMPLICIT   NONE
!
!--------------------------------------------------------------------------------
!   The purpose of this function is to calculate the permutations of n
!       things taken k at a time.
!
!   n, k          -    INTEGER, INTENT ( IN ) arguments
!
!   Combinations  -    the function name: a DOUBLE PRECISION variable
!
!   Factorial     -    a DOUBLE PRECISION function variable
!
!--------------------------------------------------------------------------------
        !   BLACK BOX:  FUNCTION  Factorial ( m )
        !
        !   The purpose of this function is to calculate m! . . . if the
        !       argument is negative or > 75, a -1 is returned as the
        !       function's value to aid error checking in any program
        !       unit that uses this function.
        !
        !   m  -  an INTEGER INTENT ( IN ) argument
        !
        !   Factorial  -  the function name: a DOUBLE PRECISION variable
        !--------------------------------------------------------------------------------
!
      INTEGER, INTENT ( IN )         ::     n, k
      DOUBLE PRECISION               ::     Combinations, Factorial
!
      Combinations=Factorial ( n )/( Factorial ( n - k )*Factorial( k ))
!
END FUNCTION  Combinations
```

*************** PROGRAM RESULTS ***************

1. Calculate Permutations.
2. Calculate Combinations.
3. Exit.

            Please enter your selection:
4   <---- user entered an invalid menu selection
      Your Menu choice was invalid
      Please try again

1. Calculate Permutations.
2. Calculate Combinations.
3. Exit.

            Please enter your selection:
2

 Please enter n:
52

 Please enter k:
53   <---- user entered an invalid value for k

 k's value is invalid... 1 <= k <= n
      Please try again.

 Please enter k:
6

      **** Combinations =   2.035852000000004E+07   <---- lottery example
1. Calculate Permutations.
2. Calculate Combinations.
3. Exit.
            Please enter your selection:
1

```
  Please enter n:
0

  n's value is invalid... 1 <= n <= 75
       Please try again.


  Please enter n:
3

  Please enter k:
2

         **** Permutations =   6.0000000000000000



1. Calculate Permutations.
2. Calculate Combinations.
3. Exit.

              Please enter your selection:
3                                          <---- user chose "Exit" option
```

It is interesting to observe how many structures in `Calculate_Permutations` and `Calculate_Combinations` are devoted to **error checking**. All that those two subroutines basically need to do is prompt for two numbers, read them and perform a calculation. The additional `DO` loops and `IF` statements are for error checking; they are used to ensure that the user doesn't enter invalid numbers. It is vital to include error checking in all programs, and it is often nontrivial to carry out as seen in this design example.

The functions `Permutations`, `Combinations` and `Factorial` will now be added to the library of useful routines created in this text. Any time in the future that a program requires one of these functions, it can simply be used as a black box. This is exactly what happened with the two subroutines:

$$\text{Print\_Blank\_Lines (n)}$$

and

$$\text{Get\_Generic\_INTEGER\_Input (Prompt, Response)}$$

These subroutines were developed in the chapter on looping, added to this text's subroutine/function library and used in program `Permutations_And_Combinations`. Don't recreate any subroutines or functions that have general use; create them once, add them to a library and use them as black boxes from then on. And remember, all that needs to be understood to use a black box is its *interface,* which is completely described in the black box's comment block.

Important Concepts Review

- **Combinations:**

$$_nC_k = \frac{n!}{(n-k)!k!}$$

- **Permutations:**

$$_nP_k = \frac{n!}{(n-k)!}$$

- **Functions on structure charts:**
    - Don't connect program unit function blocks to the main structure chart—functions are variables.
    - Program unit functions must appear on structure diagrams because the functions' algorithms need to be designed.
    - See Figure 8.6.

- **Error checking** is a vitally important feature to include in all programs and is often nontrivial to set up. Error checking ensures that users enter valid values.

- **Useful functions, subroutines and modules** should be included in libraries so they can be used as black boxes in the future.

EXERCISES

8.1 Using a loop that goes from 0 to 2π radians in increments of .33 radians, write out a table of secant and cosecant.

8.2 Write a complete Fortran function that will find the middle of three values. For example, if the actual arguments are -7, 0 and -33, the function should produce -7 as its derived value.

8.3 Design a complete Fortran function that will perform a numerical integration over interval $[a, b]$ of the mathematical function:

$$f(x) = -4x^3 + 5x^2 - 17x + 73$$

To accomplish this, sum up the areas of all the rectangles that are formed by Δx and $f(x)$ over interval $[a, b]$.

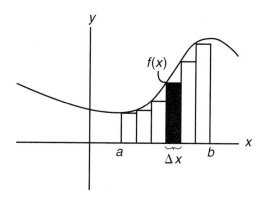

Try Δx at 0.1, 0.01 and 0.001.

8.4 Modify the function in Exercise 8.3 so that it uses the "trapezoidal" method.

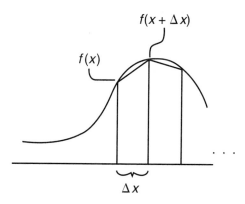

Notice how the trapezoids approximate the curve of $f(x)$ more closely.

8.5 Using the factorial function as a black box, design a function to evaluate the series for e^x to n terms. Read n and x from the user.

$$e^x = \sum_{i=0}^{n} \frac{x^i}{i!}$$

8.6 To put two character strings together, Fortran offers the concatenation operator, `//`. The following `CHARACTER` expression illustrates concatenation:

`"abc" // " " // "xyz"`

The above concatenation results in the following string:

`"abc xyz"`

Write a `CHARACTER` function that has two `INTENT ( IN )` arguments: `Title` (`INTEGER`) and `Name` (`CHARACTER`). The result of the function should be the person's name concatenated with an appropriate title such as "Mr." The valid values for `Title` are:

1. "Mr."
2. "Mrs."
3. "Ms."
4. "Hey you,"

Please use error checking to ensure a proper value for `Title` is received by the function. If the `Title` value is invalid, set the function name to `ERROR`.

8.7 Write a logical function that will return .TRUE. if its INTEGER argument is a prime number and .FALSE. if not. An integer, p, is prime if it is *evenly* divisible only by itself and one. (*Hint: use Fortran's intrinsic MOD function . . . If* MOD (a, b) = 0, a *is evenly divisible by* b; *MOD's result is the remainder of* a / b.)

8.8 The coefficients of binomial expansion are defined by combinations:

$$(x+y)^n = \binom{n}{n} x^n y^0 + \binom{n}{n-1} x^{n-1} y^1 + \cdots + \binom{n}{1} x^1 y^{n-1} + \binom{n}{0} x^0 y^n$$

$$\binom{n}{k} = \text{combinations of } n \text{ things taken } k \text{ at a time}$$

Read in an n and write out the coefficients for $(x + y)^n$. Please use the existing combinations function developed in this chapter.

8.9 In a quadratic equation, one can tell whether the roots are imaginary or not by testing the discriminant.

$$y = ax^2 + bx + c$$
$$x = \frac{-b \pm \sqrt{b^2 - 4ac}}{2a}$$

$$\text{discriminant} = b^2 - 4ac$$

Write a logical function that returns .TRUE. if $b^2 - 4ac$ is ≥ 0 (the equation has real root(s)) and .FALSE. if $b^2 - 4ac$ is < 0 (the equation has imaginary roots).

8.10 Find the maximum of ten numbers. (*Hint: look at Fortran's intrinsic functions.*)

8.11 Given a string of characters stored in variable `Text_String` (*kind*: (LEN = 80)) and another string of characters stored in variable `Sub_Text_String` (*kind*: (LEN = 5)), figure out at which character the first occurrence `Sub_Text_String` is in `Text_String`. Assume that `Sub_Text_String` **IS** contained somewhere in `Text_String`. (*Hint: look at Fortran's intrinsic functions.*)

Example:

```
Text_String:      "abcdefg79  hello there,./@@ bye"
Sub_Text_String: "79  he"
```

Answer: 8

8.12 Write a CHARACTER function with three INTENT (IN) INTEGER arguments that represent the lengths of the sides of a triangle. The function should be set to one of three values:

1. "EQUILATERAL"   —three equal sides.
2. "ISOSCELES"   —two equal sides.
3. "SCALENE"   —no equal sides.
4. "NO TRIANGLE" —no triangle possible.

One-Dimensional Arrays

Introduction

Variables that we've learned about so far are designed to hold only one value at any given time. An example of such a variable is shown below:

```
INTEGER     ::     Int_1 = 22
```

Variable `Int_1` is an `INTEGER` variable that is assigned a value of `22` at compile time because of the assignment in the declaration. If a different value is stored into `Int_1` as the program executes, the first value is "given the boot" in favor of the new value (Figure 9.1).

The fact that any simple variable can hold only one value at a time can lead to cumbersome, error-prone code if you have much data to store and manipulate. If, for example, you wanted to do two or three different analyses on 100 numbers, you would have to declare 100 variables to hold the 100 values, have 100 prompt/read statements to get those values into the variables and then create some really ugly expressions to do calculations (Figure 9.2).

To put the final nail in Figure 9.2's algorithmic coffin, imagine what it would be like to modify the program to analyze 1,000 numbers ... or 1,000,000 numbers, or worse yet, *n* numbers. Fortunately, Fortran offers data structures that simplify working with

9.1 One-dimensional arrays

Important Concepts Review

9.2 Array sections

Important Concepts Review

9.3 Array variables as arguments

9.4 Array arguments in intrinsic functions

9.5 Program design example

Important Concepts Review

9.6 DO loops for arrays

9.7 Array-masking operations —the WHERE statement

Important Concepts Review

Exercises

Figure 9.1

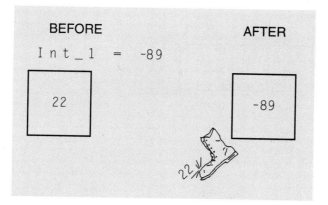

Figure 9.2

```
PROGRAM  Oh_My
            IMPLICIT   NONE
                . . .
     REAL      ::      r1,r2,r3,r4,r5,r6,r7,r8,r9,r10
     REAL      ::      r11,r12,r13,r14,r15,r16,r17,r18,r19,r20
                . . .
     REAL      ::      r91,r92,r93,r94,r95,r96,r97,r98,r99,r100
                . . .
     PRINT *, "Please enter a number"
     READ *, r1
     PRINT *, "Please enter a number"
     READ *, r2
     PRINT *, "Please enter a number"
     READ *, r3
                . . .
     PRINT *, "Please enter a number"
     READ *, r100
                . . .
! Let's calculate an average:
     Sum = r1+r2+r3+ . . . &
                 . . .      &
         & +r99+r100
     Avg = Sum / 100
! Now let's find the largest of the numbers . . . NOT ME!!
! Now let's sort and print them out . . . NOT ME!!
! Now let's pass them to a subroutine . . . NOT ME!!
!
END PROGRAM  Oh_My
```

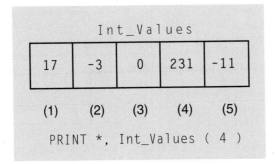

Figure 9.3

large sets of data. These structures are called **arrays**. An array is a variable that has one name, just as simple variables do, but an array variable can hold multiple values of the array's declared type. Figure 9.3 shows a simple example of an INTEGER array.

In Figure 9.3, array Int_Values can hold five distinct values. The following PRINT statement would print 231, the contents of the fourth position of array Int_Values:

PRINT *, Int_Values (4)

In the PRINT statement, (4) is a **subscript** and identifies which position of array Int_Values to print.

Fortran 90 has implemented a fantastic set of array tools including special looping structures, which aid processing the data in arrays. These looping structures are also useful in several other programming situations. Fortran's various styles of arrays, along with a host of referencing and assignment methods, address nearly every array-based computing need for program designers.

• • •

9.1 One-Dimensional Arrays

Arrays allot multiple storage locations under one variable name. In the introduction, a program was shown that required 100 numbers to be stored and analyzed in several ways. Maintaining all those variables would be a formidable task. It would be very easy to omit or mistype a variable name, operator or expression. Arrays make dealing with large volumes of data much simpler and less error prone.

Declaring Arrays

Declare array variables exactly the same way simple variables are created, with one exception: the `DIMENSION` attribute must be used to specify array variables. Figure 9.4 shows a typical array declaration. Program `Show_An_Array_Declaration` declares an `INTEGER` array variable, `Squares_Array`, with ten locations:

```
INTEGER, DIMENSION ( 1 : 10 )  ::  Squares_Array
```

Figure 9.4

```
PROGRAM   Show_An_Array_Declaration
          IMPLICIT   NONE
             . . .
!
      INTEGER, DIMENSION ( 1 : 10 )       ::      Squares_Array
      INTEGER                             ::      Ary_Counter = 1
!
      DO WHILE ( Ary_Counter <= 10 )
          Squares_Array ( Ary_Counter ) = Ary_Counter ** 2
          Ary_Counter = Ary_Counter + 1
      END DO
!
      Ary_Counter = Ary_Counter - 1    ! change from 11 to 10
!
      DO WHILE ( Ary_Counter >= 1 )
          PRINT*, "Squares_Array (", Ary_Counter," ) =", &
                                & Squares_Array ( Ary_Counter )
          Ary_Counter = Ary_Counter - 1
      END DO
!
END PROGRAM   Show_An_Array_Declaration

*************** PROGRAM RESULTS ***************

Squares_Array ( 10 ) = 100
Squares_Array (  9 ) = 81
Squares_Array (  8 ) = 64
Squares_Array (  7 ) = 49
Squares_Array (  6 ) = 36
Squares_Array (  5 ) = 25
Squares_Array (  4 ) = 16
Squares_Array (  3 ) = 9
Squares_Array (  2 ) = 4
Squares_Array (  1 ) = 1
```

The attribute, , `DIMENSION ( 1 : 10 )`, makes `Squares_Array` an **array** variable. "`( 1 : 10 )`" defines the **bounds** of `Squares_Array`, which specify how many data-holding locations an array has (`10`, in `Squares_Array`'s case).

All of an array's declared bounds together implicitly define its **dimensionality** or **shape**. Array variable `Squares_Array` is called a **one-dimensional** array because it has only *one* set of bounds. Multidimensional arrays, which have two to seven sets of bounds, are explained in Chapter 11. A pictorial representation of array `Squares_Array` is shown in Figure 9.5.

Access Array Variables' Data with Subscripts

There are a myriad of ways to reference the data in arrays. The first and most fundamental way is to reference *individual positions* of an array, such as `Squares_Array` with **subscripts**. Subscripts are placed to the right of an array variable name and consist of a set of parentheses surrounding an integer value that specifies the position of an array to be used. Subscripts must be within the limits of the declared bounds of an array. For example, `Squares_Array ( 7 )` (Figure 9.5) identifies the seventh position of array `Squares_Array` and is between `1` and `10`. When verbally describing a position of an array such as `Squares_Array ( 7 )`, one would say, "`Squares-Array` 'sub' seven."

Any individual position of an array specified by a subscript behaves *just like a simple variable* of that array's declared type. This can be seen in the `DO` loop of Figure 9.4 that is used to print all positions of array `Squares_Array`:

```
DO WHILE ( Ary_Counter >= 1 )
     PRINT *, "Squares_Array (", Ary_Counter, ") =", &
                                 & Squares_Array ( Ary_Counter )
     Ary_Counter = Ary_Counter - 1
END DO
```

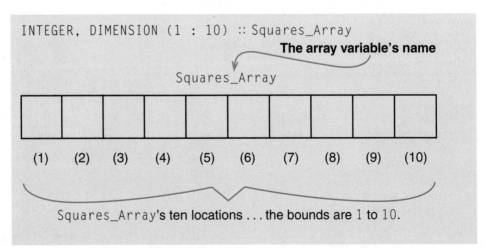

9.1 One-Dimensional Arrays

Four things are printed by the above `PRINT` statement:

1. `"Squares_Array ("` ← character string.
2. `Ary_Counter` ← variable.
3. `") ="` ← character string.
4. `Squares_Array ( Ary_Counter )` ← the `Ary_Counter` position of array variable `Squares_Array`.

The second item, `Ary_Counter`, is used as itself to display the current subscript and as an actual subscript in item four to identify which position of `Squares_Array` to print. `Squares_Array ( Ary_Counter )` is the `Ary_Counter` position of array variable `Squares_Array`, which contains one of the squares. Since each time `Squares_Array ( Ary_Counter )` is printed a single value is displayed, it is clear that `Squares_Array ( Ary_Counter )` holds only one value *just like a simple variable*. `Squares_Array ( Ary_Counter )` can also be put on the left of an assignment statement, used as the object of a read statement, included in an expression or utilized in *any* capacity that a simple variable of `Squares_Array`'s type can.

Array Assignment

In Figure 9.4, values were assigned to array positions inside a loop:

```
DO WHILE ( Ary_Counter <= 10 )
       Squares_Array ( Ary_Counter ) = Ary_Counter ** 2
       Ary_Counter = Ary_Counter + 1
END DO
```

The assignment statement in the above loop is an example of the simplest form of assignment to arrays, where one array position is set to some value. It is also possible to assign a constant or variable to an entire array. When this is done, every position of the array is assigned the value:

```
LOGICAL, DIMENSION ( 100 : 600 ) ::    Prime_Sieve
       . . .
Prime_Sieve = .FALSE.
```

In the example above, all 501 positions of array `Prime_Sieve` are set to `.FALSE.` in one assignment. This kind of mass assignment to all the positions of an array is a very powerful tool enabling entire large arrays to be set to some value without creating loops and loop counters. It is also possible to perform the above assignment in a declaration statement:

```
LOGICAL, DIMENSION ( 100 : 600 ) ::    Prime_Sieve = .FALSE.
```

Array variables can be assigned values with **array constructors,** too. Array constructors *are arrays* that are made by enclosing values in a special set of delimiters: `( /` and `/ )`. An array constructor's values are separated by commas within the delimiters and must be constants, variables or expressions. For example, an *array constructor* that has three `REAL` numbers, `-3.2`, `8.731` and `675901.0`, can be constructed and assigned to an *array variable*. Assume an array variable named `Array_Of_Reals` is declared as type `REAL` and has bounds (1 : 3). It is important that `Array_Of_Reals` and the array constructor have the same number of positions if they are to be in the same assignment statement.

```
Array_Of_Reals = (/ -3.2, 8.731, 675901.0 /)
```

The effect of the assignment above is equivalent to the following sequence of assignment statements:

```
Array_Of_Reals ( 1 ) = -3.2
Array_Of_Reals ( 2 ) = 8.731
Array_Of_Reals ( 3 ) = 675901.0
```

`Array_Of_Reals = (/ -3.2, 8.731, 675901.0 /)` assigns the *array* created by the array constructor, `(/ -3.2, 8.731, 675901.0 /)`, to array variable `Array_Of_Reals`.

There is a powerful variation on array constructors that allows large arrays to be filled efficiently with a series of values based on a linear sequence of numbers. It involves an abbreviated looping structure called an **implied DO loop**. An example of this type of array constructor is shown below:

```
INTEGER, DIMENSION ( 1 : 10 )   ::  Fill_Me
INTEGER                         ::  Fill_Index
    . . .
Fill_Me = (/ ( Fill_Index, Fill_Index = 5, 14 ) /)
```

This is equivalent to the following assignments:

```
Fill_Me ( 1 ) = 5
Fill_Me ( 2 ) = 6
Fill_Me ( 3 ) = 7
    . . .
Fill_Me ( 10 ) = 14
```

`Fill_Index` starts at 5 and is stored into `Fill_Me ( 1 )` and then changes to 6, which is stored into `Fill_Me ( 2 )`, etc. The general form of an implied DO loop in the context of array constructors is shown below:

```
( value, index = initial-value, limit, increment )
```

To help explain the way the above looping structure works, examine the equivalent `DO WHILE` in Figure 9.6. Table 9.1 shows some examples of using the "implied DO" form of array constructors (assume I is a declared `INTEGER` variable in Table 9.1). Implied DO loops are used in many other data-transferring operations such as `READ`, `WRITE` and `DATA`. These situations will be discussed as they occur.

Array Expressions

In the previous section, an array variable was assigned various values with an array constructor, which is an array in and of itself:

```
Array_Of_Reals = (/ -3.2, 8.731, 675901.0 /)
```

Figure 9.6

```
index = initial-value
DO WHILE ( index <= limit )
    put value in the next position of the array
    index = index + increment
END DO
```

Table 9.1

Implied DO Loops and Array Constructors	
Array Constructor	Created Array
(/ (I, I = 1, 1000, 2) /)	500-position array of positive ascending odd numbers.
(/ (I, I = 10, 0, -1) /)	11-position array with values $10,9,8,\ldots1,0$.
(/ (0.25 * I, I = 1, 10, 3) /)	4-position array containing $0.25,1.0,1.75,2.5$.

This is one way of getting values into an array variable and is an example of assigning *one array to another*. The assignment statement above is consistent with the general form of an assignment statement, `variable = expression`, where `variable` is an array variable and `expression` is an array constructor.

As explained in earlier chapters, expressions can be arbitrarily complicated combinations of **operators** and **operands**. This is true for array variables as well, if the operand arrays are **conformable** with each other. Arrays are said to be conformable if they are the same **shape**. For one-dimensional arrays, being the same shape means that the arrays have the same number of data-holding positions. This definition of "conformable" will be extended when multidimensional arrays are explained.

Table 9.2 (p. 270) shows some examples of conformable and nonconformable arrays. It is also important to understand that any *simple constant* or *simple variable* is conformable with any array. That is how the assignment statement seen earlier was possible:

```
LOGICAL, DIMENSION ( 100 : 600 )  ::  Prime_Sieve
        . . .
Prime_Sieve = .FALSE.
```

`.FALSE.` is, by definition, conformable with array `Prime_Sieve` and can be assigned to it, filling all positions of the array with `.FALSE.`s. If groups of arrays of the same type are conformable with each other, they can be put into expressions using any operators that are appropriate for the arrays' declared type. As with all expressions, order of precedence applies and parentheses may be used to group terms and override precedence.

When an operation is performed on two conformable arrays, the operation is executed for all corresponding positions of the operand arrays. The result is an array that is conformable with the operand arrays and contains the resulting values of the operation. Examine the example in Figure 9.7 on p. 270. From the program results in Figure 9.7, it is

Table 9.2

Conformable Arrays		
Array A	Array B	?
REAL, DIMENSION (1 : 100) :: A	REAL, DIMENSION (-50 : 49) :: B	YES
INTEGER, DIMENSION (5 : 55) :: A	INTEGER, DIMENSION (1 : 50) :: B	NO
REAL, DIMENSION (1 : 5) :: A	(/ 3.7, 4.3E-75, .0009, -2E5, 0.0, -7.9 /)	NO
REAL, DIMENSION (1 : 5) :: A	(/ 3.7, 4.3E-75, .0009, -2E5, 0.0 /)	YES
(/ 1, 1, 2, 3, 5, 8, 13, 21, 34 /)	(/ 0, 0, 0, 0, 0, -1, -1, -1, -1 /)	YES
COMPLEX, DIMENSION (-1000:1000) :: A	COMPLEX, DIMENSION (0:2000) :: B	YES

Figure 9.7

```
PROGRAM   Simple_Array_Expression
          IMPLICIT NONE
!
!------------------------------------------------------------------
! Written by      :    C. Forsythe
! Date Written    :    3/25/2004
!------------------------------------------------------------------
! This program shows an example of two conformable INTEGER arrays,
!     Operand_1 and Operand_2, being multiplied. The result of the
!     multiplication is stored into INTEGER Array_Product, which is
!     also comformable with Operand_1 and Operand_2.
!------------------------------------------------------------------
!
      INTEGER, DIMENSION( 1 : 3 ) :: Operand_1 = (/ 2, 4, 6, /)
      INTEGER, DIMENSION( 0 : 2 ) :: Operand_2 = (/ 10, 20, 30 /)
      INTEGER, DIMENSION( 1 : 3 ) :: Array_Product
!
      Array_Product = Operand_1 * Operand_2
      PRINT *, Array_Product (1), Array_Product (2), Array_Product (3)
!
END PROGRAM Simple_Array_Expression

************** PROGRAM RESULTS ****************

20 80 180
```

clear that each position of `Operand_1` is multiplied by the corresponding position of `Operand_2`, and that product is put in the corresponding position of `Array_Product`. The following assignment statement from Figure 9.7 can be thought of as three assignment statements:

$$\text{Array\_Product} = \text{Operand\_1} * \text{Operand\_2}$$

is equivalent to

```
Array_Product ( 1 ) = Operand_1 ( 1 ) * Operand_2 ( 0 )
Array_Product ( 2 ) = Operand_2 ( 2 ) * Operand_2 ( 1 )
Array_Product ( 3 ) = Operand_1 ( 3 ) * Operand_2 ( 2 )
```

Being able to operate on entire arrays as illustrated in Figure 9.7 is a very powerful computing tool. Although these examples involve small arrays, the same kinds of operations can be performed on conformable arrays of any size.

The various methods of getting values into an array are summarized in Table 9.3. (Assume that `Ary` is a three-position `INTEGER` array in Table 9.3.)

Table 9.3

Moving Values into an Array	
Method	Example
Assign a value to one array position.	`Ary (3) = -7`
Assign a value to all positions of an array.	`Ary = 22`
Assign an array made with an array constructor to an array variable.	`Ary = (/ 4, 5, -1 /)`
Assign one array to another.	`Ary = Another_Conformable_Array`
Assign an array expression to an array.	`Ary = Ary_2 / Ary_3 + Ary_4`
READ values into an array.	There are several methods of reading values into arrays; the techniques will be explained as they occur.
Set values in declarations.	`INTEGER, DIMENSION ( 1 : 3 ) :: Ary = 10`
Set values in a DATA statement.	`DATA ARY / 4, 5, -1 /`
Assign values to array sections.	See sections 9.2 and 9.3.

Important Concepts Review

- **Arrays** allot multiple storage locations under one variable name.
- Array variables are declared with the `DIMENSION` attribute.
- `DIMENSION` attributes specify array bounds.
- **Bounds** are the limits of the array.
- **Shape** is the collection of all of an array's bounds. (This will have a richer meaning when multidimensional arrays are explained in Chapter 11.)
- **Subscripts** are used to reference array positions.
- **One-dimensional arrays** only require one subscript to uniquely identify a position in an array.
- One position of an array behaves just like a simple variable of the array's type.
- **Array constructors** allow arrays with specific values to be constructed.
- **Implied DO loops** are abbreviated looping structures:

    (value, index = initial-value, limit, increment)

- **Array expressions** are meaningful combinations of conformable arrays, simple variables and/or constants, and operators.
- **Conformable arrays** have the same shape.
- One-dimensional arrays are the same shape if they have the same number of positions.
- Simple variables and constants are conformable with any array.

9.2 Array Sections

It is often necessary to work with part of an array rather than a single position or the entire array. A collection of array elements forms a subset/part of an array and is called an **array section**. Array sections are arrays in their own right and are conformable with other array sections and arrays of the same shape. As mentioned earlier in this chapter, one-dimensional arrays are the same shape if they have the same number of data-holding positions. Fortran's facilities for working with array sections are some of the language's greatest strengths.

Array Section References

Array sections are specified several different ways. The first method defines an array section by specifying a range of array elements. In Figure 9.8, two conformable arrays are added. One array, Test_Array_1, declared with bounds (1 : 5), has five positions for data. The second array, Test_Section_2, is an *array section*. It also has five positions: 17, 18, 19, 20 and 21 designated by the following array section reference:

(17 : 21)

Test_Section_2 is declared with bounds (10 : 50).

Figure 9.8

```
      PROGRAM   Array_Sections
            IMPLICIT   NONE
      !
      !-------------------------------------------------------------------
      !  Written by       :    C.  Forsythe
      !  Date Written     :    3/25/2004
      !-------------------------------------------------------------------
      !  This program demonstrates array sections.
      !
      !  Test_Array_1, Test_Section_2 - two one-dimensional INTEGER arrays
      !                                  used as array addends
      !
      !  Results_Array   -  an INTEGER array that is assigned the
      !                       sum of the Test_Array_1 and a conformable
      !                       array section of Test_Section_2
      !
      !  n                -  an INTEGER variable used as a loop counter
      !-------------------------------------------------------------------
      !
            INTEGER, DIMENSION ( 1 : 5 )        ::    Test_Array_1
            INTEGER, DIMENSION ( 10 : 50 )      ::    Test_Section_2
            INTEGER, DIMENSION ( 1 : 5 )        ::    Results_Array
            INTEGER                             ::    n = 10
      !
            DO WHILE ( n <= 50 )              ! This loop fills Test_Section_2
                Test_Section_2 (n) = n**2     ! with the squares of the numbers
                n = n + 1                     ! 10 through 50...notice that n
            END DO                            ! starts at 10 in the declaration
                                              ! above.
      !
```

continues

Figure 9.8 (continued)

```
        Test_Array_1 = (/ -7, 89, 22, 0, -32 /)    ! give Test_Array_1
                                                   ! some arbitrary values
        Results_Array = Test_Array_1 + Test_Section_2 ( 17 : 21 )
                ! an array section reference ---->  ^^^^^^^^^^
        n = 1
        DO WHILE ( n <= 5 )
            PRINT *, "Results_Array (", n, ") =", &
                    Results_Array ( n ), "... which equals:"
    !
            PRINT *, "     Test_Array_1 (", n, ") =",     &
                    &Test_Array_1 ( n ), "       +        &
                    &Test_Section_2 (", n + 16, ") =",&
                    &Test_Section_2 ( n + 16 )
            PRINT *, "-----------------------------------&
                    &----------------------------------"
            CALL Print_Blank_Lines ( 1 )
            n = n + 1
        END DO
    !
    END PROGRAM  Array_Sections

*************** PROGRAM RESULTS ***************

Results_Array  ( 1 ) is 282 . . . which equals:
     Test_Array_1 ( 1 ) = -7       +       Test_Section_2 ( 17 ) = 289
-----------------------------------------------------------------------

Results_Array  ( 2 ) is 413 . . . which equals:
     Test_Array_1 ( 2 ) = 89       +       Test_Section_2 ( 18 ) = 324
-----------------------------------------------------------------------

Results_Array  ( 3 ) is 383 . . . which equals:
     Test_Array_1 ( 3 ) = 22       +       Test_Section_2 ( 19 ) = 361
-----------------------------------------------------------------------

Results_Array  ( 4 ) is 400 . . . which equals:
     Test_Array_1 ( 4 ) = 0        +       Test_Section_2 ( 20 ) = 400
-----------------------------------------------------------------------

Results_Array  ( 5 ) is 409 . . . which equals:
     Test_Array_1 ( 5 ) = -32      +       Test_Section_2 ( 21 ) = 441
-----------------------------------------------------------------------
```

The fact that Test_Section_2 is declared (10 : 50) is unimportant to the calculation in Figure 9.8. What IS important is that Test_Section_2 is used with an **array section reference**: (17 : 21). This limits the used portion of Test_Section_2 to five positions (17, 18, 19, 20 and 21), making it conformable with array Test_Array_1. Figure 9.9 shows a pictorial representation of how the array addition in Figure 9.8 works.

Besides having array sections as operands in expressions (as seen in Figure 9.9), array section references can also be used on the left side of an assignment statement. By doing so, any subset of array elements in an array can be set to new values in a single assignment statement.

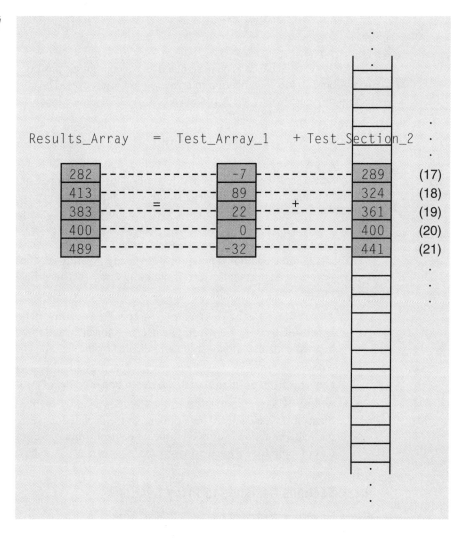

Figure 9.9

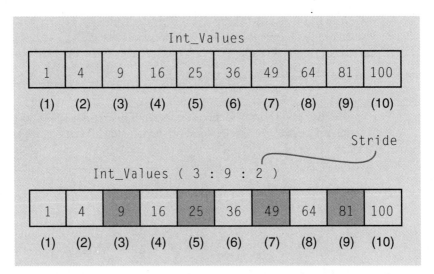

Figure 9.10

Array sections can also be designated as noncontiguous positions of an array by using the optional **stride** specification in array section references. The general form of an array section reference with a stride is shown below:

```
Array_Name ( lower : upper [: stride] )
Array_Name ( upper : lower [: -stride] ) ! negative stride
```

Stride defines an even separation between array elements. Note that when a negative stride is specified, the upper and lower bounds are swapped and that both the upper and lower bounds must be specified. Figure 9.10 shows an array section with a stride of 2.

The array section shown in Figure 9.10 is a four-position array and can be used in any context where a four-position INTEGER array is appropriate. The array section reference is equivalent to the following array constructor: (/ 9, 25, 49, 81 /).

There are three additional notations for specifying array sections. Basically, the three new styles are just shorthand ways of denoting array sections. Each can be used with a stride as well:

1. Array (: n [: stride])—specifies an array section that contains all elements of array Array from the declared lower bound through position n. It can be used with an *optional* stride as indicated by the [: stride] notation above.
2. Array (n : [: stride])—specifies an array section that contains the elements from position n to the declared upper bound of Array.
3. Array (: [: stride])—specifies an array section that includes the elements from the declared lower bound to the declared upper bound of Array. This particular notation is only of value when used with a *positive* stride or in the context of multidimensional arrays.

Array Sections Assigned to Array Sections

A very interesting feature of manipulating array sections is that overlapping array sections of the same array can be on opposite sides of an assignment statement. This is useful for

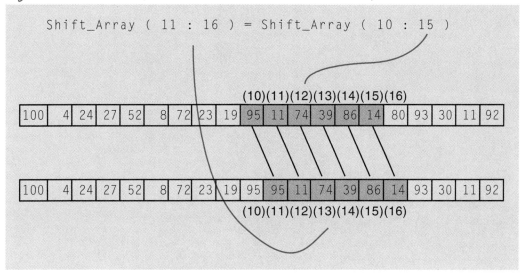

Figure 9.11

shifting array elements so that new data can be inserted. Figure 9.11 shows a graphic representation of this technique and is followed by a program that uses it in Figure 9.12 (p. 278). The array sections are underlined for clarity in Figure 9.12. The first line of output represents the array before the assignment, and the second line shows the shifted section.

Vector Subscripts

There are many ways to specify and manipulate array sections, as seen in the previous sections. The final and most powerful technique is **vector subscripts**. A vector subscript is an array or array section of integer values whose elements specify which positions of the subscripted array are to be collected together as an array section. The advantage of creating an array section with this method is that ANY set of elements, contiguous or otherwise, can be brought together as an array section. This is illustrated in Figure 9.13 (p. 279), where a vector subscript is used to create an array section.

In program Vector_Subscripts of Figure 9.13, the PRINT statement prints an array section that is defined by a vector subscript: (/ 2, 3, 5, 8, 9 /). The A in the *resulting* array section comes from the 2 in the vector subscript: (/ 2, 3, 5, 8, 9 /), which specifies the second position of array Vowels. "E" is in the second position of the resulting array section and is obtained from the third position of array Vowels. This is defined by the second value in the vector subscript: (/ 2, 3, 5, 8, 9 /). The last three values in the resulting array section are I, O and U, which come from array Vowels and are designated by the last three values in the vector subscript: (/ 2, 3, 5, 8, 9 /).

It is also acceptable to have a vector subscript reference of an array that is receiving data. The following assignment, which uses the same vector subscript as above, replaces the vowels in array Vowels with digits:

Vowels ((/ 2, 3, 5, 8, 9 /)) = (/ "1", "2", "3", "4", "5" /)
. . .

Figure 9.12

```
PROGRAM Shift_Array_Sections
          IMPLICIT NONE
!
!----------------------------------------------------------------
!  Written by      :     C. Forsythe
!  Date Written    :     3/25/2004
!----------------------------------------------------------------
!  This program takes an array section and shifts it one position by
!     assigning it to a conformable array section in the same array
!     that is one position from the original array section.
!
!  Shift_Array    - an INTEGER array that holds some arbitrary integers
!----------------------------------------------------------------
!
      INTEGER, DIMENSION ( 1 : 20 )      ::    Shift_Array
!
      DATA Shift_Array / 100, 4, 24, 27, 52, 8, 72, 23, 19, 95, &
                    & 39, 86, 14, 80, 93, 30, 11, 92, 11, 74 /
!
      CALL Print_Array ( Shift_Array )
      Shift_Array ( 11 : 16 ) = Shift_Array ( 10 : 15 )
      CALL Print_Array ( Shift_Array )
!
END PROGRAM Shift_Array_Sections
!================================================================
SUBROUTINE  Print_Array ( Shift_Array )
            IMPLICIT NONE
!
!----------------------------------------------------------------
!  This subroutine prints a 20-position INTEGER array.
!
!  Shift_Array   - an INTENT ( IN ) INTEGER array
!
!  I             - an INTEGER loop index variable
!----------------------------------------------------------------
!
      INTEGER, INTENT ( IN ), DIMENSION ( 1 : 20 )   ::   Shift_Array
      INTEGER                                         ::   I
!
      WRITE ( UNIT = *, FMT = 57 ) Shift_Array
      57 FORMAT ( 20I3 )
!
RETURN
END SUBROUTINE  Print_Array

*************** PROGRAM RESULTS ***************

100  4 24 27 52  8 72 23 19 95 39 86 14 80 93 30 11 92 11 74
100  4 24 27 52  8 72 23 19 95 95 39 86 14 80 93 11 92 11 74
```

Figure 9.13

```
Program Vector_Subscripts
          IMPLICIT NONE
!
!--------------------------------------------------------------
!
! This program shows a simple vector subscript.
!
! Vowels - a CHARACTER ( LEN = 1 ) one-dimensional array that
!          holds letters
!
!--------------------------------------------------------------
!
   CHARACTER ( LEN = 1 ), DIMENSION ( 1 : 10 ) :: VOWELS
   DATA Vowels / "B", "A", "E", "J", "I", "M", "Z", "O", "U", "K" /
!
   PRINT *, Vowels ( (/ 2, 3, 5, 8, 9 /) )
!
END PROGRAM Vector_Subscripts

************** PROGRAM RESULTS **************
AEIOU
```

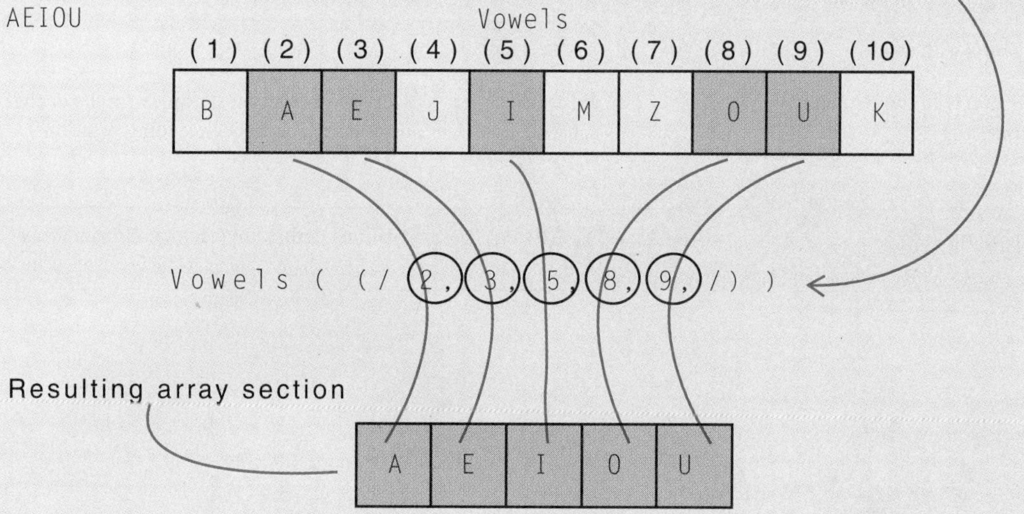

Vowels

(1)	(2)	(3)	(4)	(5)	(6)	(7)	(8)	(9)	(10)
B	A	E	J	I	M	Z	O	U	K

Vowels ((/ 2, 3, 5, 8, 9 /))

Resulting array section

A	E	I	O	U

Array Vowels in Figure 9.13 will contain the following after the assignment above:

B12J3MZ45K

There are three caveats when working with vector subscripts:

1. It is illegal to have two or more equal values in a vector subscript if it is used to define an array section that is receiving data. In the following assignment, two different values, 7 and -3, are both mapped to position 2 of array Illegal_Vector ... NO GOOD!

```
Illegal_Vector ( (/ 2, 6, 2 /) ) = (/ 7, 6, -3 /)
```

2. Array positions specified by a vector subscript must be within the subscripted array's declared bounds.
3. Array sections defined by vector subscripts can *only* be passed to other program units as `INTENT ( IN )` arguments.

In all the examples above, array constructors were used as the vector subscripts; this was done for clarity. It is *legal* to use any one-dimensional `INTEGER` array or `INTEGER` array section as a vector subscript.

There are *many* ways to reference arrays. Table 9.4 summarizes them.

Table 9.4

	Array Reference Methods	
Reference Method	Example	Array Position(s) Referenced
Subscripted array position	`A ( 4 )`	4
Array section	`A ( 3 : 7 )`	3, 4, 5, 6, 7
Array section with stride	`A ( 2 : 9 : 3 )`	2, 5, 8
Array section with implied bounds and a stride	`A ( : 9 : 3 )`	Every third position starting with the declared lower bound and not exceeding position 9
Entire array	`A`	All
Vector subscript	`A ( Vec_Array )`	Positions defined by vector subscript array `Vec_Array`

Important Concepts Review

- An **array section** is a collection of array elements forming a subset/part of an array.
- Array sections are specified with **array section references or vector subscripts**.
- **Array section reference:**

    ```
    ( lower : upper [ : stride ] )
    ( upper : lower [ : -stride] )   ! negative stride
    ```

- `Stride` defines an even separation between array elements that form an array section.

- A **vector subscript** is an `INTEGER` array used as a subscript. Its elements specify positions in the subscripted array. The collection of the specified elements form an array section.

- Array sections may be assigned to other array sections if they are conformable.

9.3 Array Variables As Arguments

In Figure 9.12, array `Shift_Array` was passed to subroutine `Print_Array` as an **actual argument** so that `Shift_Array`'s contents could be printed. Arrays are variables and can be passed to other program units. The only difference between passing simple variables and array variables as arguments is that array arguments must be declared with the `DIMENSION` attribute.

Passing Arrays and Their Bounds to Program Units

When an array is passed to a program unit, its bounds must be known to the called program so that the array formal argument can be declared correctly. Figure 9.14 (p. 282) shows an example of passing an array to a subroutine and a function.

In addition to the `DIMENSION` attribute appearing in the formal and actual argument declarations, Figure 9.14 shows several other noteworthy features of program `Passing_An_Array_As_An_Argument`. First, the declaration of array `To_Be_Averaged`'s bounds is controlled by `INTEGER PARAMETER How_Big`:

```
INTEGER, PARAMETER                  ::   How_Big = 5
REAL, DIMENSION ( 1 : How_Big )     ::   To_Be_Averaged
```

`How_Big` helps generalize program `Passing_An_Array_As_An_Argument` because it is used in the main program as the upper bound for array `To_Be_Averaged` and as the limit for the loop. It is also passed to other program units for similar purposes. By simply changing `How_Big`'s value in its declaration, array sizes and loop limits are changed. If, for example, it is necessary to change program `Passing_An_Array_As_An_Argument` so that it calculates the average of a hundred numbers, only one line needs to be changed:

```
INTEGER, PARAMETER                  ::   How_Big = 100
```

With the change above, all arrays are declared with bounds (1 : 100) and loop limits automatically change to 100 in all program units. There are even better ways to accomplish this sort of array size generalization; see Chapter 10.

Figure 9.14

```
        PROGRAM  Passing_An_Array_As_An_Argument
                IMPLICIT   NONE
!
!-----------------------------------------------------------------
!  Written by     :   C. Forsythe
!  Date Written   :   3/25/2004
!-----------------------------------------------------------------
!  This program demonstrates using an array as an argument.  The
!     program units that have the array arguments are 1) a SUBROUTINE
!     that reads values into the array and 2) a FUNCTION that
!     calculates the average of all the elements in the array.
!
!  To_Be_Averaged - a one-dimensional REAL array that holds values
!                   to be averaged
!
!  How_Big        - an integer PARAMETER that defines the size of array
!                   To_Be_Averaged
!
!  Average        - a REAL FUNCTION variable that calculates the
!                   average of all elements in array To_Be_Averaged
!
!  Loop_Counter   - an INTEGER variable used as a loop counter
!-----------------------------------------------------------------
      !  BLACK BOX:  SUBROUTINE  Print_Blank_Lines ( n )
      !
      !  The purpose of this subroutine is to print n blank lines.
      !
      !  n          - an INTENT ( IN ) INTEGER argument that defines how
      !               many blank lines to print
      !-----------------------------------------------------------------
!
        INTEGER, PARAMETER                  ::   How_Big = 5
        REAL, DIMENSION ( 1 : How_Big )     ::   To_Be_Averaged
        REAL                                ::   Average
        INTEGER                             ::   Loop_Counter = 1
!
        CALL  Fill_Array ( To_Be_Averaged, How_Big )
        CALL  Print_Blank_Lines ( 1 )
```

Figure 9.14 (continued)

```fortran
        DO WHILE ( Loop_Counter <= How_Big )
           PRINT *, "Value", Loop_Counter, "=", &
                                         To_Be_Averaged ( Loop_Counter )
           Loop_Counter = Loop_Counter + 1
        END DO
        CALL Print_Blank_Lines ( 3 )
        PRINT *, "   Average:", Average ( To_be_Averaged, How_Big )
!
END PROGRAM   Passing_An_Array_As_An_Argument
!================================================================================
SUBROUTINE   Fill_Array ( Input_Array, Array_Size )
             IMPLICIT    NONE
   !
   !--------------------------------------------------------------------
   ! This subroutine reads values into REAL INTENT ( OUT ) array argument
   !    Input_Array whose size is defined by INTENT ( IN ) INTEGER
   !    argument Array_Size.
   !
   ! Loop_Counter  - a local INTEGER variable used as a loop counter
   !--------------------------------------------------------------------
   !
        INTEGER, INTENT ( IN )                              ::   Array_Size
        REAL, DIMENSION ( 1 : Array_Size ), INTENT ( OUT )  ::   Input_Array
        INTEGER                                             ::   Loop_Counter
   !
        Loop_Counter = 1
        DO WHILE ( Loop_Counter <= Array_Size )
           WRITE ( UNIT = *, FMT = 35, ADVANCE = "NO" ) &
                                         & "Please enter a value: "
   !
                     ! When the ADVANCE = "NO" specifier is used,
                     ! it can only be used in a WRITE statement and the
                     ! WRITE statement must be used with a FORMAT.
   !
35         FORMAT ( A )   ! The "A" FORMAT descriptor will
                          ! print a CHARACTER string of any
                          ! length.
   !
           READ  *, Input_Array ( Loop_Counter )
           Loop_Counter = Loop_Counter + 1
        END DO
   !
END SUBROUTINE   Fill_Array
```

continues

Figure 9.14 (continued)

```
!==========================================================================
      FUNCTION  Average ( To_Be_Averaged, How_Big )
                IMPLICIT   NONE
!
!--------------------------------------------------------------------------
!     This function calculates the average of all of the elements of REAL
!        INTENT ( IN ) array To_Be_Averaged.
!
!     Average    - the function's name - a REAL variable
!
!     How_Big    - an INTENT ( IN ) INTEGER variable that represents the
!                  number of positions/elements in To_Be_Averaged
!--------------------------------------------------------------------------
!
      INTEGER,                              INTENT ( IN ) :: How_Big
      REAL, DIMENSION ( 1 : How_Big ), INTENT ( IN ) :: To_Be_Averaged
!
      REAL                         ::    Average
!
      Average = SUM ( To_Be_Averaged ) / SIZE ( To_Be_Averaged )

         ! SUM is an intrinsic Fortran function that sums the
         ! elements of its array argument.
         !
         ! SIZE is an intrinsic Fortran function that returns
         ! an INTEGER value that represents how many
         ! array positions/elements are in SIZE's array argument.
!
```

The second feature that should be noted is the array declarations in the subroutine and function. Examine the declarations in subroutine `Fill_Array`:

```
SUBROUTINE  Fill_Array ( Input_Array, Array_Size )
            IMPLICIT   NONE
                . . .
INTEGER, INTENT ( IN )                              :: Array_Size
REAL, DIMENSION ( 1 : Array_Size ), INTENT ( OUT ) :: Input_Array
                . . .
```

In the declarations above, `Input_Array` is given an upper bound of `Array_Size`. `Array_Size` is an integer value passed from the calling program unit that represents the size of `Input_Array`. Declaring arrays by passing bound values is perfectly acceptable as long as care is taken to declare bound values (e.g., `Array_Size`) BEFORE declaring arrays that use the bound values. Variable upper bound(s) and/or lower bound(s) must be "known" to a subroutine before any array declarations use them.

Figure 9.14 (continued)

```
      END FUNCTION  Average

*************** PROGRAM RESULTS ***************

Please enter a value: 7.89e+2
Please enter a value: 4.900132e+2
Please enter a value: 400.0
Please enter a value: 617.098997
Please enter a value: 1.341231e+2

  Value 1 =   7.8900000E+02
  Value 2 =   4.9001321E+02
  Value 3 =   4.0000000E+02
  Value 4 =   6.1709900E+02
  Value 5 =   1.3412309E+02


      Average:   4.8604706E+02
```

The final interesting point of this program is the use of intrinsic functions to perform the average calculation. Two functions, SUM and SIZE, are used:

```
Average = SUM ( To_Be_Averaged ) / SIZE ( To_Be_Averaged )
```

SUM adds up all the values in the array, and SIZE calculates how many positions are in the array. These powerful intrinsic functions allow the function variable, Average, to be given its value in a single assignment statement. No loops, loop counters or accumulation variables are necessary.

Passing Array Sections and Individual Array Positions

In addition to passing entire arrays to program units, array sections or individual array positions can be passed:

```
    . . .
    LOGICAL, DIMENSION ( 1 : 10 )      ::     Truth_or_Else
    . . .
    CALL  Some_Sub(Truth_or_Else,Truth_or_Else( 3 : 5 ),Truth_or_Else( 7 ))
    . . .
    SUBROUTINE  Some_Sub ( TE_Array, TE_Section, TE_Element )
    . . .
        LOGICAL, DIMENSION ( 1 : 10 ), INTENT ( IN )   ::    TE_Array
        LOGICAL, DIMENSION ( 1 : 3 ),  INTENT ( OUT )  ::    TE_Section
        LOGICAL, INTENT ( INOUT )                      ::    TE_Element
    . . .
```

In the program fragment above, three actual arguments are passed to subroutine Some_Sub. They are shown below:

1. Truth_or_Else
2. Truth_or_Else (3 : 5)
3. Truth_or_Else (7)

Truth_or_Else, the first argument, is an array argument. It is passed to formal argument TE_Array, which is declared exactly the same as Truth_or_Else. The entire array will be handed off to Some_Sub as an INTENT (IN) argument and used as an array inside Some_Sub.

Truth_or_Else (3 : 5) is the second argument and is an array section. Array sections are arrays and must be passed to conformable formal arguments. TE_Section is actual argument Truth_or_Else (3 : 5)'s corresponding formal argument and is declared as a three-position conformable array and used as such:

```
LOGICAL, DIMENSION ( 1 : 3 ), INTENT ( OUT ) ::     TE_Section
```

The third argument is Truth_or_Else (7), which is one position of array Truth_or_Else. Individual array positions behave **just like simple variables** so this argument is passed to a formal argument that is declared *without* the DIMENSION attribute. TE_Element is that argument and will be used as a *nonsubscripted* variable within Some_Sub:

```
LOGICAL, INTENT ( INOUT )             ::     TE_Element
```

If TE_Element's value is changed within Some_Sub, its corresponding actual argument will be changed in the calling program unit because TE_Element is INTENT (INOUT). Since the actual argument is Truth_or_Else (7), changes are reflected in the seventh position of array Truth_or_Else.

9.4 Array Arguments in Intrinsic Functions

Many intrinsic functions are able to take array variables as actual arguments. When an array variable is passed to an intrinsic function, the function variable result (the value placed in the function's name) is an array that represents the function's algorithm operating on **each position** of the array argument. The array function result is implicitly *conformable* with the array actual argument. To see how this works, examine the program in Figure 9.15.

After the first loop fills array Numbers with the values 1.0 through 10.0, the following assignment statement is executed:

```
Square_Roots = SQRT ( Numbers )
```

This is one elegant piece of code ... high-level languages at their best. The square root of each position of array Numbers is put into the corresponding position of conformable array Square_Roots. Imagine how much work a compiler has to do to convert the simple assignment statement above into a set of appropriate machine language instructions.

Figure 9.15

```
PROGRAM   Array_As_Intrinsic_FUNCTION_Arg
          IMPLICIT   NONE
!
!----------------------------------------------------------------
! Written by      :   C. Forsythe
! Date Written    :   3/25/2004
!----------------------------------------------------------------
! This program demonstrates an intrinsic FUNCTION with an array
!    argument.
!
! Numbers, Square_Roots - one-dimensional REAL arrays
!
! Loop_Counter          - an INTEGER loop counter variable
!----------------------------------------------------------------
!
    REAL, DIMENSION ( 1 : 10 )    ::   Numbers, Square_Roots
    INTEGER                       ::   Loop_Counter
!
    Loop_Counter = 1
    DO WHILE ( Loop_Counter <= 10 )
        Numbers ( Loop_Counter ) = Loop_Counter
        Loop_Counter = Loop_Counter + 1
    END DO
!
    Square_Roots = SQRT ( Numbers )
!
    Loop_Counter = 1
    DO WHILE ( Loop_Counter <= 10 )
        PRINT *, Numbers( Loop_Counter ), Square_Roots ( Loop_Counter )
        Loop_Counter = Loop_Counter + 1
    END DO
!
END PROGRAM   Array_As_Intrinsic_FUNCTION_Arg

*************** PROGRAM RESULTS ***************

    1.0000000      1.0000000
    2.0000000      1.4142135
    3.0000000      1.7320508
    4.0000000      2.0000000
    5.0000000      2.2360680
    6.0000000      2.4494898
    7.0000000      2.6457512
    8.0000000      2.8284271
    9.0000000      3.0000000
   10.0000000      3.1622777
```

9.5 Program Design Example

In the "real world" (whatever that means) there are countless examples of large sets of data arranged so they can be searched conveniently. This is called **sorted** data. One example of sorted data is a phone book, where numbers can be looked up quickly because they are sorted by name.

This has enormous implications in the computing industry, as data is sorted all the time, and it is a time-consuming process from a computing point of view. There are many sort algorithms and books on the subject.

In the following design example, data is sorted using the Forsythe lickety-split **binary insertion sort**. After data is put into a one-dimensional array, each position is compared to previous positions to find its "sorted" place in the array. All intervening array contents are shifted and the value is inserted where it belongs. The basic idea is shown in Figure 9.16.

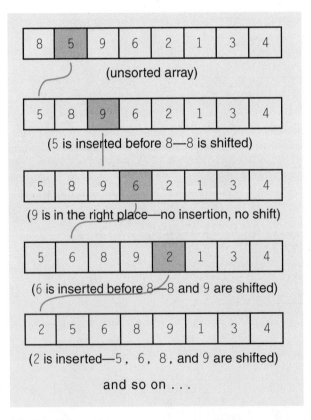

Figure 9.16

Program Design

STATE THE PROBLEM:

In this program, I want to sort data using an insertion sort. I'll store the data in a one-dimensional array. Each "unsorted" array position must be fit into the already-sorted part of the array. This will be done by searching the array to figure out where the value belongs and then shifting array contents to make room for it. I will need to create a subroutine to read numbers into the array and it will probably be useful in other programs, so I'll put it in my library for further use. Since the idea of a sort is to get an "ultimate value" (a sorted array variable), I'll make it a function and also save it in my library. In addition, I'll need a search function to figure out where a given array value fits in the sorted part of the array.

STRUCTURE CHART:

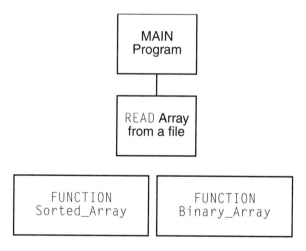

(Remember: Functions are specialized variables . . . don't connect their blocks to the structure chart.)

MAIN PROGRAM:

Purpose:

As usual, the main program drives the whole program by calling its subordinate subroutines. The main program also prints the unsorted array and its sorted counterpart.

List to do:

1. Call a subroutine to read values into the array.
2. Write out the array.
3. Write out the sorted array.

Data to and from calling program unit:

None. The main program isn't called.

English Fortran:

Not necessary . . . it's a simple sequential structure.

READ VALUES INTO THE ARRAY FROM A FILE:

Purpose:

This subroutine reads in "Upper_Bound" integers into an INTENT (OUT) array from file "INTEGERS.DAT". (Reading from a file is explained after "The Complete Program" below.)

List to do:

1. Open the file.
2. Read a value into the next available position of the input array. ("Next available" means the first position of the array for the first read.)
3. Increment the subscript.
4. Repeat steps 2 and 3.
5. Stop after "Upper_Bound" integer values are read.
6. Close file "INTEGERS.DAT".

Data to and from calling program unit:

The unsorted array INTENT (OUT) and its upper bound INTENT (IN).

English Fortran:

```
OPEN INTEGERS.DAT
Subscript = 1
DO WHILE ( Subscript <= Upper_Bound )
   Read Int_Array (Subscript)
   Subscript = Subscript + 1
END DO
CLOSE INTEGERS.DAT
```

SORT FUNCTION:

Purpose:

The purpose of this function is to sort its INTENT (IN) INTEGER array argument.

List to do:

1. Save a copy of the value currently being sorted.
2. Figure out where it fits in the already-sorted part of the array.
3. Shift array values.
4. Store the value currently being sorted in the "opened up" array position.
5. Repeat the steps above for each position of the array.

Data to and from calling program unit:

The unsorted array and its upper bound . . . both, INTENT (IN).

English Fortran:

```
Subscript = 2 ! Start sorting from position 2
DO WHILE ( Subscript <= Upper_Bound )
   Insert_Me = Array ( Subscript )
   Insert = Where Value Fits ! Find the position with
                               a binary search
   Array ( Insert+1 : Subscript ) = Array ( Insert :
                                             Subscript-1 )
   Array ( Subscript ) = Insert_Me
   Subscript = Subscript +1
END DO
```

BINARY SEARCH FUNCTION:

Purpose:

This function is a supplied black box.

```
FUNCTION  Binary_Search ( Int_Array, Search_Value, Array_Size )
          IMPLICIT NONE
!
!-------------------------------------------------------------------
! This subroutine performs a binary search on Int_Array.
!
!  Int_Array              - an INTEGER INTENT ( IN ) array that holds
!                           sorted integers
!
!  Array_Size             - an INTENT ( IN ) INTEGER variable used
!                           to declare Int_Array
!
!  Top, Bottom, Middle    - INTEGER variables used in the binary search
!
!  Search_Value           - used to hold the search value
!
!  Binary_Search          - FUNCTION variable . . . for this FUNCTION
!-------------------------------------------------------------------
!
```

```
            INTEGER, INTENT ( IN )                          :: Array_Size
            INTEGER, DIMENSION ( 1 : Array_Size ), INTENT ( IN ) :: Int_Array
            INTEGER, INTENT ( IN )        ::   Search_Value
            INTEGER                       ::   Binary_Search  !  FUNCTION variable
            INTEGER                       ::   Top, Bottom, Middle
        !
            Top    = 1
            Bottom = Array_Size
            Middle = ( Top + Bottom ) / 2
        !
            DO WHILE (Int_Array ( Middle ) /= Search_Value.AND.Top <= Bottom)
                IF ( Search_Value < Int_Array ( Middle ) ) THEN
                    Bottom = Middle - 1
                ELSE
                    Top    = Middle + 1
                END IF
                Middle = ( Top + Bottom ) / 2
            END DO
        !
            IF (Int_Array ( Middle ) == Search_Value) THEN
                Binary_Search = Middle
            ELSE
                Binary_Search = Middle + 1
            ENDIF
        !
        END FUNCTION  Binary_Search
```

The Complete Program

```
PROGRAM Forsythes_Binary_Insertion_Sort
         IMPLICIT NONE
    !
    !-----------------------------------------------------------------
    !  Written by     :   C. Forsythe
    !  Date Written   :   3/25/2004
    !-----------------------------------------------------------------
    !  The purpose of this program is to sort an array of values using a
    !     binary insertion sort.
    !
    !  Upper_Bound   - an INTEGER parameter that defines the upper bound
    !                    of array To_Be_Sorted.
    !
    !  To_Be_Sorted  - an INTEGER array with bounds ( 1 : Upper_Bound )
    !
    !  Sorted_Array  - an INTEGER array function variable whose result is
    !                    a sorted version of its array argument.
    !-----------------------------------------------------------------
    !
```

```
      INTEGER, PARAMETER                          ::  Upper_Bound = 250
      INTEGER, DIMENSION ( 1 : Upper_Bound )  ::  To_Be_Sorted
!
      INTERFACE
          FUNCTION Sorted_Array ( Unsorted_Array, Upper_Bound )
                      IMPLICIT NONE
              INTEGER, INTENT ( IN )            ::  Upper_Bound
              INTEGER, DIMENSION ( 1: Upper_Bound), INTENT ( INOUT ) :: &
                                                         Unsorted_Array
              INTEGER, DIMENSION ( 1 : Upper_Bound )           :: &
                                                         Sorted_Array
          END FUNCTION  Sorted_Array
      END INTERFACE
!
! INTERFACES and why they are needed in certain situations is explained
! in detail with assumed shape arrays in Chapter 10.
!
    CALL   Input_INTEGER_Array_Values ( To_Be_Sorted, Upper_Bound )
    WRITE (UNIT=*, FMT=22) To_Be_Sorted,      &
                           Sorted_Array ( To_Be_Sorted, Upper_Bound )
 22 FORMAT ( 1X, 25I3 ) ! will print 25 three-digit integers
!
END PROGRAM Forsythes_Binary_Insertion_Sort
!==================================================================
SUBROUTINE  Input_INTEGER_Array_Values ( Input_Array, Upper_Bound )
              IMPLICIT NONE
!
!------------------------------------------------------------------
! This subroutine puts values into one-dimensional INTEGER array
!    Input_Array.  The values are read from a file whose name is
!    INTEGERS.DAT.  The values in file INTEGERS.DAT are stored one
!    per line.
!
! Upper_Bound   - an INTENT ( IN ) INTEGER variable used to declare
!                 Input_Array and control the input loop
!
! Input_Array   - an INTENT ( OUT ) INTEGER array that holds the
!                 incoming integers
!
! Loop_Counter - an INTEGER variable used control the input loop
!------------------------------------------------------------------
!
      INTEGER, INTENT ( IN )                            ::  Upper_Bound
      INTEGER, DIMENSION (1:Upper_Bound), INTENT ( OUT ) ::  Input_Array
      INTEGER                                           ::  Loop_Counter
!
```

```
        Loop_Counter = 1
        OPEN ( UNIT = 37, FILE = "INTEGERS.DAT", STATUS = "OLD" )
        DO WHILE ( Loop_Counter <= Upper_Bound )
            READ ( UNIT = 37, FMT = * ) Input_Array ( Loop_Counter )
            Loop_Counter = Loop_Counter + 1
        END DO
        CLOSE ( UNIT = 37 )
!
RETURN
END SUBROUTINE   Input_INTEGER_Array_Values
!==============================================================================
FUNCTION Sorted_Array ( Unsorted_Array, Upper_Bound )
             IMPLICIT NONE
!
!-------------------------------------------------------------------------------
! This function sorts the contents of Unsorted_Array in ascending
!    order using an insertion sort. A binary search function is used
!    to quickly figure out where a given value should be inserted.
!
! Upper_Bound           - an , INTENT ( IN ) INTEGER variable used to
!                           declare Input_Array and control the input loop
!
! Unsorted_Array        - INTENT ( IN ) array of values to be sorted
!
! Current_Position      - Current_Position is an INTEGER variable used
!                           as the subscript to identify which array
!                           position is currently being sorted
!
! Insert_Here           - INTEGER variable that is set to where to
!                           insert the currently-being-sorted array
!                           position
!
! Insert_Me             - an INTEGER variable that holds the current
!                           value to be inserted
!
! Binary_Search         - INTEGER array FUNCTION variable that returns
!                           a sorted version of Unsorted_Array
!
! Sorted_Array          - FUNCTION variable ... this function
!
! Hold_Array            - local INTEGER array used to hold
!                           Unsorted_Array because Unsorted_Array is
!                           INTENT ( IN ) and can't be changed
!-------------------------------------------------------------------------------
!
```

```
        INTEGER, INTENT ( IN )                              :: Upper_Bound
        INTEGER, DIMENSION ( 1 : Upper_Bound ), INTENT ( IN ) :: Unsorted_Array
!
        INTEGER  ::   Current_Position, Insert_Here, Insert_Me
        INTEGER  ::   Binary_Search  !  search FUNCTION variable
        INTEGER, DIMENSION ( 1 : Upper_Bound )  :: Sorted_Array ! function
        INTEGER, DIMENSION ( 1 : Upper_Bound )  :: Hold_Array
!
        Hold_Array = Unsorted_Array
        Current_Position = 2
        DO WHILE ( Current_Position <= Upper_Bound )
            Insert_Me = Hold_Array ( Current_Position )
            Insert_Here = &
            Binary_Search ( Hold_Array ( 1 : Current_Position - 1 ),     &
                                 &  Insert_Me, Current_Position - 1 )
            Hold_Array ( Insert_Here + 1 : Current_Position ) =          &
                    &  Hold_Array ( Insert_Here : Current_Position - 1 )
            Hold_Array ( Insert_Here ) = Insert_Me
            Current_Position = Current_Position + 1
        END DO
        Sorted_Array = Hold_Array       !  set the function variable
!
END FUNCTION  Sorted_Array
!========================================================================
FUNCTION  Binary_Search ( Int_Array, Search_Value, Array_Size )
          IMPLICIT NONE
    !
    !-----------------------------------------------------------------
    !  This subroutine performs a binary search on Int_Array.
    !
    !  Int_Array            - an INTEGER INTENT ( IN ) array that holds
    !                         sorted integers
    !
    !  Array_Size           - an INTENT ( IN ) INTEGER variable used
    !                         to declare Int_Array
    !
    !  Top, Bottom, Middle  - INTEGER variables used in the binary search
    !
    !  Search_Value         - used to hold the search value
    !
    !  Binary_Search        - FUNCTION variable ... for this FUNCTION
    !-----------------------------------------------------------------
    !
```

```
        INTEGER, INTENT ( IN )                              :: Array_Size
        INTEGER, DIMENSION ( 1 : Array_Size ), INTENT ( IN ) :: Int_Array
        INTEGER                         :: Binary_Search  ! FUNCTION variable
        INTEGER, INTENT ( IN )          :: Search_Value
        INTEGER                         :: Top, Bottom, Middle
!
        Top    = 1
        Bottom = Array_Size
        Middle = ( Top + Bottom ) / 2
!
        DO WHILE ( Int_Array ( Middle ) /= Search_Value .AND. Top <= Bottom )
            IF ( Search_Value < Int_Array ( Middle ) ) THEN
                Bottom = Middle - 1
            ELSE
                Top    = Middle + 1
            END IF
            Middle = ( Top + Bottom ) / 2
        END DO
!
        IF (Int_Array ( Middle ) == Search_Value) THEN
            Binary_Search = Middle
        ELSE
            Binary_Search = Middle + 1
        ENDIF
!
    END FUNCTION Binary_Search

*************** PROGRAM RESULTS ***************
```

```
   4  50   6  19   2  11   4  50   6  14  21  31  30  87  99 100 101 101 100  99  98  97  96  95  94
  93  92  91  90  89  88  87  86  85  84  83  82  81  82  83  84  85  86  87  88  89  89  90  91  22
  18  61   4  32  11   4  50   6  14  21  31  30  17  19   2  11   4  50   6  14  21  31  30  17  18
  61  14  21  31  30  17  18  61  26  19   2  11   7   9   2   9   8   7  22   4  33  31  30  22  18
  61   4  32  11   4  50   6  14  21  31  30  17  19   2  11   4  50   6  19   2  11   4  50   6  14
  21  31  30  22  18  61   4  32  11   4  50   6  14  21  31  30  17  19   2  11   4  50   6  14  21
  31  30  17  18  61  14  21  31  30  17  18  61  26  19   2  11   4  50   6  14  21  31  30  17  50
 113  54  50  50  50  71  61  60  63   7   9   2   9   8   7  22   4  33  26  19   2  11   4  50   6
  14  21  31  30  22  18  61   4  32  11   4  50   6  14  21  31  30   4  50   6  19   2  11   4  50
   6  14  21  31  30  22  18  61   4  32  11   4  50   6  14  21  31  30  17  19   2  11   4  50   6
   2   2   2   2   2   2   2   2   2   4   4   4   4   4   4   4   4   4   4   4
   4   4   4   4   4   4   4   4   6   6   6   6   6   6   6   6   6   6   6
   6   7   7   7   8   8   9   9   9   9  11  11  11  11  11  11  11  11  11  11  11
  11  14  14  14  14  14  14  14  14  14  14  14  17  17  17  17  17  17  17  17  18
  18  18  18  18  18  18  18  19  19  19  19  19  19  19  19  21  21  21  21  21  21
  21  21  21  21  21  21  22  22  22  22  22  22  22  26  26  26  30  30  30  30  30  30  30
  30  30  30  30  30  30  31  31  31  31  31  31  31  31  31  31  31  31  31  31  32  32  32
  32  32  33  33  50  50  50  50  50  50  50  50  50  50  50  50  50  50  50  50  50  50  54
  60  61  61  61  61  61  61  61  61  61  63  71  81  82  82  83  83  84  84  85  85  86  86  87
```

Unsorted

Sorted

In the main program, a very powerful WRITE statement is used:

WRITE (UNIT=*, FMT=22) To_Be_Sorted, Sorted_Array (To_Be_Sorted, Upper_Bound)

Two array variables are written: To_Be_Sorted and Sorted_Array. Using a function variable invokes its algorithm so Sorted_Array's appearance in the WRITE statement causes the sort algorithm and its associated binary search function to be executed, producing the output above.

In subroutine Input_INTEGER_Array_Values, integers are read from a file. This topic will be thoroughly covered in Chapter 12, but until that chapter is met, Figure 9.17 will help explain the basics.

An OPEN statement prepares a file of data for use and associates it with a UNIT number. INTEGERS.DAT's UNIT number is 41 in Figure 9.17. Whenever UNIT = 41 is used in a READ statement, data is taken from the source that UNIT = 41 represents, INTEGERS.DAT; data is read from this file. Use a CLOSE (UNIT = . . .) statement after all the necessary data is read from the file to release it to the system; files should not be left open any longer than necessary to help protect them from becoming corrupted.

In function Sorted_Array, a **binary search** is used to figure out where to insert a given datum in the already sorted part of the array. A binary search is a very efficient method of searching *sorted* data. It works as follows: find the middle point of the sorted data and then determine which half should contain the searched-for value. Repeat the process for that half of the data: find the middle and figure out which half should contain the datum. Cutting the sorted list in half repeatedly finds a datum, or very quickly confirms that it doesn't exist. For example, finding any value in a sorted list of 4,000,000,000 items would take 33 comparisons *at most* (the mind reels).

Figure 9.17

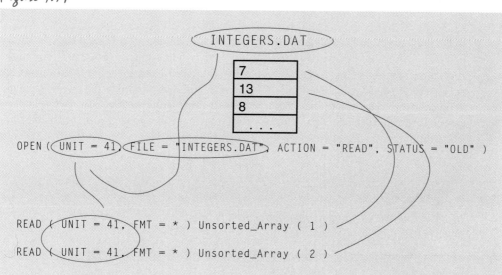

Important Concepts Review

- **Formal arguments** that are arrays must be declared with the `DIMENSION` attribute.
- Lower and/or upper bounds for an array argument may be passed as arguments and used to declare an array argument. When this is done, the bounds must be declared BEFORE the array is declared.
- Array sections and individual array positions may be passed as arguments to other program units.
- When an array variable is passed to an intrinsic function, the resulting function variable is an array that represents the function's algorithm operating on **each position** of the array argument.
- **Sorted data** are arranged in some order that facilitates the data being searched.
- **Binary search** is an efficient way to search *sorted* data; data are found by repeatedly reducing the sorted list by half.

9.6 DO Loops for Arrays

In the previous sections, standard `DO WHILE` conditional loops were used to access array positions. Although this works fine, it is unwieldy when a specific number of iterations is required. Fortran offers a special loop called the **fixed-limit** `DO` loop. It eliminates the necessity of manually incrementing a loop counter variable, as well as having to figure out a logical expression to terminate the loop. **Fixed-limit** loops are very easy to use and are especially useful in the context of arrays. The following is the general form of a fixed-limit `DO` loop:

```
DO   index-variable = initial-value, limit [, increment]
        Fortran statements
END DO
```

(Note: recall that anything in square brackets is optional. If the "increment" is omitted, it defaults to 1.)

Conditional loops that are functionally identical to the fixed-limit `DO` above are shown below to help clarify fixed-limit loops.

increment > 0

```
index-variable = initial-value
DO WHILE ( index-variable <= limit )
      Fortran statements
      index-variable = index-variable + increment
END DO
```

increment < 0

```
index-variable = initial-value
DO WHILE ( index-variable >= limit )
      Fortran statements
      index-variable = index-variable + increment  ! Adding a
END DO                                             ! negative
```

If you need to do a looping operation *n* times, why would you choose a conditional loop? The conditional loop has more statements, requires more typing and forces a programmer to think about controlling a loop counter variable.

Please don't misunderstand. Conditional loops are widely used and essential for constructing certain looping structures. Any fixed-limit loop can be written as a conditional loop (the reverse is not true). But when it comes to working with arrays, the array bounds are always known or can be figured out, which makes fixed-limit loops the way to go. Be sure *not to change* the `index-variable` implicitly or explicitly within the scope of a fixed-limit DO loop; Fortran controls the `index-variable` and expects it to be unchanged by the programmer while the loop is iterating.

There is a variant of fixed-limit loops that can be implemented on a *single line* called an **implied** DO **loop**. Implied loops were first introduced with array constructors earlier in this chapter. These looping structures use an abbreviated version of the fixed-limit DO and may be used in a variety of *data transfer* situations. Implied DO loops have the following form:

```
( value-list, index-variable = initial value, limit [, increment] )
```

`value-list` is the data being transferred, perhaps in a `READ`, `WRITE` or `DATA` statement. Similarities between fixed-limit DOs and implied DOs are fairly obvious. Putting the two structures together sets off those similarities:

Implied DO:

```
( value-list, index-variable = initial value, limit [, increment] )
```

Fixed-limit DO:

```
DO   index-variable = initial-value, limit [, increment]
        Fortran statements
END DO
```

Implied DO loops always have something to do with moving data from one place to another and have the advantage of offering an *on-the-same-line* looping structure suitable for working with `PRINT` or `DATA`, etc. Fixed-limit DOs, on the other hand, are of more general use as they are able to control a block of statements. Table 9.5 (p. 300) lists the circumstances under which implied DO loops may be used, and Table 9.6 shows a set of examples.

Fixed-limit DOs and implied DOs will be used when appropriate from this point on.

Table 9.5

Where to Use Implied DO Loops
WRITE
READ
DATA
Array Constructors
PRINT

Table 9.6

Implied DO Loops	
Example	Effect
`DATA (ARY(I), I=1,10)/ 1,2,3,4,5,6,7,8,9,10 /`	Fills positions 1 through 10 of array ARY with the integers 1 to 10.
`DATA (B(k), k=5,8,3) / 2 * -3/`	Puts -3 in positions 5 and 8 of array B.
`WRITE(UNIT=*,FMT=*)("HELLO", J=1,3)`	Writes: HELLO HELLO HELLO on the screen ... on the same line.
`WRITE(UNIT=*,FMT=*)(J, J**2,"YEA!", J=1,2)`	Writes: 1 1 YEA! 2 4 YEA! on the screen ... on the same line.
`Ary = (/ ( I, I=1, 100 ) /)`	Fills Ary with the counting numbers from 1 to 100 ... Ary MUST be declared with 100 positions.
`READ (UNIT=32, FMT=*)(A(I),B(I),I=1,1000)`	Reads 2000 values from a storage device named "32". A(1) and B(1) each get a value, then A(2) and B(2) and so on.
`PRINT*, ("a(",I,")=",a(I),I=1,5)`	Prints a(1) = followed by a(1)'s value then prints a(2) = followed by a(2)'s value and repeats this process five times.
`V(5:24)=(/(I**3,i=24,5,-1)/)`	Fills the array section v(5:24) with 24**3 in position v(5), 23**3 in position v(6), etc.

9.7 Array-Masking Operations— The WHERE Statement

A **masking operation** is where only certain elements of an array are acted on based on an *array condition.* Fortran's WHERE statement allows an entire array's elements to be exam-

Figure 9.18

```
WHERE ( array-logical-expression )
       array-assignments-1
[ ELSEWHERE
       array-assignments-2 ]
END WHERE
```

(Note that anything in square brackets is optional.)

ined one by one with other arrays' corresponding positions in an array logical expression. If a corresponding position of the arrays in the array logical expression conforms to the WHERE's condition, that element will be involved in some kind of *array assignment*. Figure 9.18 shows the general format of the WHERE statement.

This is a very powerful high-level language selection structure. To see how it works, examine the example in Figure 9.19 (p. 302). The left column of numbers in the program results displays the original values put into array Test in the DATA statement. Corresponding numbers on the right are from array S_Root. Notice how negative values in Test are transformed to -1.0 in S_Root and all other values in S_Root are the square roots of the corresponding positions in Test.

Examine the WRITE statement's implied DO loop and FORMAT, which write the program results out. All the values are created by **ONE** WRITE statement. For each I, two values are printed: Test (I) and S_Root (I). *Each time the FORMAT is used up, the next two values are printed on a new line* giving the two columns.

There is one tricky concern about using the WHERE statement. If the statement is being used to avoid program-crashing calculations such as the WHERE statement in Figure 9.19 (where taking the square root of a negative number is scrupulously avoided), one can still get into trouble. Consider the following WHERE statement.

```
WHERE ( Test >= 0.0 )
      Test = Test + PRODUCT ( SQRT ( Test ) )
END WHERE
```

If a given position of Test is ≥ 0.0, then the array assignment Test = Test + PRODUCT (SQRT (Test)) will be executed. The operand PRODUCT (SQRT (Test)) gives a single numeric result, which is the product of the square roots of *every* position of array Test, so even the negative positions of the array will be square rooted and included in the function's result. Therefore, if *any* of Test's values are negative, the program will crash even though the WHERE appears to be avoiding such situations. The key here is that PRODUCT has a single-value result, not an array result.

Figure 9.19

```
PROGRAM WHERE_Test
          IMPLICIT    NONE
!
!-------------------------------------------------------------------
! Written by:    C. Forsythe
! Date written:  3/25/2004
!-------------------------------------------------------------------
! This program demonstrates the WHERE array-masking statement.  Array
!    Test is filled with some arbitrary real numbers and array S_Root
!    is set to various values determined by the WHERE selection
!    structure. If a position of Test >= 0.0, the corresponding
!    position of S_Root is set to SQRT (Test). If Test's position
!    is negative, a -1.0 is put in the corresponding position of
!    S_Root indicating an imaginary result.
!
! Test    - a six-position one-dimensional REAL array
!
! S_Root  - object of the WHERE statement; a six-position one-
!            dimensional REAL array
!
! I       - an index variable for a fixed-limit loop
!-------------------------------------------------------------------
!
      INTEGER                      ::   I
      REAL, DIMENSION ( 1 : 6 )    ::   Test, S_Root
      DATA  Test / -72.4, 100.0, 0.0, -2.0, 2 * 5.0 /
!
      WHERE ( Test >= 0.0 )
          S_Root = SQRT ( Test )
      ELSEWHERE
          S_Root = -1.0
      END WHERE
!
      WRITE ( UNIT = *, FMT = 199 ) ( Test ( I ), S_Root (I), I = 1, 6 )
199   FORMAT ( 5X, F10.5, 5X, F10.5 )
!
END PROGRAM WHERE_Test

************** PROGRAM RESULTS *****************

         -72.40000         -1.00000
         100.00000         10.00000
           0.00000         -0.00000
          -2.00000         -1.00000
           5.00000          2.23607
           5.00000          2.23607
```

Important Concepts Review

- **Fixed-limit** DO loops eliminate the necessity of manually incrementing a loop counter variable and having to figure out a logical expression to terminate the loop.
  ```
  DO    index-variable = initial-value, limit [, increment]
         Fortran statements
  END DO
  ```

- The index-variable changes by increment each time a fixed-limit loop iterates. DON'T change index-variable within a fixed-limit DO loop!

- An initial-value is index-variable's first value.

- **Limit** is the maximum value for index-variable.

- **Increment** is the amount added to index-variable each time the loop iterates.

- **Implied** DO **loops** are abbreviated one-line looping structures used in data-transferring statements such as PRINT and READ.

- A **masking operation** is a high-level language facility where only certain elements of an array are acted on based on an *array logical expression*.

- **Array logical expressions** are formed with relational operators, logical operators and conformable operands.

- WHERE [ELSEWHERE] END WHERE is Fortran's array-masking statement.

EXERCISES

9.1 Read 10 REAL values into an array and print them out. Use fixed-limit DO loops.

9.2 Explain why arrays are necessary as data structures.

9.3 Write a program that will declare an array called Alphabet that has 26 positions. Use a fixed-limit DO loop to fill the array with the alphabet. (Hint: use the ACHAR function.) With another fixed-limit loop, copy array Alphabet into a second array in reverse order. Both arrays should be type CHARACTER (LEN = 1).

9.4 Define shape and conformability.

9.5 Write an algorithm that will determine whether a word is a palindrome or not. A palindrome is a word that reads the same from left to right as it does from right to left, e.g., "eye." Read a word from the keyboard by putting each character of it in one location of a CHARACTER (LEN = 1) array and displaying an appropriate message.

9.6 What gets printed?
```
          INTEGER, DIMENSION ( 1 : 10 )      ::     A, B
          DATA      A, B  / 5, -3, 7*8, 4*-5, 9, -10, 4*11, 18 /
     a) PRINT *, A ( 6 : 9 )
     b) PRINT *, B ( 4 )
     c) PRINT *, ( B ( J ), J = 10, 3, -2 )
     d) PRINT *, ( A ( J * 2 ), B ( J ), J = 1, 5 )
     e) PRINT *, B ( 2 : 8 : 3 )
     f) PRINT *, A ( 3 : 5 ) + (/ 9, -32, 0 /)
     g) PRINT *, (/ 22, 104, 14 /) ** (/ 1, 2, 0 /)
     h) PRINT *, A ( 1 : 10 : 4 ) * A ( 5 : 10 : 2 )
     i) PRINT *, A ( B ( 4 ) )
     j) PRINT *, B ( (/ 3, 4, 8, 10 /) )
```

9.7 Give an example of an array section being assigned to another array section in the same array; use overlapping sections. Test the answer on a computer.

9.8 Write a complete Fortran program that will examine the contents of a 100-position CHARACTER (LEN = 1) array and find out which element of the array occurs most often. As an example, the following five-position array's element that occurs most often is j.

j	f	A	j	$

9.9 Write a complete Fortran program that uses a LOGICAL array to find all the prime numbers between 2 and 1,000 using a "prime number sieve." The algorithm works as follows:
   1. Declare a LOGICAL array with bounds, 2 : 1000.
   2. Fill the array with .TRUE.s.
   3. Change every array position that is a multiple of the position that contains the current .TRUE. to .FALSE.
   4. Find the next .TRUE. in the array.
   5. Repeat steps 3 and 4 until you reach array position 500. The subscripts of .TRUE. positions represent the prime numbers.

9.10 Write a program that will alphabetize 10 names held in a CHARACTER (LEN = 15) array.

9.11 Write a program that will convert Roman numerals to decimal numbers. Read each character of the Roman numeral into one position of a CHARACTER array.

9.12 Write a program that will add 100 digit numbers. (*Hint: store each digit in a separate position of a one-dimensional array.*)

9.13 Write a program that will find the smallest of *n* values stored in an array.

9.14 Write a fixed-limit DO loop that functions the same as the following conditional loop:

```
N = -5
DO WHILE ( L < N )
    PRINT *, "WHOLE"
    N = N - 2
END DO
```

9.15 Write an assignment statement that will ensure that all elements of an array are nonnegative without changing the magnitude of the elements.

9.16 Replace a REAL array's values with their natural logarithms by using a WHERE statement; be sure not to take the logarithm of any array position whose value is less than or equal to zero.

9.17 Assume two INTEGER arrays are filled with arbitrary integers. Write a single WHERE statement that will fill a conformable LOGICAL array such that a position is .TRUE. if the sum of the corresponding integer array positions is greater than zero, and .FALSE. otherwise.

9.18 Write a program that will reverse the elements of an INTEGER array in one assignment statement. Print the array before the assignment and after.

Styles of Arrays

10

Introduction

In Chapter 9 you learned about one-dimensional arrays and the various ways they can be referenced and manipulated. A new, simplified looping structure called the fixed-limit DO loop and a new selection structure that operates on entire arrays, the WHERE statement, were introduced as well. That chapter offered a very large body of information, but the power of arrays has just begun to be revealed. This chapter explains five new styles of arrays:

1. Assumed-shape arrays.
2. Allocatable arrays.
3. Automatic arrays.
4. Parallel arrays.
5. Derived data type arrays.

Each of these styles of arrays has its own valuable features. For example, **allocatable arrays** can be created while a program is running. This is very useful because often, when you want to fill an array, you don't know exactly how much data you have. With allocatable arrays, first use an algorithm to figure out how much data you will be processing, and then you can allocate an array to the appropriate size. When you're done with the array, it can be deallocated and then subsequently reallocated as a different size array for other data.

Another very useful style of array is the **assumed-shape array**. Assumed-shape arrays are *formal arguments* that don't have specific declared bounds. Their bounds are defined by the

10.1 Assumed-shape arrays

10.2 Allocatable arrays

10.3 Automatic arrays

Important Concepts Review

10.4 Parallel arrays

10.5 Derived data type arrays

10.6 Program Design Example

Important Concepts Review

Exercises

actual argument they receive when the routine they're in is invoked. This is a very desirable feature because it reduces the number of arguments that must be specified in a subroutine or function definition; the array bounds don't have to be sent as actual arguments. This saves time by reducing the possibility of introducing errors into your program designs while typing arguments for the bounds of arrays.

All of the various styles of arrays that are introduced in the next few sections can be used with the diverse array reference techniques shown in Chapter 9, such as array section references, vector subscripts, etc.

● ● ●

10.1 Assumed-Shape Arrays

Assumed-shape arrays are *formal arguments* whose bounds may change each time their program unit is used. Rather than passing variables or constants as arguments for array bound declarations, the bounds are *not specified* except as an optional lower bound. An assumed-shape array declaration has the following general form:

```
type, DIMENSION ( [lower-bound] : ) :: formal-argument-array-names
```

When an array actual argument is passed to a program unit whose corresponding formal argument is an assumed-shape array, it *adopts* the array bounds of the actual argument. The calling program *implicitly* sends the bounds information. Figure 10.1 (p. 308) shows two different-sized arrays being passed to an assumed-shape array formal argument.

There are some points of interest in program `Assumed_Shape_Arrays` of Figure 10.1. First, consider the `INTERFACE` block in the main program. When a called program unit, such as `Print_The_Array` is going to use assumed-shape arrays, *the calling program needs to know that* because the called program is going to require extra information (the bounds of the actual argument) when the called program unit is invoked.

`INTERFACE` blocks solve this problem by giving calling program units a "peek" at the style of called program units, `SUBROUTINE`s or `FUNCTION`s and the nature of their arguments. The general form of an `INTERFACE` block is shown below:

```
INTERFACE
     interface-body
END INTERFACE
```

Figure 10.1

```
      PROGRAM Assumed_Shape_Arrays
            IMPLICIT  NONE
      !
      !-------------------------------------------------------------------
      ! Written by    :   C. Forsythe
      ! Date Written  :   3/25/2004
      !-------------------------------------------------------------------
      ! This program passes two different-shaped one-dimensional arrays to
      !    a subroutine whose purpose is simply to print the array out. The
      !    "shape" of a one-dimensional array is its size. Shape will have
      !    a more involved meaning when multidimensional arrays are
      !    explained.
      !-------------------------------------------------------------------
            ! BLACK BOX:   SUBROUTINE Print_Blank_Lines ( n )
            !
            ! The purpose of this subroutine is to print n blank lines.
            !
            ! n              an INTENT ( IN ) INTEGER argument that defines how
            !                many blank lines to print
            !-------------------------------------------------------------------
      !
            INTEGER, DIMENSION ( 1 : 3 )      ::      Three_Position_Array
            INTEGER, DIMENSION ( 1 : 5 )      ::      Five_Position_Array
      !
            INTERFACE
                SUBROUTINE  Print_The_Array ( Array_To_Be_Printed )
                    INTEGER, DIMENSION ( : ), INTENT ( IN ) :: &
                                                    Array_To_Be_Printed
                END SUBROUTINE  Print_The_Array
            END INTERFACE
      !
            DATA Three_Position_Array / 2, 4, 6 /
            DATA Five_Position_Array / 1, 3, 5, 7, 9 /
      !
            CALL  Print_The_Array ( Three_Position_Array )
            CALL  Print_Blank_Lines ( 2 )
            CALL  Print_The_Array ( Five_Position_Array )
      !
      END PROGRAM  Assumed_Shape_Arrays
```

Figure 10.1 (continued)

```
    SUBROUTINE  Print_The_Array ( Array_To_Be_Printed )
            IMPLICIT   NONE
    !
    !--------------------------------------------------------------------
    !  This subroutine prints whatever INTEGER array is passed to it.
    !
    !  Array_To_Be_Printed   - an INTENT ( IN ) INTEGER array that is
    !                          used as an assumed-shape array
    !
    !  Size_Of_Array         - a local INTEGER variable that is set to
    !                          the size of the array: the declared size of
    !                          actual argument Array_To_Be_Printed
    !
    !  Loop_Counter          - a local INTEGER variable used as a loop
    !                          counter
    !--------------------------------------------------------------------
    !
        INTEGER, DIMENSION ( : ), INTENT ( IN )  ::  Array_To_Be_Printed
        INTEGER                        ::   Size_Of_Array, Loop_Counter
    !
        Size_Of_Array = SIZE ( Array_To_Be_Printed )
                                        ! SIZE is an intrinsic
                                        ! Fortran function
                                        ! that returns the number
                                        ! of position in an array.
        Loop_Counter = 1
        DO WHILE ( Loop_Counter <= Size_Of_Array )
            PRINT *, "Array(", Loop_Counter, ") =", &
                                Array_To_Be_Printed ( Loop_Counter )
            Loop_Counter = Loop_Counter + 1
        END DO
    !
    RETURN
    END SUBROUTINE Print_The_Array
```

continues

Interface-body is similar to a *black box* description in that it supplies the style of program unit and a syntactical description of all of the arguments. In Figure 10.1, the interface block is as follows:

```
INTERFACE
    SUBROUTINE Print_The_Array ( Array_To_Be_Printed )
        INTEGER, DIMENSION ( : ), INTENT ( IN ) :: Array_To_Be_Printed
    END SUBROUTINE Print_The_Array
END INTERFACE
```

Figure 10.1 (continued)

```
*************** PROGRAM RESULTS ***************

Array(  1  ) = 2
Array(  2  ) = 4
Array(  3  ) = 6


Array(  1  ) = 1
Array(  2  ) = 3
Array(  3  ) = 5
Array(  4  ) = 7
Array(  5  ) = 9
```

With this information, the main program knows all it needs to know about subroutine Print_The_Array. It knows that subroutine Print_The_Array's formal argument is an assumed-shape array, because it is not declared with bounds. Therefore, Assumed_Shape_Arrays will send essential array bounds information to Print_The_Array, in addition to the actual argument, each time the subroutine is called.

A final point about INTERFACE blocks is that when they are created, they can be collected and stored in a MODULE. This keeps them out of the way so they don't clutter up program units. The INTERFACE blocks can then be easily accessed with a USE interface-module-name statement.

The next interesting feature of program Assumed_Shape_Arrays from Figure 10.1 is the assumed-shape array declaration:

INTEGER, DIMENSION (:), INTENT (IN) :: Array_To_Be_Printed

Assumed-shape arrays specify no bounds in the DIMENSION attribute. This is because the formal argument, Array_To_Be_Printed, is going to *assume* the bounds of the actual argument. There is one caveat when declaring assumed-shape arrays: when the lower bound of an actual argument being passed to an assumed-shape formal argument doesn't have 1 as its declared lower bound, the lower bound of the assumed-shape array must be specified. The default lower bound(s) of assumed-shape arrays is 1. If the actual argument being passed to the formal argument assumed-shape array has a lower bound other than 1, the arrays won't correspond in a natural way. See Figure 10.2.

In subroutine Test_Bound_1, the lower bound is not specified in the assumed-shape array declaration. Therefore, the assumed-shape array's first position defaults to 1. Test_Formal still has 11 positions, but the bounds are assumed to be (1 : 11) instead of (-5 : 5). Subroutine Test_Bound_2 uses the lower bound feature of assumed-shape arrays and therefore, Test_Formal has bounds (-5 : 5) inside the subroutine just as its actual argument does in the main program. Both methods work the same, but it is important that programmers understand the difference so subscripting errors aren't made.

Figure 10.2

```
PROGRAM Assumed_Shape_Array_Lower_Bound
         IMPLICIT NONE
!
!-----------------------------------------------------------------
! Written by      :  C. Forsythe
! Date Written    :  3/25/2004
!-----------------------------------------------------------------
! This program shows what happens if an actual argument with lower
!   bounds other than 1 is passed to an assumed-shape formal argument.
!
! Test     - an 11-position REAL array with bounds -5 : 5
!
! I        - an INTEGER index variable
!-----------------------------------------------------------------
!
      INTEGER                              ::  I
      REAL, DIMENSION ( -5 : 5 )           ::  &
                                  Test = (/ ( I / 2.0, I = 1, 11 ) /)
!     fills Test with 0.5, 1.0, ...,5.0, 5.5 ^^^^^^^^^^^^^^^^^^^^^^^^
!
      INTERFACE
         SUBROUTINE Test_Bound_1 ( Test_Formal )
            REAL, INTENT ( IN ), DIMENSION ( : ) ::  Test_Formal
         END SUBROUTINE Test_Bound_1
      END INTERFACE
!
      INTERFACE
         SUBROUTINE Test_Bound_2 ( Test_Formal, Lower_B )
            INTEGER, INTENT ( IN )                ::  Lower_B
            REAL, INTENT ( IN ), DIMENSION ( : )  ::  Test_Formal
         END SUBROUTINE Test_Bound_2
      END INTERFACE
!
      CALL Test_Bound_1 ( Test )
      CALL Test_Bound_2 ( Test, -5 )
!
END PROGRAM Assumed_Shape_Array_Lower_Bound
```

continues

A final important feature of program `Assumed_Shape_Arrays` in Figure 10.1 is subroutine `Print_The_Array`'s use of the `SIZE` intrinsic function. This function provides a convenient way to figure out how many positions are in assumed-shape array `Array_To_Be_Printed`. The result of the function, the "size" of

Figure 10.2 (continued)

```
      SUBROUTINE  Test_Bound_1 ( Test_Formal )
                IMPLICIT NONE
      !
      !-------------------------------------------------------------------
      !  This subroutine does NOT use the lower-bound feature of assumed-
      !     shape array declarations.
      !
      ! Test_Formal - an assumed shape INTENT ( IN ) REAL array
      !-------------------------------------------------------------------
      !
          REAL, INTENT ( IN ), DIMENSION ( : )  ::  Test_Formal
      !
          PRINT *, Test_Formal ( 1 ), Test_Formal ( 11 )
      !
      RETURN
      END SUBROUTINE Test_Bound_1
!==================================================================
      SUBROUTINE  Test_Bound_2 ( Test_Formal, Lower_B )
                IMPLICIT NONE
      !
      !-------------------------------------------------------------------
      ! This subroutine DOES use the lower bound feature of assumed-
      !     shape array declarations.
      !
      ! Test_Formal  - an assumed-shape INTENT ( IN ) REAL array
      !
      ! Lower_B      - the lower bound for array Test_Formal
      !-------------------------------------------------------------------
      !
          INTEGER, INTENT ( IN )                      :: Lower_B
          REAL, INTENT ( IN ), DIMENSION ( Lower_B : )  :: Test_Formal
      !
          PRINT *, Test_Formal ( -5 ), Test_Formal ( 5 )
      !
      RETURN
      END SUBROUTINE Test_Bound_2

*************** PROGRAM RESULTS ***************

  0.5000000   5.5000000
  0.5000000   5.5000000
```

`Array_To_Be_Printed`, is stored into `Size_Of_Array`, which in turn is used as a loop limit.

10.2 Allocatable Arrays

Allocatable arrays are arrays whose size and dimensionality can be defined and redefined as a program runs. This is known as **dynamic allocation**. When an allocatable array is allocated, it is given its bounds and created while the program is running. The necessary memory is given to the program to create the array, assuming the computer has enough memory. If the computer doesn't have enough memory, the program crashes with a run-time error.

Allocatable arrays are useful because the size of an array often is not known at compile time. Perhaps, for example, the size of an array depends on some read-in or calculated value. Without allocatable arrays, standard arrays have to be declared with bounds that are large enough to accommodate any situation. This usually results in declaring an array that is too large and wastes memory. This is an inefficient, error-prone way of using arrays.

In the example program in Figure 9.14, an array's size is based on the value of a named constant called `How_Big`. (Figure 9.14's original main program is reproduced on p. 314 as Figure 10.3.) This was a reasonable generalization of the program because it only required a modification of one line to make it possible to calculate the average of a different-sized array. The problem with that generalization is that each time a different number of values need to be averaged, surgery must be performed on the program to change the value of `How_Big`. A better solution is to use allocatable arrays. By asking the user how many values to average, an array of the appropriate size can be dynamically created (Figure 10.4, p. 315).

There are several differences between the main programs in Figures 10.3 and 10.4:

- `How_Big` is used as a variable instead of a named constant in Figure 10.4. `How_Big` is given a value by the user that is used to `ALLOCATE` array `To_Be_Averaged` with the right number of positions.

- `How_Big`, the size of the array, is determined by input from the user rather than editing the program source code (which would require recompiling and linking the program).

- Array `To_Be_Averaged` is declared as follows:

    `REAL, DIMENSION ( : ), ALLOCATABLE  :: To_Be_Averaged`

- No bounds are specified; the bounds are created dynamically with the `ALLOCATE()` statement.

- The `ALLOCATABLE` attribute is added to the declaration statement, allowing bounds to be given to `To_Be_Averaged` while the program runs.

- `ALLOCATE ( To_Be_Averaged ( 1 : How_Big ) )` is used to create array `To_Be_Averaged`.

- `DEALLOCATE ( To_Be_Averaged )` is used to return `To_Be_Averaged`'s memory to the computer's memory pool. This is a very important thing to

Figure 10.3

```
PROGRAM Passing_An_Array_As_An_Argument
          IMPLICIT NONE
!
!------------------------------------------------------------------
!
!  Written by       :     C. Forsythe
!  Date Written     :     3/25/2004
!------------------------------------------------------------------
!  This program demonstrates using an array as an argument.  The
!     program units that have the array arguments are 1) a SUBROUTINE
!     that reads values into the array and 2) a FUNCTION that
!     calculates the average of all the elements in the array.
!
!  To_Be_Averaged - a one-dimensional REAL array that holds values
!                   to be averaged
!
!  How_Big         - an INTEGER PARAMETER that defines the size of
!                    array To_Be_Averaged
!
!  Average         - a REAL FUNCTION variable that calculates the
!                    average of all elements in array To_Be_Averaged
!
!  Loop_Counter    - an INTEGER variable used as a loop counter
!------------------------------------------------------------------
      !  BLACK BOX:  SUBROUTINE  Print_Blank_Lines (n)
      !
      !  The purpose of this subroutine is to print n blank lines.
      !
      !  n           - an INTENT ( IN ) INTEGER argument that defines how
      !                many blank lines to print
      !------------------------------------------------------------------
!
      INTEGER, PARAMETER             ::    How_Big = 5
      REAL, DIMENSION ( 1 : How_Big ) ::   To_Be_Averaged
      REAL                           ::    Average
      INTEGER                        ::    Loop_Counter = 1
!
      CALL Fill_Array ( To_Be_Averaged, How_Big )
      CALL Print_Blank_Lines ( 1 )
      DO WHILE ( Loop_Counter <= How_Big )
          PRINT *, "Value", Loop_Counter, "=", &
                                   & To_Be_Averaged ( Loop_Counter )
          Loop_Counter = Loop_Counter + 1
      END DO
      CALL Print_Blank_Lines ( 3 )
      PRINT *, "   Average:", Average ( To_Be_Averaged, How_Big )
!
END PROGRAM  Passing_An_Array_As_An_Argument
```

Figure 10.4

```
PROGRAM Passing_An_Array_As_An_Argument
          IMPLICIT NONE
!
!--------------------------------------------------------------------
! Written by       :    C. Forsythe
! Date Written     :    3/25/2004
!--------------------------------------------------------------------
! This program demonstrates using an array as an argument.  The
!    program units that have the array argument are 1) a SUBROUTINE
!    that reads values into the array and 2) a FUNCTION that
!    calculates the average of all the elements of the array.
!
! To_Be_Averaged  - a one-dimensional ALLOCATABLE REAL array that holds
!                   the values to be averaged
!
! How_Big         - an INTEGER variable that defines the size of array
!                   To_Be_Averaged
!
! Average         - a FUNCTION variable that calculates the average of
!                   all the elements of array To_Be_Averaged
!
! Loop_Counter    - an INTEGER variable used as a loop counter
!--------------------------------------------------------------------
!    BLACK BOX:   SUBROUTINE Print_Blank_Lines (n)
!
!    The purpose of this subroutine is to print n blank lines.
!
!    n            - an INTENT ( IN ) INTEGER argument that defines how
!                   many blank lines to print
!    !-----------------------------------------------------------------
!
    INTEGER                                   ::   How_Big
    REAL, DIMENSION ( : ), ALLOCATABLE        ::   To_Be_Averaged
    REAL                                      ::   Average
    INTEGER                                   ::   Loop_Counter = 1
!
    PRINT *, "How many numbers to average?"
    READ *,  How_Big
    ALLOCATE ( To_Be_Averaged ( 1 : How_Big ) )
    CALL Fill_Array ( To_Be_Averaged, How_Big )
    CALL Print_Blank_Lines ( 1 )
    DO WHILE ( Loop_Counter <= How_Big )
         PRINT *, "Value" , Loop_Counter, "=", &
                                        To_Be_Averaged ( Loop_Counter )
         Loop_Counter = Loop_Counter + 1
    END DO
    DEALLOCATE ( To_Be_Averaged )
    CALL Print_Blank_Lines ( 3 )
    PRINT *, "    Average:", Average ( To_Be_Averaged, How_Big )
!
END PROGRAM   Passing_An_Array_As_An_Argument
```

do—especially in called program units. Allocatable arrays should *always be deallocated* within the program unit that creates them.

A final version of the program from Figure 10.4 that uses both allocatable arrays and assumed-shape arrays is shown in Figure 10.5.

Figure 10.5

```fortran
PROGRAM Passing_An_Array_As_An_Argument
         IMPLICIT  NONE
!
!------------------------------------------------------------------
! Written by      :  C. Forsythe
! Date Written    :  3/25/2004
!------------------------------------------------------------------
! This program demonstrates using an array as an argument. The
!    program units that have the array argument are 1) a SUBROUTINE
!    that reads values into the array and 2) a FUNCTION that
!    calculates the average of all the elements of the array.
!
! To_Be_Averaged  - a one-dimensional ALLOCATABLE REAL array that holds
!                     the values to be averaged
!
! How_Big         - an INTEGER variable that defines the size of array
!                     To_Be_Averaged
!
! Average         - a FUNCTION variable that calculates the average of
!                     all the elements of array To_Be_Averaged
!------------------------------------------------------------------
      ! BLACK BOX: SUBROUTINE Print_Blank_Lines ( n )
      !
      ! The purpose of this subroutine is to print n blank lines.
      !
      ! n - an INTENT ( IN ) INTEGER argument: how many blank lines
      !
      !---------------------------------------------------------------
!
      INTEGER                                 ::   How_Big
      REAL, DIMENSION ( : ), ALLOCATABLE      ::   To_Be_Averaged
      REAL                                    ::   Average   ! Function
!
      INTERFACE   ! so Input_Array is known as an assumed-shape array
         SUBROUTINE Fill_Array ( Input_Array )
            REAL, DIMENSION ( : ), INTENT ( OUT ) :: Input_Array
         END SUBROUTINE Fill_Array
      END INTERFACE
!
```

Figure 10.5 (continued)

```
          INTERFACE   ! so To_Be_Averaged is known as an assumed-shape array
             FUNCTION Average ( To_Be_Averaged )
                 REAL, DIMENSION ( : ), INTENT ( IN ) :: To_Be_Averaged
                 REAL                                 :: Average
             END FUNCTION Average
          END INTERFACE
             ! The above INTERFACE blocks are necessary because the
             ! formal arguments are declared as assumed-shape arrays.
             ! Extra info, the array bounds, will need to be
             ! sent to the program units and the above INTERFACE blocks
             ! tell this program unit about the situation.
    !
          PRINT *, "How many numbers to average?"
          READ *,  How_Big
          ALLOCATE ( To_Be_Averaged ( 1 : How_Big ) ) ! Give To_Be_Averaged
                                                      ! bounds
          CALL Fill_Array ( To_Be_Averaged )
          CALL Print_Blank_Lines ( 3 )
          PRINT *, "   Average:", Average ( To_Be_Averaged )
          DEALLOCATE ( To_Be_Averaged )
    !
    END PROGRAM Passing_An_Array_As_An_Argument
!================================================================================
    SUBROUTINE Fill_Array ( Input_Array )
               IMPLICIT   NONE
    !
    !--------------------------------------------------------------------------------
    !  This subroutine reads values into REAL INTENT ( OUT ) assumed-shape
    !    array argument Input_Array.
    !
    !  Loop_Counter  - an INTEGER variable used as a loop counter
    !--------------------------------------------------------------------------------
    !
          REAL, DIMENSION ( : ), INTENT ( OUT ) :: Input_Array
          INTEGER                               :: Loop_Counter
    !
          Loop_Counter = 1
          DO WHILE ( Loop_Counter <= SIZE ( Input_Array ) )
              WRITE (UNIT=*, FMT=35, ADVANCE="NO") " Please enter a value: "
35            FORMAT ( A )
              READ *, Input_Array ( Loop_Counter )
              Loop_Counter = Loop_Counter + 1
          END DO
    !
    END SUBROUTINE Fill_Array
!================================================================================
```

continues

Figure 10.5 (continued)

```
FUNCTION Average ( To_Be_Averaged )
          IMPLICIT  NONE
!
!-----------------------------------------------------------------
! This function calculates the average of all of the elements of REAL
!    INTENT ( IN ) assumed-shape array To_Be_Averaged.
!
! Average    - the function's name - a REAL variable
!-----------------------------------------------------------------
!
     REAL, DIMENSION ( : ), INTENT ( IN ) :: To_Be_Averaged
     REAL                                 :: Average
!
     Average = SUM ( To_Be_Averaged ) / SIZE ( To_Be_Averaged )
!
END FUNCTION Average

*************** PROGRAM RESULTS ***************

 How many numbers to average?
4
 Please enter a value: 9.769
 Please enter a value: 2.435
 Please enter a value: 9.400221
 Please enter a value: 6.7767077


     Average:   7.0952320
```

10.3 Automatic Arrays

Automatic arrays are arrays created to be *conformable* with other arrays. An automatic array is *never* a formal argument; it is a conformable clone of another array's shape that is used as a local variable. This is a useful feature when arrays of different shapes are passed as arguments, and a conformable "work" array is required in subsequent array expressions. Figure 10.6 is an illustration of just such a situation.

In program `Passing_An_Array_As_An_Argument`, an `INTERFACE` block lets the main program know that calling subroutine `Reverse_Em` involves an assumed-shape array. So that array `Palindrome` can be a general one-dimensional array, it is declared with the `ALLOCATABLE` attribute. The array is allocated, which means it is dynamically declared, to a size that is defined by user input. Then, so arguments representing

Figure 10.6

```fortran
    PROGRAM Reverse_Array
        USE Generic_Input
        IMPLICIT   NONE
!
!-----------------------------------------------------------------------
!   Written by    :  C. Forsythe
!   Date Written  :  3/25/2004
!-----------------------------------------------------------------------
!   The purpose of this program is to reverse the values in a one-
!      dimensional CHARACTER ( LEN = 1 ) array. If the array contains:
!      "a" "b" "c" "d" "e" it will contain "e" "d" "c" "b" "a" after
!      subroutine Reverse_Em is called.  Notice the use of assumed-shape
!      arrays, allocatable arrays, and, of course, the automatic array
!      in subroutine Reverse_Em.
!
!   Palindrome   -  a CHARACTER ( LEN = 1 ), ALLOCATABLE array
!
!   What_Size    -  an INTEGER variable that defines the shape/size
!                   of Palindrome
!
!   Loop_Counter -  an INTEGER variable used to control loops
!-----------------------------------------------------------------------
        !  BLACK BOX:
        !      SUBROUTINE Get_Generic_Numeric_Input ( Prompt, Response )
        !
        !  Prompt - CHARACTER INTENT ( IN ) argument that holds the prompt
        !
        !  Response - INTEGER
        !-------------------------------------------------------------------
!
        CHARACTER ( LEN = 1 ), DIMENSION ( : ), ALLOCATABLE :: Palindrome
        INTEGER              :: What_Size, Loop_Counter
!
        INTERFACE
            SUBROUTINE Reverse_Em ( Palindrome )
                CHARACTER ( LEN = 1 ), DIMENSION ( : ), &
                                        INTENT ( INOUT ) :: Palindrome
            END SUBROUTINE Reverse_Em
        END INTERFACE
            ! The above INTERFACE block is necessary because the
            ! formal argument Palindrome in subroutine Reverse_Em
            ! is declared as an assumed-shape array. When Reverse_Em
            ! is called, extra info—the array bounds—will need to be
            ! sent to the subroutine and the above INTERFACE block
            ! tells this program unit about the situation.
```

continues

Figure 10.6 (continued)

```
!
      CALL Get_Generic_Numeric_Input ( "Array size? ", What_Size )
      ALLOCATE ( Palindrome ( 1 : What_Size ) )
      Loop_Counter = 1
      DO WHILE ( Loop_Counter <= What_Size )
          WRITE ( UNIT = *, FMT = 46, ADVANCE = "NO" ) "Enter a value: "
46        FORMAT ( 5X, A )
          READ *, Palindrome ( Loop_Counter )
          Loop_Counter = Loop_Counter + 1
      END DO
      PRINT *, Palindrome
      CALL Reverse_Em ( Palindrome )
      PRINT *, Palindrome
!
END PROGRAM Reverse_Array
!==============================================================================
SUBROUTINE Reverse_Em ( Palindrome )
            IMPLICIT   NONE
!
!------------------------------------------------------------------------------
!  This subroutine reverses the positions of array Palindrome.
!
!  Palindrome       - a CHARACTER ( LEN = 1 ) INTENT ( INOUT ) one-
!                     dimensional array
!
!  Loop_Counter     - an INTEGER variable used to control the loop
!
!  Size_Palindrome  - an INTEGER variable that holds the size
!                     of Palindrome
!
!  Work             - a CHARACTER ( LEN = 1 ) local automatic array
!------------------------------------------------------------------------------
!
      CHARACTER (LEN=1), DIMENSION ( : ), INTENT ( INOUT ):: Palindrome
      CHARACTER (LEN=1), DIMENSION ( 1 : SIZE ( Palindrome )) :: Work
            ! The above array is an automatic array. It is NOT an
            ! argument, but rather a local array that is automatically
            ! made conformable with array Palindrome through the
            ! use of the intrinsic SIZE function in Work's declaration.
      INTEGER          ::    Loop_Counter, Size_Palindrome
!
      Loop_Counter = 0
      Size_Palindrome = SIZE ( Palindrome )
```

Figure 10.6 (continued)

```
        DO WHILE ( Loop_Counter < Size_Palindrome )
            Work(Size_Palindrome-Loop_Counter)=Palindrome(Loop_Counter+1)
            Loop_Counter = Loop_Counter + 1
        END DO
        Palindrome = Work
    !
    END SUBROUTINE Reverse_Em

*************** PROGRAM RESULTS ***************

Array size? 5
   Enter a value: A
   Enter a value: B
   Enter a value: C
   Enter a value: D
   Enter a value: E
ABCDE
EDCBA
```

Palindrome's bounds don't have to be sent to subroutine Reverse_Em, an assumed-shape array is used for the formal argument. Finally, an automatic array called Work is created that is conformable with assumed-shape array Palindrome. Examine the automatic array declaration in the subroutine fragment below:

```
SUBROUTINE  Reverse_Em ( Palindrome )
         . . .
    CHARACTER (LEN=1), DIMENSION ( : ), INTENT ( INOUT ):: Palindrome
    CHARACTER (LEN=1), DIMENSION ( 1 : SIZE ( Palindrome ) ) :: Work
         . . .
    Loop_Counter = 0
    Size_Palindrome = SIZE ( Palindrome )
    DO WHILE ( Loop_Counter < Size_Palindrome )
        Work(Size_Palindrome-Loop_Counter)=Palindrome(Loop_Counter+1)
        Loop_Counter = Loop_Counter + 1
    END DO
    Palindrome = Work
         . . .
```

Work is an automatic array. When a nonformal argument array (a **local array variable**) is created with the intrinsic SIZE function as seen above, the array is automatically created so that it is conformable with SIZE's array argument.

    CHARACTER (LEN=1), DIMENSION (1 : SIZE (Palindrome)) :: Work

The key to this declaration is (1 : SIZE (Palindrome)). Work's bounds are declared 1 to SIZE (Palindrome). This implicitly/automatically makes Work conformable with Palindrome, making Work the right size/shape to be filled with the reverse of Palindrome. Finally, Work is stored back into Palindrome to complete the reversal:

```
Palindrome = Work
```

Important Concepts Review

- **Assumed-shape arrays** are *formal arguments* whose bounds may change each time their program unit is used; assumed-shape arrays adopt the bounds of their corresponding actual arguments.

- **Allocatable arrays** are arrays whose size and dimensionality are defined and redefined as a program runs.

- **Dynamic allocation** is memory being allocated to a program's variables as it runs.

- **Deallocate** means to recover allocated memory from a program and return that memory to the computer's memory pool.

- **Automatic arrays** are arrays created to be conformable with other arrays. An automatic array is never a formal argument but a conformable clone of another array.

10.4 Parallel Arrays

Parallel arrays is a method of working with related data that is stored in different arrays. A simple example of this is associating an array of student names with an array that contains their final grades. For each name in array Name, there is a corresponding grade in array Grade. Array Grade and array Name work in *parallel* with each other. Figure 10.7 shows how parallel arrays work.

Virginia Rye's grade is 97; Linda Forsythe's is 94. Array Name's data corresponds to array Grade's. When a subscript is used to identify a student by her name, that same subscript can be used to find her grade in array Grade.

Figure 10.7

Name		Grade
Rye, Olaf	----------	91
Burns, Diana	----------	78
Forsythe, Linda	----------	94
Churchill, Winston	----------	84
Rye, Virginia	----------	97
. . .	----------	. . .

There are no declaration considerations for one-dimensional parallel arrays other than ensuring that the arrays are the same size. Parallel arrays are only a convenient convention for relating two or more arrays. There are situations when parallel arrays must be used, but when possible, derived data type arrays should be used instead of parallel arrays.

10.5 Derived Data Type Arrays

When a collection of aggregate data records is needed in a program such as the student name/grade example in the last section, a derived data type should be created. Then, an array variable of that derived type can be declared to hold the data. Figure 10.8 (p. 324) shows how this is accomplished.

To help make array `All_Grades` in Figure 10.8 clearer, Figure 10.10 (p. 325) shows a programming example that finds the student with the highest grade in the class. But before perusing the example in Figure 10.10, look at Figure 10.9 (p. 324), which shows an English Fortran algorithm for finding the maximum value in an array of positive `INTEGER` data. The key is maintaining a "current largest value" that is updated whenever a larger value is met.

`Current Largest` is set to the first value in `array` and becomes the `Current Largest`. As each array position is examined, `Current Largest` is changed if the value in that array position is larger than `Current Largest`. On to Figure 10.10 . . .

Program `Derived_Data_Type_Array` produced one very reuseable function: `Upper_Case`. Function `Upper_Case` can be used whenever a string needs to be capitalized. Its worth is obvious in program `Derived_Data_Type_Array` by allowing a user to type any case variation of `END` such as `EnD`, `enD`, etc., without affecting how the program works.

Figure 10.8

```
TYPE Name_And_Grade
     CHARACTER ( LEN = 20 )     ::     Name
     REAL                       ::     Grade
END TYPE Name_And_Grade
!
TYPE ( Name_And_Grade ), DIMENSION ( 1 : 500 ) ::   All_ Grades
```

Each of the five hundred positions of array All_Grades contains Name and Grade.

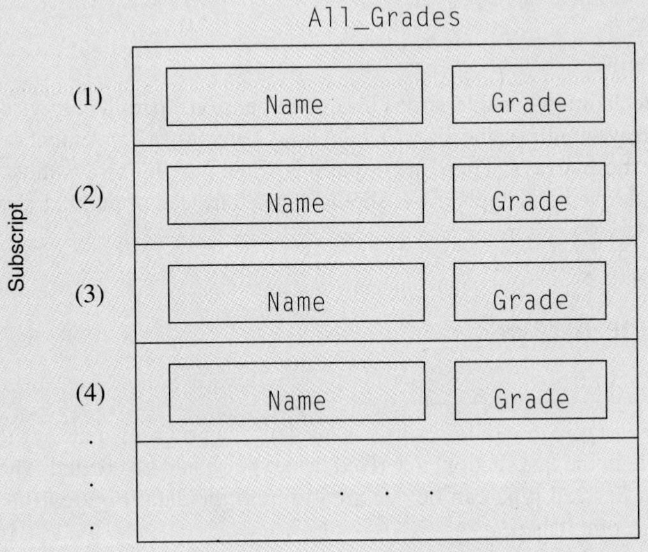

All_Grades

Subscript		Name	Grade
(1)		Name	Grade
(2)		Name	Grade
(3)		Name	Grade
(4)		Name	Grade
. . .			

Figure 10.9

```
Current largest = array ( 1 )
DO I = 2, number of array positions
    IF array ( I ) > Current largest THEN
        Current largest = array ( I )
    END IF
END DO
```

Figure 10.10

```
        MODULE Students_And_Grades
               IMPLICIT NONE
        !
        !----------------------------------------------------------------
        ! This module contains the derived data type for the names and grades
        !   of students.
        !
        ! Name_And_Grade - a derived data type that contains two variables
        !
        ! Name           - a CHARACTER ( LEN = 20 ) variable that is a
        !                    subobject of derived data type Name_And_Grade
        !
        ! Grade          - an INTEGER variable that is a subobject
        !                    of derived data type Name_And_Grade
        !----------------------------------------------------------------
        !
            TYPE Name_And_Grade
                CHARACTER ( LEN = 20 ) ::   Name
                INTEGER                ::   Grade
            END TYPE Name_And_Grade
        !
        END MODULE Students_And_Grades
!================================================================================
        PROGRAM Derived_Data_Type_Array
               USE Students_And_Grades
               IMPLICIT NONE
        !
        !----------------------------------------------------------------
        ! Written by     :  C. Forsythe
        ! Date Written   :  3/25/2004
        !----------------------------------------------------------------
        ! The purpose of this program is to find the highest grade contained
        !    in array All_Grades.
        !
        ! All_Grades       - an array of type Name_And_Grade
        !
        ! How_Many         - INTEGER variable that holds a value representing
        !                      how many grades are in array All_Grades
        !
        ! Valedictorian    - type Name_And_Grade variable that holds
        !                      the student with the highest grade
        !
        ! Find_Highest_Grade - a FUNCTION variable, type Name_And_Grade
        !
        ! Length_Of_Name   - an INTEGER variable used to hold the length, in
        !                      characters, of the valedictorian's name so
        !                      the answers will have proper spacing
        !----------------------------------------------------------------
```

continues

Figure 10.10 (continued)

```
            !   BLACK BOX:   SUBROUTINE Print_Blank_Lines ( n )
            !
            !   The purpose of this subroutine is to print n blank lines.
            !
            !   n             an INTENT ( IN ) INTEGER argument that defines how
            !                       many blank lines to print
            !------------------------------------------------------------------
!
            TYPE ( Name_And_Grade ), DIMENSION ( 1 : 500 ) :: All_Grades
            TYPE ( Name_And_Grade )  ::   Valedictorian, Find_Highest_Grade
            INTEGER                  ::   How_Many, Length_Of_Name
!
            CALL  Fill_The_Array ( All_Grades, How_Many )
            Valedictorian = Find_Highest_Grade ( All_Grades, How_Many )
            Length_Of_Name = LEN_TRIM ( Valedictorian % Name ) ! the length of
                                        ! Valedictorian % Name not including
                                        ! trailing blanks
            CALL  Print_Blank_Lines ( 2 )
            PRINT *, "Valedictorian: ",                                     &
                  & Valedictorian % Name ( 1 : Length_Of_Name ),", Grade:",&
                  & Valedictorian % Grade
!
      END PROGRAM Derived_Data_Type_Array
!==================================================================================
      SUBROUTINE Fill_The_Array ( All_Grades, How_Many )
                  USE Students_And_Grades
                  IMPLICIT NONE
            !
            !-----------------------------------------------------------------
            ! This subroutine fills derived data type array All_Grades with
            !    student information.
            !
            ! All_Grades      - an INTENT ( OUT ) array of type Name_And_Grade
            !
            ! How_Many        - an INTENT ( OUT ) INTEGER variable that holds a
            !                     value representing how many grades are in array
            !                     All_Grades
            !
            ! No_More_Students - a CHARACTER PARAMETER that contains the
            !                     end-of-data flag
            !
            ! Upper_Case      - a FUNCTION variable whose value is its argument with
            !                     all alphabetic characters capitalized
            !
            ! Local_Name      - a CHARACTER that represents Name and protects
            !                     All_Grades ( How_Many ) % Name from being
            !                     changed to all capitals
            !-----------------------------------------------------------------
            !
```

Figure 10.10 (continued)

```
            TYPE ( Name_And_Grade ), DIMENSION ( 1 : 500 ), INTENT ( OUT ) ::&
                             All_Grades
            INTEGER, INTENT ( OUT )                 ::    How_Many
            CHARACTER ( LEN = 3 ), PARAMETER        ::    No_More_Students = "END"
            CHARACTER ( LEN = 20 )                  ::    Upper_Case, Local_Name
    !
            How_Many = 1
            PRINT *, "Please enter the Student's name and grade"
            READ *, All_Grades ( How_Many )
            Local_Name = Upper_Case ( All_Grades ( How_Many ) % Name )
            DO WHILE ( Local_Name /= No_More_Students )
                How_Many = How_Many + 1
                PRINT *, "Please enter the Student's name and grade"
                READ *, All_Grades ( How_Many )
                Local_Name = Upper_Case ( All_Grades ( How_Many ) % Name )
            END DO
            How_Many = How_Many - 1
    !
        RETURN
        END SUBROUTINE Fill_The_Array
!================================================================================
        FUNCTION Upper_Case ( String )
                IMPLICIT NONE
    !
    !--------------------------------------------------------------------------------
    !   This FUNCTION is used to convert the all alphabetic characters in
    !     argument String to capitals.
    !
    !   String        - a CHARACTER INTENT ( IN ) variable that contains the
    !                     string to be capitalized
    !
    !   Upper_Case    - a CHARACTER variable - the FUNCTION name
    !
    !   S             - an INTEGER variable that "points" to the current
    !                     character in Local_String
    !
    !   Local_String  - a CHARACTER variable that holds a copy of String.
    !                     String must not be changed as it is INTENT ( IN )
    !--------------------------------------------------------------------------------
    !
            CHARACTER ( LEN = * ), INTENT ( IN )  ::  String
            CHARACTER ( LEN = * )                 ::  Upper_Case
            CHARACTER ( LEN = 1000 )              ::  Local_String
            INTEGER                               ::  S
    !
            Local_String = String
            S = LEN_TRIM ( Local_String )  ! the length of Local_String not
                                           ! including trailing blanks
```

continues

Figure 10.10 (continued)

```
            DO WHILE ( S >= 1 )
                SELECT CASE ( Local_String ( S : S ) )
                    CASE( "a" : "z" )
                        Local_String ( S : S ) = &
                            & ACHAR ( ICHAR ( Local_String ( S : S ) ) - 32 )
                END SELECT
                S = S - 1
            END DO
            Upper_Case = Local_String
    !
        END FUNCTION Upper_Case
    !===============================================================
        FUNCTION Find_Highest_Grade ( All_Grades, How_Many )
            USE Students_And_Grades
            IMPLICIT NONE
    !
    !-------------------------------------------------------------
    !
    !   This FUNCTION finds the highest grade in array All_Grades.
    !
    !   All_Grades           - an INTENT ( IN ) array of type Name_And_Grade
    !
    !   How_Many             - an INTENT ( IN ) INTEGER variable that holds a
    !                          value representing how many grades are in array
    !                          All_Grades
    !
    !   Find_Highest_Grade   - a TYPE ( Name_And_Grade ) variable - the
    !                          FUNCTION name
    !
    !   Current_Highest_Grade - an INTEGER that always has the current
    !                           highest grade
    !
    !   Index                - an INTEGER loop counter
    !-------------------------------------------------------------
    !
            TYPE ( Name_And_Grade ), DIMENSION ( 1 : 500 ), INTENT ( IN ) :: &
                                    All_Grades
            TYPE ( Name_And_Grade )      :: Find_Highest_Grade ! FUNCTION name
            INTEGER                      :: Current_Highest_Grade, Index
            INTEGER, INTENT ( IN )       :: How_Many
    !
            Current_Highest_Grade = All_Grades ( 1 ) % Grade
                                    ! initialized Current_Highest_Grade
                                    ! to the first grade in the array
```

Figure 10.10 (continued)

```
            DO Index = 2, How_Many
                IF(All_Grades ( Index ) % Grade > Current_Highest_Grade) THEN
                    Current_Highest_Grade = All_Grades ( Index ) % Grade
                    Find_Highest_Grade % Name = All_Grades ( Index ) % Name
                END IF
            END DO
            Find_Highest_Grade % Grade = Current_Highest_Grade
        !
        END FUNCTION Find_Highest_Grade

*************** PROGRAM RESULTS ***************

 Please enter the Student's name and grade
 Alex, 94                                      <- first student record
 Please enter the Student's name and grade
 Natasha 90                                    <- data separated by a blank
 Please enter the Student's name and grade        instead of a comma
 Matt                                          <- data entered on two
 96                                               separate lines
 Please enter the Student's name and grade
 Chester    85
 Please enter the Student's name and grade
 Linda,95
 Please enter the Student's name and grade
 eNd,0                                         <- the flag value that ends input
                                                  uses an irregular case . . .
                                                  FUNCTION Upper_Case solves
                                                  the problem

 Valedictorian: Matt, Grade: 96
```

10.6 Program Design Example

In this design, a set of crops will be analyzed to figure out which of them will grow in a given geographical region. Characteristics of the crops and regions are shown below:

1. Name of the crop.
2. Minimum number of days in the growing season. (INTEGER)
3. Minimum average temperature (in °F) during the growing season. (INTEGER)
4. Maximum average temperature (in °F) during the growing season. (INTEGER)
5. Minimum precipitation (in inches) during the growing season. (REAL)
6. Maximum precipitation (in inches) during the growing season. (REAL)

Region data contains the following:

1. Name of the region.
2. Number of days in the growing season.   (INTEGER)
3. Average temperature (in °F) during the growing season. (INTEGER)
4. Precipitation (in inches) during the growing season.   (REAL)

The crop names with their characteristics will be stored in a file called CROPS.DAT.

Program Design

STATE THE PROBLEM:

For this program I will need to create two derived data types: one for crops and the other for regions. I will need to create an array to hold all the various crops. I will put a crop that has "END" for its name as the last crop in the file so I can use a conditional loop to read an unknown number of crops into an array. Once the crops are read in, I'll ask the user to enter data for a region and then I'll cycle through the crops to figure out which will grow in that region.

STRUCTURE CHART:

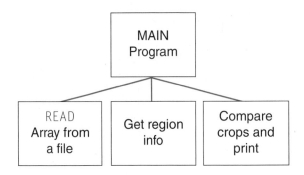

MAIN PROGRAM:

Purpose:

As usual, the main program drives the whole program by calling its subordinate subroutines.

List to do:

1. Call a subroutine to read crop data into the crops array.
2. Get regional information.
3. Compare all crops to the regional information and print any crops that will grow in the region.
4. Repeat steps 2 and 3 until the user enters QUIT for a regional name.

Data to and from calling program unit:

None. The main program isn't called.

English Fortran:

```
CALL Read_Crop_Data ( Crops_Array )
CALL Read_Region_Info   ! Read first region
DO WHILE Region_Name /= "QUIT"
    CALL Find_Acceptable_Crops ( Crops_Array )
    CALL Read Region Info   ! Get next region
END DO
```

READ VALUES INTO THE CROP ARRAY FROM CROPS.DAT:

Purpose:

This subroutine reads in an unknown number of crop records. Input stops when a crop bearing the name "END" is met. For simplicity, it is assumed that there will never be more than 100 crops.

List to do:

1. Set a subscript variable to 1.
2. Open CROPS.DAT (explained briefly at the end of the design example in Chapter 9. In addition, Chapter 12 is dedicated to the study of files).
3. Read a crop record into the next available position of the crops array. ("Next available" means the first position of the array for the first read.)
4. Increment the subscript.
5. Repeat steps 3 and 4. This will be organized as the "standard input algorithm."
6. Stop when a crop named "END" is read.
7. Close CROPS.DAT.

English Fortran:

```
i = 1
OPEN CROPS.DAT
Read a crop record into crops array (i)    ! First crop
DO WHILE ( Crops_Array (i)'s Name /= "END" )
   i = i + 1
   Read a crop record into crops array (i) ! Next crop
END DO
CLOSE CROPS.DAT
```

GET REGIONAL INFORMATION:

Purpose:

The purpose of this subroutine is to acquire regional information from a user.

List to do:

1. Prompt the user for the regional information.
2. Read the values.

Data to and from calling program unit:

Regional info to the calling program, INTENT (OUT).

English Fortran:

```
Print "Please enter region name"
Read Region % Name
IF Region % Name /= "QUIT" THEN
    Read Region % Growing_Days
    Read Region % Growing_Temp
    Read Region % Precipitation
END IF
```

CHECK CROPS AGAINST REGIONAL CONSTRAINTS:

Purpose:

The purpose of this subroutine is to check each crop to see if it will grow in the region.

List to do:

1. Set a subscript variable, i, to 1.
2. Compare crops array (i) to the regional info.
3. Print the crop name if it will grow in the region.
4. Increment i.
5. Repeat steps 2 through 4 until the crop whose name is "END" is encountered.

Data to and from calling program unit:

Regional info and the crops array come from the calling program unit; they're INTENT (IN).

English Fortran:

```
i = 1
DO WHILE Crops_Array (i) % name /= "END"
    IF
        Region % Growing_Days >= Crops_Array (i) % Min_Growing_Days
                            .AND.
        Region % Avg_Growing_Temp >= Crops_Array (i) % Min_Growing_Temp
                            .AND.
        Region % Avg_Growing_Temp <= Crops_Array (i) % Max_Growing_Temp
                            .AND.
        Region % Precipitation >= Crops_Array (i) % Min_Precipitation
                            .AND.
        Region % Precipitation <= Crops_Array (i) % Max_Precipitation
                            THEN
            Print Crops_Array (i) % Name
    END IF
    i = i + 1
END DO
```

The Complete Program

```
      MODULE Crop_And_Region_TYPEs
              IMPLICIT NONE
      !
      !-----------------------------------------------------------------
      !  This module defines the derived data types for crops and regions
      !     used in program What_Crops_Grow_In_A_Region.
      !
      !  The variables in the derived types below (the subobjects) are well
      !     named and are self-explanatory.
      !-----------------------------------------------------------------
      !
          TYPE Crops
              CHARACTER ( LEN = 10 )  ::   Name
              INTEGER                 ::   Min_Growing_Days
              INTEGER                 ::   Min_Avg_Temp
              INTEGER                 ::   Max_Avg_Temp
              REAL                    ::   Min_Avg_Precip
              REAL                    ::   Max_Avg_Precip
          END TYPE Crops
      !
```

```
        TYPE Region
            CHARACTER ( LEN = 10 ) ::    Name
            INTEGER                ::    Growing_Days
            INTEGER                ::    Average_Temp
            REAL                   ::    Precipitation
        END TYPE Region
    !
    END MODULE   Crop_And_Region_TYPEs
!=================================================================================
PROGRAM What_Crops_Grow_In_A_Region
              USE Crop_And_Region_TYPEs
              IMPLICIT NONE
    !
    !---------------------------------------------------------------------------
    ! Written by      :   C. Forsythe
    ! Date Written    :   3/25/2004
    !---------------------------------------------------------------------------
    ! This program reads crop data into an array and compares them to user
    !    supplied regional data to determine which crops will grow in that
    !    region.
    !
    ! Crops_Array       - a TYPE ( Crops ) array variable used to hold the
    !                        crop data
    !
    ! Region_From_User  - TYPE ( Region ) derived data type variable
    !
    ! No_More_Regions   - a named constant that represents the flag "QUIT"
    !                        CHARACTER ( LEN = 4 )
    !---------------------------------------------------------------------------
        ! BLACK BOX: FUNCTION Upper_Case ( String )
        !
        ! This FUNCTION is used to convert all the alphabetic characters
        !    in argument String to uppercase. This function was developed
        !    earlier in this chapter.
        !
        ! String           - a CHARACTER variable that contains the string
        !                       to be capitalized
        !
        ! Upper_Case       - function variable CHARACTER ( LEN = * )
        !---------------------------------------------------------------------
    !
```

```
        TYPE ( Crops ), DIMENSION ( 1 : 100 ) ::   Crops_Array
        TYPE ( Region )                       ::   Region_From_User
        CHARACTER ( LEN = 4 ), PARAMETER      ::   No_More_Regions = "QUIT"
        CHARACTER ( LEN = 4 )                 ::   Upper_Case
!
                    ! The following interface blocks are used
                    ! because the formal argument, Crops_Array,
                    ! is an assumed shape array.
!
        INTERFACE
            SUBROUTINE Read_Crops ( Crops_Array )
                USE Crop_And_Region_TYPEs
                TYPE ( Crops ), DIMENSION ( : ), INTENT ( OUT ) :: &
                                                    Crops_Array
            END SUBROUTINE Read_Crops
        END INTERFACE
!
        INTERFACE
            SUBROUTINE Print_Crops_That_Will_Grow &
                            ( Crops_Array, Region_From_User )
                USE Crop_And_Region_TYPEs
                TYPE ( Crops ), DIMENSION ( : ), INTENT ( IN ) :: &
                                                    Crops_Array
                TYPE ( Region ), INTENT ( IN )                  :: &
                                                    Region_from_User
            END SUBROUTINE Print_Crops_That_Will_Grow
        END INTERFACE
!
        CALL Read_Crops ( Crops_Array )
        CALL Get_Region_Info ( Region_From_User )
        DO WHILE(Upper_Case (Region_From_User % Name) /= No_More_Regions )
            CALL Print_Crops_That_Will_Grow(Crops_Array,Region_From_User)
            CALL Get_Region_Info ( Region_From_User )
        END DO
!
END PROGRAM   What_Crops_Grow_In_A_Region
!=================================================================
```

```fortran
      SUBROUTINE Read_Crops ( Crops_Array )
            USE Crop_And_Region_TYPEs
            IMPLICIT NONE
!
!-------------------------------------------------------------------
! This subroutine gets crop data from file CROPS.DAT and stores them
!    in INTENT ( OUT ) array argument Crops_Array.
!
! Crops_Array      - a TYPE ( Crops ) array variable
!
! Out_Of_Data      - a named constant that represents the flag "END"
!                    CHARACTER ( LEN = 3 )
!
! i                - local INTEGER variable used as an array index
!-------------------------------------------------------------------
      !  BLACK BOX: FUNCTION Upper_Case ( String )
      !
      !  This function is used to convert the all alphabetic characters
      !     in argument String to capitals.
      !
      !  String      - a CHARACTER variable that contains the string
      !                  to be capitalized
      !
      !  Upper_Case  - function variable, CHARACTER ( LEN = * )
      !-------------------------------------------------------------------
!
      TYPE ( Crops ), DIMENSION ( : ), INTENT ( OUT ) ::  Crops_Array
      INTEGER                                         ::  i
      CHARACTER ( LEN = 3 ), PARAMETER   ::  Out_Of_Data = "END"
      CHARACTER ( LEN = 3 )              ::  Upper_Case
!
      OPEN (UNIT=63, FILE = "CROPS.DAT", STATUS = "OLD", ACTION = "READ")
      i = 1
      READ ( UNIT = 63, FMT = * ) Crops_Array ( i )
      DO WHILE ( Upper_Case ( Crops_Array ( i ) % Name ) /= Out_Of_Data)
         PRINT *, Crops_Array ( i )   ! Echo the data file as read
         i = i + 1
         READ ( UNIT = 63, FMT = * ) Crops_Array ( i )
      END DO
      CLOSE ( UNIT = 63 )
!
      RETURN
      END SUBROUTINE  Read_Crops
!===================================================================
```

```
      SUBROUTINE Get_Region_Info ( Region_From_User )
                 USE Crop_And_Region_TYPEs
                 IMPLICIT NONE
!
!------------------------------------------------------------------------
!   This subroutine gets regional data from the user and stores them in
!      INTENT ( OUT ) argument Region.
!
!   Region_From_User - a TYPE ( Region ) variable
!
!   No_More_Regions  - a named constant that represents the flag "QUIT"
!                      CHARACTER ( LEN = 4 )
!------------------------------------------------------------------------
       !  BLACK BOX:   FUNCTION Upper_Case ( String )
       !
       !  This function is used to convert the all alphabetic characters
       !     in argument String to capitals.
       !
       !  String       - a CHARACTER variable that contains the string
       !                  to be capitalized
       !
       !  Upper_Case   - function variable, CHARACTER ( LEN = * )
       !-----------------------------------------------------------------
!
      TYPE ( Region ), INTENT ( OUT )   ::   Region_From_User
      CHARACTER ( LEN = 4 ), PARAMETER  ::   No_More_Regions = "QUIT"
      CHARACTER ( LEN = 4 )             ::   Upper_Case
!
      WRITE ( UNIT = *, FMT = 784, ADVANCE = "NO" )
 784  FORMAT ( 5X, "Please Enter a region name or 'QUIT' to exit: " )
      READ *, Region_From_User % Name
!
      SELECT CASE ( Upper_Case ( Region_From_User % Name ) )
          CASE ( No_More_Regions )
              ! Do nothing...this value will end the loop
              ! in the main program.
```

```
              CASE DEFAULT   ! Gets the rest of the region's data
                    WRITE ( UNIT = *, FMT = 785, ADVANCE = "NO" )
785                 FORMAT (5X,"Please Enter the number of growing days: " )
                    READ *, Region_From_User % Growing_Days
                    WRITE ( UNIT = *, FMT = 786, ADVANCE = "NO" )
786                 FORMAT (5X,"Please Enter the average temperature:   " )
                    READ *, Region_From_User % Average_Temp
                    WRITE ( UNIT = *, FMT = 787, ADVANCE = "NO" )
787                 FORMAT (5X,"Please Enter the precipitation:       " )
                    READ *, Region_From_User % Precipitation
         END SELECT
!
      RETURN
      END SUBROUTINE Get_Region_Info
!==================================================================
      SUBROUTINE Print_Crops_That_Will_Grow ( Crops_Array, Region_From_User )
            USE Crop_And_Region_TYPEs
            IMPLICIT NONE
      !
      !------------------------------------------------------------------
      !  This subroutine compares crop data to regional data and prints out
      !     the crops that will grow in the region.
      !
      !  Crops_Array        - a TYPE ( Crops ) array variable, INTENT ( IN )
      !
      !  Region_From_User   - a TYPE ( Crops ) variable, INTENT ( IN )
      !
      !  Out_Of_Data        - a named constant that represents the flag "END"
      !                          CHARACTER ( LEN = 3 )
      !
      ! i                   - local INTEGER variable used as an array index
      !------------------------------------------------------------------
            !  BLACK BOX:   FUNCTION Upper_Case ( String )
            !
            !  This function is used to convert the all alphabetic characters
            !     in argument String to capitals.
            !
            !  String        - a CHARACTER variable that contains the string
            !                     to be capitalized
            !
            !  Upper_Case    - function variable CHARACTER ( LEN = * )
            !------------------------------------------------------------------
      !
```

```fortran
        TYPE ( Crops ), DIMENSION ( : ), INTENT ( IN ) ::  Crops_Array
        TYPE ( Region ), INTENT ( IN )       ::      Region_From_User
        INTEGER                              ::      i
        CHARACTER ( LEN = 3 ), PARAMETER     ::      Out_Of_Data = "END"
        CHARACTER ( LEN = 3 )                ::      Upper_Case
!
        i = 1
        DO WHILE ( Upper_Case ( Crops_Array ( i ) % Name ) /= Out_Of_Data )
            IF ( Region_From_User % Growing_Days >=                        &
                                 & Crops_Array ( i ) % Min_Growing_Days &
          & .AND.                                                          &
                 Region_From_User % Average_Temp >=                        &
                                 & Crops_Array ( i ) % Min_Avg_Temp      &
          & .AND.                                                          &
                 Region_From_User % Average_Temp <=                        &
                                 & Crops_Array ( i ) % Max_Avg_Temp      &
          & .AND.                                                          &
                 Region_From_User % Precipitation >=                       &
                                 & Crops_Array ( i ) % Min_Avg_Precip    &
          & .AND.                                                          &
                 Region_From_User % Precipitation <=                       &
                                 & Crops_Array ( i ) % Max_Avg_Precip ) &
          & THEN
                 Print *, Crops_Array ( i ) % name, "Will grow in ",       &
                                 & Region_From_User % Name
            END IF
            i = i + 1
        END DO
!
        RETURN
        END SUBROUTINE   Print_Crops_That_Will_Grow
```

10 Styles of Arrays

```
*************** PROGRAM RESULTS ***************

Wheat          130 38 94    8.0000000   25.0000000
Barley         110 38 86   10.0000000   36.0000000
Cotton         140 60 120  15.0000000   35.0000000
Oats            95 50 90   10.0000000   27.0000000
Soybeans       100 54 88   12.0000000   28.0000000    The input subroutine
Corn           120 46 80    8.0000000   30.0000000    prints the file as it is read.
Peanuts        120 48 90   12.0000000   24.0000000
Sugarbeet      140 33 97   13.0000000   32.0000000
Potatoes        90 20 80    5.0000000   15.0000000
Beans           90 32 70   14.0000000   24.0000000
    Please Enter a region name or 'QUIT' to exit:   California
    Please Enter the number of growing days:    135
    Please Enter the average temperature:       75
    Please Enter the precipitation:             10.5
Wheat     Will grow in California
Barley    Will grow in California
Oats      Will grow in California
Corn      Will grow in California
Potatoes  Will grow in California
    Please Enter a region name or 'QUIT' to exit:   Michigan
    Please Enter the number of growing days:    98
    Please Enter the average temperature:       71
    Please Enter the precipitation:             22.7
Oats      Will grow in Michigan
    Please Enter a region name or 'QUIT' to exit: QuIt
```

The irregular capitalization of `QuIt` is handled by function `Upper_Case`.

`Upper_Case` was certainly useful in this program to compensate for the user's total disregard for the directions that asked her to type `QUIT` or `END` in capitals. As always, save ALL useful program units for future use.

Important Concepts Review

- **Parallel arrays** are a method of working with related data that is stored in different arrays.
- **Derived data type arrays** are arrays whose elements are of some derived type.
- When possible, use derived data type arrays instead of parallel arrays.

EXERCISES

10.1 Give a detailed description of each of the five styles of arrays introduced in this chapter.
   a) Assumed-shape arrays.
   b) Allocatable arrays.
   c) Automatic arrays.
   d) Parallel arrays.
   e) Derived data type arrays.

10.2 Write declarations for the following array descriptions:
   a) An assumed-shape array that can contain values that have decimal parts and only supplies values to a calling program.
   b) An automatic array that will be a clone for the following array:

   `REAL, DIMENSION ( : ), INTENT ( INOUT ) :: Z`

   c) An allocatable `REAL` array called `Chi_Square`.
   d) Allocate `Chi_Square` from the last question with bounds that go from -72 to 72.
   e) An assumed-shape array whose corresponding actual argument is declared as follows:

   `LOGICAL, DIMENSION ( -7 : -1 ) :: Boolean`

   f) Create any derived data type. Declare an allocatable array of that `TYPE`.

10.3 Write a complete program that does the following:
   - Prompts for and reads a positive integer, n, where $1 \leq n \leq 50$.
   - Allocates a `DOUBLE PRECISION` array to have n positions.
   - Fills the array with whole numbers from `1.0D0` to n.
   - Calculates $n!$ by using Fortran's `PRODUCT` function.
   - Deallocates the array.
   - Repeats the above steps until the user types a –1 for n.
   - Uses error checking whenever possible.

10.4 Write a complete program that will calculate the dot product of two vectors. To calculate the dot product, add up the products of corresponding positions of two parallel arrays where the arrays represent the vectors. Use `DOUBLE PRECISION` arrays whose allocated size is n, a value read from a user. Also use assumed-shape arrays whenever an array is passed to another program unit. Use error checking and don't forget to deallocate any allocated arrays.

10.5 Write a complete program that merges two `REAL` arrays. Again, get values from a user for each array size (they may be different lengths) and allocate the arrays. Values for the arrays are input from the keyboard. Sort each array using a bubble sort and then perform the merge. The resulting merged array's size should be allocated so that it can accommodate all the merged values. A merge works as seen in the illustration below:

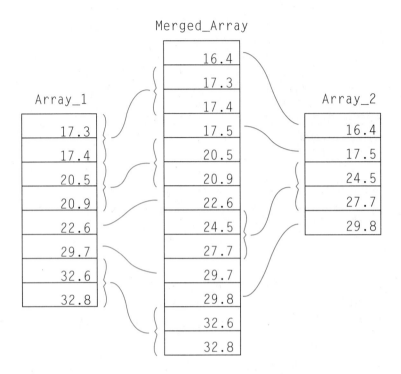

The bubble sort routine is a supplied black box. An example program that uses the bubble sort function is shown below; include the function in the merge program.

```
PROGRAM Test_The_Bubble_Sort
        IMPLICIT    NONE
!
!-------------------------------------------------------------------
! Written by      :  C. Forsythe
! Date Written    :  3/25/2004
!-------------------------------------------------------------------
! This program tests FUNCTION Bubble_Sort.
!
! n, i            - Two INTEGER variables: n is used to allocate
!                   Test_Array and as a loop limit; i is the loop
!                   index variable.
!
! Test_Array      - an allocatable REAL array
!-------------------------------------------------------------------
!
```

```
        INTEGER      ::  n, i
        REAL, DIMENSION ( : ), ALLOCATABLE  :: Test_Array
  !
  !  The following interface block lets this program unit know that
  !  Bubble_Sort's argument is an assumed-shape array. It also serves
  !  as a declaration of the function variable Bubble_Sort, which is,
  !  interestingly, an automatic array.
  !
        INTERFACE
           FUNCTION  Bubble_Sort ( UnSorted )
              REAL, DIMENSION ( : ), INTENT ( IN )       :: UnSorted
              REAL, DIMENSION ( 1 : SIZE ( UnSorted ) ) :: Bubble_Sort
           END FUNCTION
        END INTERFACE_Sort
  !
        PRINT *, "Please enter an array size"
        READ *, n
        ALLOCATE ( Test_Array ( 1 : n ) )
        DO i = 1, n
           PRINT *, "Please enter a value"
           READ *, Test_Array ( i )
        END DO
        PRINT *, Test_Array
        PRINT *, Bubble_Sort ( Test_Array )
        DEALLOCATE ( Test_Array )
  !
     END PROGRAM   Test_The_Bubble_Sort
!==============================================================================
     FUNCTION Bubble_Sort ( UnSorted_Array )
                IMPLICIT   NONE
  !
  !------------------------------------------------------------------
  !  This function performs a bubble sort on its assumed-shape array
  !     argument UnSorted_Array, which is type REAL, INTENT ( IN ).
  !
  !  UnSorted_Array     - a REAL, INTENT ( IN ) assumed-shape array
  !
  !  Bubble_Sort        - the function variable...an automatic array
  !                       that adopts the size of UnSorted_Array
  !
  !  UnSorted           - an automatic array to clone UnSorted_Array...
  !                       UnSorted_Array can't be changed because it is
  !                       INTENT ( IN ).
  !
  !  Size_Of_Array, i   - INTEGER variables used in the loops
  !
  !  Temp               - used in switching array positions that are
  !                       out of order
  !
  !  Sorted             - a LOGICAL variable that flags when the sort
  !                       is done
  !------------------------------------------------------------------
```

```
        !
                REAL, DIMENSION ( : ), INTENT ( IN )         :: UnSorted_Array
                REAL, DIMENSION ( 1 : SIZE ( UnSorted_Array ) ) :: Bubble_Sort
                REAL, DIMENSION ( 1 : SIZE ( UnSorted_Array ) ) :: UnSorted
                INTEGER           :: Size_Of_Array, i
                REAL              :: Temp
                LOGICAL           :: Sorted
        !
                UnSorted = UnSorted_Array  ! Get a copy of UnSorted_Array into
                                           ! automatic array UnSorted.
                Size_Of_Array = SIZE ( UnSorted )
                Sorted = .FALSE. ! Sort will end when no adjacent positions
                                 ! of array UnSorted are switched.
                DO WHILE ( .NOT. Sorted )
                    Sorted = .TRUE.     ! Assume the array is sorted.
                    DO i = 1, Size_of_Array - 1
                        IF ( UnSorted ( i ) > UnSorted ( i + 1 ) ) THEN
                            Temp              = UnSorted ( i )
                            UnSorted ( i )    = UnSorted ( i + 1 )
                            UnSorted ( i + 1 ) = Temp
                            Sorted            = .FALSE. ! Can't be sure the
                                                        ! array is sorted
                                                        ! on this pass.
                        END IF
                    END DO
                    Size_Of_Array = Size_Of_Array - 1 ! Don't need to check the
                                                      ! last value bubbled to
                                                      ! the end on the next pass.
                END DO
        !
                Bubble_Sort = UnSorted
        !
        END FUNCTION   Bubble_Sort
*************** PROGRAM RESULTS ***************
 Please enter an array size
5
 Please enter a value
-17.2
 Please enter a value
0
 Please enter a value
3.5
 Please enter a value
1.6
 Please enter a value
-4.7
  -17.2000008   0.0000000E+00   3.5000000   1.6000000  -4.6999998
  -17.2000008  -4.6999998   0.0000000E+00   1.6000000   3.5000000
```

10.6 Write a subroutine that will shift an allocatable array in a left-circular fashion. A circular shift is done by moving all array elements to the left one position, and then the value that was originally in the leftmost position is assigned to the rightmost array position. Use assumed-shape and automatic arrays whenever appropriate, and don't forget to include an interface block for any assumed-shape arrays.

10.7 Solve Exercise 10.4 using Fortran's DOT_PRODUCT function.

10.8 Write a program that calculates standard deviation. Standard deviation is a statistic used as a measure of the dispersion or variation in a distribution. Standard deviation is equal to the square root of the average of the squares of the deviation from the arithmetic mean. This can be expressed mathematically by the following equation:

$$\text{standard deviation} = \sqrt{\frac{1}{n}\sum_{i=1}^{n}(\bar{x}-x_i)^2}$$

Open a file called STATS.DAT using the OPEN statement below. Read the first value into an INTEGER variable named n. n represents how many numbers are in the file; there is one number per line.

```
OPEN ( UNIT = 43, FILE = "STATS.DAT", STATUS = "OLD", ACTION = "READ" )
```

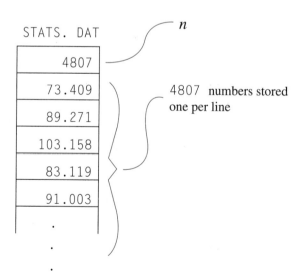

Allocate an appropriate REAL array and read in the values. Calculate the standard deviation using assumed-shape and automatic arrays as appropriate, and error check whenever possible.

10.9 Modify the Bubble_Sort function from Exercise 10.5 so that it will alphabetize CHARACTER data.

10.10 Generate 100-digit Fibonacci numbers. A Fibonacci number sequence is where each member of the sequence is the sum of the previous two, e.g.:

$$1\ 1\ 2\ 3\ 5\ 8\ 13\ 21\ 34\ldots$$

(*Hint: use the algorithm created in Exercise 9.12.*)

10.11 Generate 100-digit factorials.

10.12 Given the following derived data type, create an array called Class that is allocated to the size of an input value from the first line of a file. Calculate the class average on each assignment and each students' average on all assignments.

```
TYPE ( Student )
    CHARACTER ( LEN = 20 )      :: Name
    CHARACTER ( LEN = 11 )      :: Student_Number
    REAL, DIMENSION ( 1 : 4 )   :: Assignment_Grades
    REAL                        :: Average
END TYPE Student
```

10.13 Write a complete program that offers the following menu until option 5 is selected:

1. Enter Data.

2. Circular Shift Left.

3. Circular Shift Right.

4. Print Array.

5. Quit.

Option 1 should ask a user for a value for the size of an array and then allocate and fill the array with user input. Option 2 should do the same thing that Exercise 10.6 did, and option 3 should do the reverse. Options 4 and 5 have obvious purposes.

There is a caveat with this program: if one tries to use options 2, 3 or 4 before using option 1, the program will crash. The way to avoid this is by using Fortran's array inquiry LOGICAL function ALLOCATED. ALLOCATED will return a value of .TRUE. if the array is allocated and .FALSE. if the array has not been allocated or has been deallocated. As always, use assumed-shape, automatic, and allocatable arrays whenever appropriate. Also be sure to use error checking and don't forget to deallocate allocated arrays when they are no longer necessary.

Multidimensional Arrays

Introduction

As this is the third chapter on arrays, you can understand the importance of these data structures. The rich collection of features presented in Chapter 9 and 10 allow you to represent data in scores of ways. The only limitation in those chapter is that all the arrays had only one dimension. Often, data in the real world is more easily depicted by arrays that have multiple dimensions.

Let's say, for instance, that you want to do some research on forest fires and the methods used to extinguish them. First, you need to represent a forest that includes trees, rivers and fires. Using a CHARACTER two-dimensional array, you can create a facsimile of a forest by filling the array with symbols that represent the three elements. Let river water be represented by "■," trees by "T" and forest fires by "✖." A map can then be created that shows a given region of a forest. Algorithms can be created to cause the fires to expand over time in the array, and other algorithms can define what puts the fire out, such as human fire fighters, helicopters carrying water, heavy equipment to create fire stopping trenches, etc. The simulation can be used to study resource allocation and, hopefully, lead to more efficient ways of controlling forest fire disasters (Figure 11.1).

11.1 Multidimensional arrays

11.2 Assignment and expressions

Important Concepts Review

11.3 Array sections

11.4 Multidimensional arrays and loops

11.5 Program Design Example

Program Design

Important Concepts Review

Exercises

11 Multidimensional Arrays

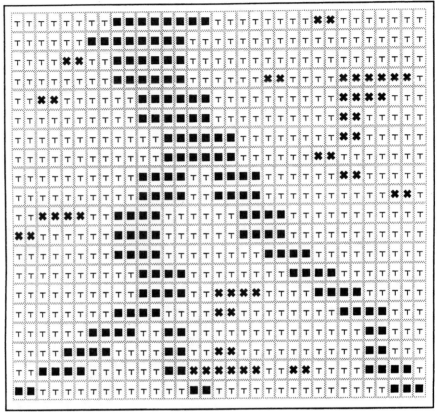

Figure 11.1

One-dimensional arrays are extremely useful, but Figure 11.1 demonstrates why programs need the facilities of multidimensional arrays. As you look at the map in Figure 11.1, you can think of it as a stack of one-dimensional array rows or a collection of adjacent one-dimensional array columns . . . multidimensional arrays are a natural outgrowth of the one-dimensional array model.

As you will find in the following sections, all the techniques described for one-dimensional arrays in Chapters 9 and 10, such as array section references, vector subscripts, etc., can be applied to multidimensional arrays as well.

11.1 Multidimensional Arrays

Multidimensional arrays come in all the same forms as one-dimensional arrays; there are just more sets of bounds. When one-dimensional arrays were introduced, they were likened to a row of data-holding boxes (Figure 11.2). Individual values in the boxes are accessed with subscripts. For one-dimensional arrays, one subscript is required to uniquely identify any position in an array. In Figure 11.2, `Int_Values ( 4 )` is printed. `Int_Values`'s fourth position is uniquely specified by the subscript, (4).

Take a first step into using multidimensional arrays by examining the two-dimensional array in Figure 11.3.

Figure 11.2

```
           Int_Values
      ┌────┬────┬────┬─────┬─────┐
      │ 17 │ -3 │  0 │ 231 │ -11 │
      └────┴────┴────┴─────┴─────┘
       (1)  (2)  (3)  (4)   (5)

      PRINT *, Int_Values ( 4 )
```

Figure 11.3

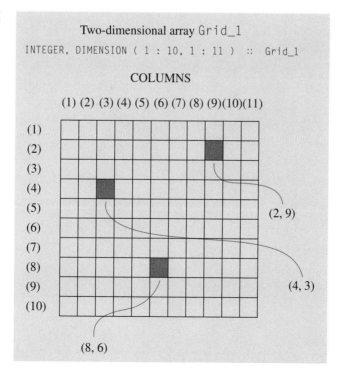

One can think of a two-dimensional array as a grid of boxes such as those on a piece of graph paper. To access a single array position, two subscripts are required. In Figure 11.3, the topmost shaded array location is in row *two*, column *nine*; the appropriate subscript to access that array position is (2, 9).

This "visualization" idea in which arrays are thought of as spacial objects is going to be short lived because Fortran permits arrays of up to seven dimensions—not so easy to visualize. For example, a cube could be used to represent a three-dimensional array, but even that very comprehendible shape is hard to work with when one gets into its middle positions (Figure 11.4).

The way to think of multidimensional arrays is to ascribe meaning to the subscripts of arrays that have more than two dimensions. Think of the subscripts as *classifying information*. For example, consider the following array reference:

```
Grades ( College, Course, Students, Assignment_Grades )
```

Grades is a four-dimensional array that holds all the assignment grades for every student in all courses among all colleges at some university. This works by forming an association between the declared bounds of the array and the actual data that it represents. Declaring the array above might take the form seen in Figure 11.5.

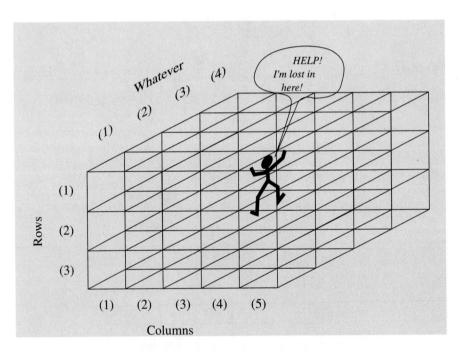

Figure 11.5

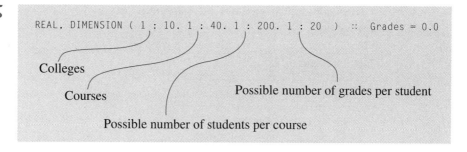

Each declared set of bounds in Figure 11.5 has meaning. If *conventions* such as the ones below are followed, all grades can be stored in array Grades in a clear, unambiguous way.

Colleges	
Subscript	College
1	Computer Science
2	English
3	Fine arts
...	

Each set of courses within a college is then associated with a subscript. An example of how this might be organized for the College of English is shown below:

Courses	
Subscript	College
1	Composition 101
2	Composition 102
3	Literature 240
...	

Subscript three in the array definition is for students, which can be arranged by using the convention that each subscript represents the next student in alphabetical order. The example below is for the Literature 240 course:

Literature 240	
Subscript	Student
1	Adams, John
2	Cleopatra
3	The Great, Alexander
...	

Finally, the assignments for Literature 240 might have the following subscript association:

Assignments	
Subscript	Assignment
1	Quizzes
2	Three-page paper
3	Term project
...	

Once the conventions above are accepted, the array can be used to hold all grades in a nonconfusing way. Remember that all the aforementioned choices are purely conceptual; another set of conventions could just have easily been chosen. It doesn't matter as long as all the programs that use array Grades view the data as being stored the same way.

If Cleopatra received a 3.75 on her quizzes in Literature 240, her grade would be stored as follows:

$$\text{Grades (2, 3, 2, 1)} = 3.75$$

The subscripts from left to right specify the College of English, Literature 240, Cleopatra, Quizzes. With all the permutations of all subscripts, the array holds all the grades for every student in all courses among all colleges at the university—*no visualizing the fourth dimension.*

It is also noteworthy that a multidimensional array's **size** is the product of its declared bounds. Thinking of that with respect to array Grades makes one realize that multidimensional arrays can be very memory hungry, but then Grades does represent a plethora of scores. Array Grades has $10 \times 40 \times 200 \times 20 = 1,600,000$ positions. Given that each grade is probably represented by four bytes of memory, the entire array would need 6,400,000 bytes of RAM. Beware of memory-hungry multidimensional arrays when using a computer with limited resources.

The declaration for array Grades also has the compile time assignment, Grades = 0.0, as shown below. This assignment will cause all 6,400,000 bytes of memory to be set to zero ... a very powerful assignment statement.

```
REAL, DIMENSION ( 1 : 10, 1 : 40, 1 : 200, 1 : 20 ) :: Grades = 0.0
```

The number of declared bounds for a multidimensional array define its **dimensionality**: Grid_1 from Figure 11.3 is a *two-dimensional* array and Grades from Figure 11.5 is a *four-dimensional* array. Every multidimensional array has **size** (which is the total number of positions in the array) and **shape** (which is the collection of the array's bounds).

Fortran represents the **shape** of a given array as a *one-dimensional INTEGER array* whose size is equal to the dimensionality of the array in question, and whose elements are the sizes of each dimension in the given array. For example, array Grades' shape is defined by the following one-dimensional array:

The shape of array Grades is (/ 10, 40, 200, 20 /)

The array above is one dimensional, its size is 4, which is the same as the number of dimensions in array Grades, and each position contains a value that is the number of position in each of the four dimensions: 10 in the shape array corresponds to Grades' first set of declared bounds, 1 : 10; 40 in the shape array corresponds to Grades' second set of declared bounds, 1 : 40; and so on. Another example of the shape of an array is (/ 10, 11 /), which is the shape of array Grid_1 in Figure 11.3.

Arrays that have the same shape are **conformable** and may be used in any expression that is appropriate for the operand arrays' declared type. Table 11.1 shows various arrays and whether they are conformable with array Grid_1 from Figure 11.3.

Even though some of the declared types of the arrays in the examples in Table 11.1 differ, they are all numeric and therefore can be used in expressions together. Although such expressions can be constructed, don't forget that all the warnings for mixed-mode expressions and integer division apply. To make the conformability issues in Table 11.1 clearer, Table 11.2 corresponds to Table 11.1 and shows the shape of each array as a one-dimensional INTEGER array.

Table 11.1

INTEGER, DIMENSION (1 : 10, 1 : 11) :: Grid_1	
Array	**Conformable?**
INTEGER, DIMENSION (-9 : 0, 11 : 21)...	YES
REAL, DIMENSION (0 : 9, -70 : -60) ...	YES
INTEGER, DIMENSION (1 : 10)...	NO
DOUBLE PRECISION, DIMENSION (14 : 23, -5 : 5)...	YES
INTEGER, DIMENSION (-9 : 0, 11 : 21, 1 : 7)...	NO
INTEGER, DIMENSION (300 : 310, -11 : -1)...	NO

Table 11.2

The Shape of Grid_1 is (/ 10, 11 /)	
Shape	**Conformable?**
(/ 10, 11 /)	YES
(/ 10, 11 /)	YES
(/ 10 /)	NO
(/ 10, 11 /)	YES
(/ 10, 11, 7 /)	NO
(/ 11, 11 /)	NO

11.2 Assignment and Expressions

When one-dimensional arrays were explained, several features of expressions, assignment and conformability were explained. They all apply to multidimensional arrays as well and are summarized in Table 11.3.

When a multidimensional array expression is evaluated and assigned to a conformable multidimensional array or array section, each position of the evaluated array expression is assigned to the corresponding array position in the array variable on the left side of the assignment. Figure 11.6 demonstrates how this works.

Every position of `Array_1` in Figure 11.6 is assigned the concatenation of corresponding positions of arrays `Ar_A` and `Ar_B`. The assignment statement below, from

Table 11.3

Feature	Assignment and Expressions	
	1-D	Multi-D
Any simple variable or constant is conformable with any array.	Yes	Yes
If an array is sent to one of Fortran's intrinsic functions, the result of it will be an array of the same shape as the argument, where each position of the *result array* contains the value of the function acting on the corresponding position of the *argument array*.	Yes	Yes
Conformable array operands may be combined with operators appropriate for the operands' type to form array expressions.	Yes	Yes
An array expression may be assigned to an array or an array section that is conformable with the result of that expression.	Yes	Yes

Figure 11.6

```
PROGRAM Conformable_2D_Array
        IMPLICIT NONE
!
!---------------------------------------------------------------
! Written by      :   C. Forsythe
! Date Written    :   3/25/2004
!---------------------------------------------------------------
!   Array_1    - a CHARACTER array that holds the results of a
!                concatenation operation
!
!   Ar_A, Ar_B - CHARACTER arrays used in the concatenation expression
!---------------------------------------------------------------
!
```

```
      CHARACTER ( LEN = 2 ), DIMENSION ( 1 : 5, 1 : 3 ) :: Array_1
      CHARACTER ( LEN = 1 ), DIMENSION ( 5 : 9, 2 : 4 ) :: Ar_A, Ar_B
      INTEGER            :: row, col
!
      Ar_A = "G"
      Ar_B = "O"
      Array_1 = Ar_A // Ar_B   ! Concatenate Ar_A and Ar_B and store
                               ! the resulting array in Array_1.
!
      DO row = 1, 5
          WRITE ( UNIT = *, FMT = 662 )        &
                          ( Array_1 ( row, col ), col = 1, 3 )
      END DO
!
 662  FORMAT ( 3( 1X, A2 ) )      ! Write 3 array positions per
                                  ! line separated by blanks.
!
END PROGRAM Conformable_2D_Array

*************** PROGRAM RESULTS ***************

 GO GO GO
 GO GO GO
 GO GO GO
 GO GO GO
 GO GO GO
```

program `Conformable_2D_Array`, is equivalent to the 15 assignment statements that follow.

$$Array\_1 = Ar\_A // Ar\_B$$
$$\equiv$$

```
Array_1 ( 1, 1 ) = Ar_A ( 5, 2 ) // Ar_B ( 5, 2 )
Array_1 ( 2, 1 ) = Ar_A ( 6, 2 ) // Ar_B ( 6, 2 )
Array_1 ( 3, 1 ) = Ar_A ( 7, 2 ) // Ar_B ( 7, 2 )
Array_1 ( 4, 1 ) = Ar_A ( 8, 2 ) // Ar_B ( 8, 2 )
Array_1 ( 5, 1 ) = Ar_A ( 9, 2 ) // Ar_B ( 9, 2 )
Array_1 ( 1, 2 ) = Ar_A ( 5, 3 ) // Ar_B ( 5, 3 )
Array_1 ( 2, 2 ) = Ar_A ( 6, 3 ) // Ar_B ( 6, 3 )
Array_1 ( 3, 2 ) = Ar_A ( 7, 3 ) // Ar_B ( 7, 3 )
Array_1 ( 4, 2 ) = Ar_A ( 8, 3 ) // Ar_B ( 8, 3 )
Array_1 ( 5, 2 ) = Ar_A ( 9, 3 ) // Ar_B ( 9, 3 )
Array_1 ( 1, 3 ) = Ar_A ( 5, 4 ) // Ar_B ( 5, 4 )
Array_1 ( 2, 3 ) = Ar_A ( 6, 4 ) // Ar_B ( 6, 4 )
Array_1 ( 3, 3 ) = Ar_A ( 7, 4 ) // Ar_B ( 7, 4 )
Array_1 ( 4, 3 ) = Ar_A ( 8, 4 ) // Ar_B ( 8, 4 )
Array_1 ( 5, 3 ) = Ar_A ( 9, 4 ) // Ar_B ( 9, 4 )
```

The way this works is exactly the same as with one-dimensional arrays; there are just more sets of bounds in multidimensional arrays.

Important Concepts Review

- **Multidimensional arrays** are arrays that have more than one set of declared bounds.

- Don't try to visualize 3- to 7-dimensional arrays.

- **Ascribe meaning** to the subscripts of multidimensional arrays . . . think of the subscripts as classifying information.

- **Size** is the product of the declared bounds of an array; it determines how many array positions there are in an array.

- **Dimensionality** is the number of declared sets of bounds of an array.
  `INTEGER, DIMENSION ( 1 : 4, 6 : 9 )` . . . is a 2D array.

- **Shape** is the collection of an array's bounds. Shape is represented by a one-dimensional `INTEGER` array.

- **Conformable arrays** have the same shape and may be used in expressions where all operands are of compatible types.

11.3 Array Sections

Sections of multidimensional arrays can be specified by using section references for any set of declared bounds. Figure 11.7 shows a program that uses array section references. As always, any array section reference can include a **stride** element, and strides may be negative (which specifies elements in reverse order).

There is one interesting feature of specifying array section references in multidimensional arrays: the bounds don't *all* have to be section references; one or more subscripts can be regular subscript references. An easy way to understand the impact of this idea is to take a two-dimensional array and think of it as a collection of adjacent one-dimensional arrays (Figure 11.8, p. 358).

Figure 11.7

```
PROGRAM show_2D_Section_Reference
          IMPLICIT NONE
!
!------------------------------------------------------------
! Written by    :  C. Forsythe
! Date Written  :  3/25/2004
!------------------------------------------------------------
! This program puts some values in two-dimensional INTEGER array
!    Test_Array using array section references. r and c are simple
!    INTEGER variables used as loop indexes.
!------------------------------------------------------------
!
      INTEGER, DIMENSION ( 1 : 10, 5 : 9 ) :: Test_Array = 0
      INTEGER                              :: r, c
!
      Test_Array ( 6 : 10, 7 : 9 ) = -7
      Test_Array ( 1 : 5,  5 : 6 ) = 3
!
      WRITE ( UNIT = *, FMT = 702 )   &
                   & ( ( Test_Array ( r, c ), c = 5, 9), r = 1, 10 )
                         ! The above implied DO loop is a nested
                         ! implied DO loop. It writes the array
                         ! by rows. This sort of nested loop is
                         ! explained in the next section.
 702 FORMAT ( 5I3 )
!
END PROGRAM   Show_2D_Section_Reference

*************** PROGRAM RESULTS ***************

  3    3    0    0    0
  3    3    0    0    0
  3    3    0    0    0
  3    3    0    0    0
  3    3    0    0    0
  0    0   -7   -7   -7
  0    0   -7   -7   -7
  0    0   -7   -7   -7
  0    0   -7   -7   -7
  0    0   -7   -7   -7
```

The array section references specify the two-dimensional array sections indicated on the left.

11 Multidimensional Arrays

At the top of the Figure 11.8, Ary is represented as four *rows* of one-dimensional arrays stacked on top of each other. The lower depiction of Ary shows its *columns* as a collection of one-dimensional arrays. Individual one-dimensional array references in Figure 11.8 are specified as follows:

Ary's rows as 1D Arrays	Ary's columns as 1D arrays
Ary (1, 1 : 7)	Ary (1 : 4, 1)
Ary (2, 1 : 7)	Ary (1 : 4, 2)
Ary (3, 1 : 7)	Ary (1 : 4, 3)
Ary (4, 1 : 7)	Ary (1 : 4, 4)
	Ary (1 : 4, 5)
	Ary (1 : 4, 6)
	Ary (1 : 4, 7)

Figure 11.8

. . . , DIMENSION (1 : 4, 1 : 7) :: Ary

	(1)	(2)	(3)	(4)	(5)	(6)	(7)
(1)		Ary (1, 1 : 7)					
(2)		Ary (2, 1 : 7)					
(3)		Ary (3, 1 : 7)					
(4)		Ary (4, 1 : 7)					

or

. . . , DIMENSION (1 : 4, 1 : 7) :: Ary

	(1)	(2)	(3)	(4)	(5)	(6)	(7)
(1)	Ary (1 : 4, 1)	Ary (1 : 4, 2)	Ary (1 : 4, 3)	Ary (1 : 4, 4)	Ary (1 : 4, 5)	Ary (1 : 4, 6)	Ary (1 : 4, 7)
(2)							
(3)							
(4)							

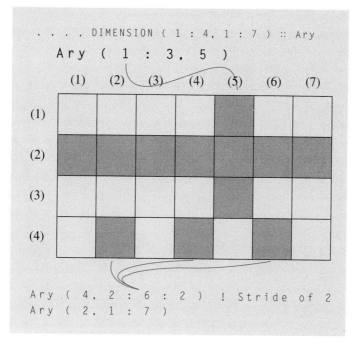

Using the various section references, multidimensional arrays can be accessed in many powerful ways, as seen in Figure 11.9.

It is also acceptable to use a vector subscript for any subscript in a multidimensional array reference. **Vector subscripts** are one-dimensional INTEGER arrays whose elements specify various array positions along one dimension of the subscripted array. Vector subscripts were thoroughly explained in Chapter 9.

11.4 Multidimensional Arrays and Loops

Much data can be stored in the various dimensions of a multidimensional array. Those data will need to be accessed, processed through expressions, read in, written out, etc., and there need to be ways to do such operations efficiently. Cycling through an array's data is primarily accomplished by using loops. Figure 11.10 (p. 360) shows a program that reads values into a 4 × 3 two-dimensional array and then writes it out using *nested* looping structures.

Figure 11.10

```
      PROGRAM  Read_And_Write_A_2D_Array
                IMPLICIT  NONE
      !
      !---------------------------------------------------------------
      !  The purpose of this program is to read values into a 4 X 3
      !     two-dimensional array and write it out to experiment with
      !     nested loops.
      !
      !  Grid       - a 4 X 3 2D CHARACTER ( LEN = 1 ) array
      !
      !  r, c       - INTEGER loop indexes
      !---------------------------------------------------------------
      !
            CHARACTER ( LEN = 1 ), DIMENSION ( 1 : 4, 1 : 3 ) ::    Grid
            INTEGER                                           ::    r, c
      !
            DO r = 1, 4
               DO c = 1, 3
                  WRITE ( UNIT = *, FMT = 56, ADVANCE = "NO" )
       56         FORMAT (4X, "Please enter one character:  " )
                  READ  ( UNIT = *, FMT = 29 ) Grid ( r, c )
       29         FORMAT ( A1 )
                              !
                              ! This set of loops will read
                              ! twelve characters filling the top
                              ! row first from left to right,
                              ! then the second row, and so on.
               END DO
            END DO
      !
            DO r = 1, 4
               WRITE ( UNIT = *, FMT = 66 ) ( Grid ( r, c ), c = 1, 3 )
       66      FORMAT ( 3( 1X, A1 ) ) ! equivalent to: 1X, A1 1X, A1 1X, A1
            END DO
      !
      END PROGRAM  Read_And_Write_A_2D_Array

     *************** PROGRAM RESULTS ***************

     Please enter one character: u
     Please enter one character: z
     Please enter one character: #
     Please enter one character: $
     Please enter one character: Q
     Please enter one character: t
     Please enter one character: 7
     Please enter one character: -
     Please enter one character: @
     Please enter one character: J
     Please enter one character: 4
     Please enter one character: *
  u z #
  $ Q t
  7 - @
  J 4 *
```

11.4 Multidimensional Arrays and Loops

Figure 11.10 uses a **nested looping** structure to fill the rows of array `Grid`:

```
DO r = 1, 4
    DO c = 1, 3
        READ ( UNIT = *, FMT = 29 ) Grid ( r, c )
    END DO
END DO
```

Although the structure above may appear foreboding and fill one with apprehension, it isn't that difficult to understand. In nested looping structures, the innermost loops iterate fastest. The nested loops below help explain how this process works and are basically the same as the ones above.

```
DO r = 1, 4
    DO c = 1, 3
        PRINT *, "r =", r, "   c =", c
    END DO
END DO
```

```
r = 1       c = 1
r = 1       c = 2
r = 1       c = 3
r = 2       c = 1
r = 2       c = 2
r = 2       c = 3
r = 3       c = 1
r = 3       c = 2
r = 3       c = 3
r = 4       c = 1
r = 4       c = 2
r = 4       c = 3
```

Notice that `r`'s value remains constant at 1 while `c`'s value cycles from 1 to 3. Then `r` changes to 2 and stays at 2 while `c` again cycles from 1 to 3, and so on. This process of analogous to an automobile odometer in which the innermost wheel (the "tenths" wheel) changes fastest. When the "tenths" wheel reaches 9 (its "loop limit"), it moves back to zero and the next wheel to the left is incremented (the "outer loop").

Data can also be transferred into or out of multidimensional arrays by `PRINT`, `READ`, `WRITE` or `DATA` statements with nested implied `DO` loops. The general form of a nested implied `DO` loop is displayed below:

```
[(]...(list,ix=init,lmt,[incr])[,ix-2=init-2,lmt-2,[incr-2])]...
```

Parentheses indicated by "`[(]`..." on the left above are used to balance any parentheses that are introduced on the right when adding additional "nests" to the implied `DO`. Examples of nested implied `DO` loops are shown in Figure 11.11 (p. 362). When using nested implied `DO` loops to write a two-dimensional array as seen in Figure 11.11, the `FORMAT` controls separating the array rows onto different lines. If there are more items to be written than a `FORMAT` can accommodate, the `FORMAT` is reused and a new line is begun. See Appendix D.

Figure 11.11

```
PROGRAM   Read_And_Write_A_2D_Array
           IMPLICIT  NONE
!
!-----------------------------------------------------------------
! The purpose of this program is to read in a 4 X 3 two-dimensional array
!     and write it out to experiment with nested implied DO loops.
!
! Grid       - a 4 X 3 2D CHARACTER ( LEN = 1 ) array
!
! r, c       - INTEGER loop indexes
!-----------------------------------------------------------------
!
     CHARACTER ( LEN = 1 ), DIMENSION ( 1 : 4, 1 : 3 )  ::   Grid
     INTEGER                                            ::   r, c
!
     WRITE ( UNIT = *, FMT = 56, ADVANCE = "NO" )
56   FORMAT (4X, "Please enter twelve characters: " )
     READ ( UNIT = *, FMT = 29 )( ( Grid ( r, c ), c = 1,3 ), r = 1,4 )
29   FORMAT ( 12(1X, A1 ) )
                  !
                  ! The above nested implied DO
                  ! will get all the characters
                  ! from ONE line of input.
                  !
!
     WRITE ( UNIT = *, FMT = 66 )( ( Grid ( r, c ), c = 1,3 ), r = 1,4 )
66   FORMAT ( 3( 1X, A1 ) )
                  !
                  ! Each time the FORMAT is used up, it is reused
                  ! and writing starts on a new line.
                  !
END PROGRAM    Read_And_Write_A_2D_Array

*************** PROGRAM RESULTS ***************

     Please enter twelve characters:    u z # $ Q t 7 - @ J 4 *
 u z #
 $ Q t
 7 - @
 J 4 *
```

11.5 Program Design Example

In this design example, a program will be created to operate on sets. The sets will be stored as CHARACTER data in two-dimensional arrays. Set operations included in the program will be union, intersection and difference. Figure 11.12 reviews how these set operations work.

Program Design

STATE THE PROBLEM

In this program I want to fill two two-dimensional arrays with characters to represent the sets; each array position will hold only one character. All set elements will be designated by "*", and the rest of the array will be filled with blanks. The two sets will be stored in separate arrays, and both sets will have the same number of rows and columns. I will write a subroutine that will read in one set from a disk file and use that same subroutine to read in the other set as well. I will also create a subroutine to print out a two-dimensional CHARACTER array. This subroutine will be used often because I want to display the sets after they are read in and, of course, after each set operation is completed. The result of the operation will be put into a conformable third array. The easiest way for a user to operate the program will be with a menu such as the one below:

```
1. Union.
2. Intersection.
3. A - B.
4. B - A.
5. Quit.
```

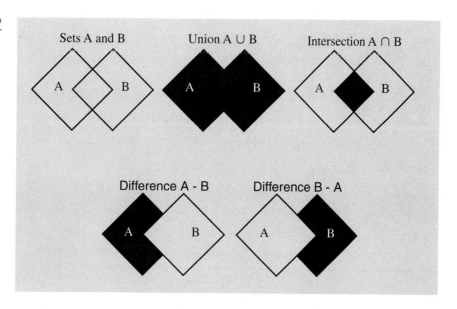

Figure 11.12

I will limit the array size to 20 × 75 so that any of the printed arrays will fit on a computer screen.

STRUCTURE CHART:

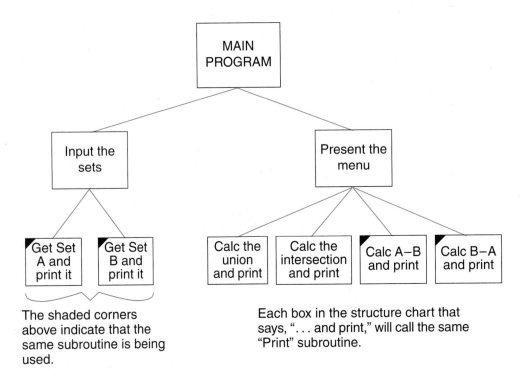

The shaded corners above indicate that the same subroutine is being used.

Each box in the structure chart that says, "... and print," will call the same "Print" subroutine.

MAIN PROGRAM:

Purpose:

MAIN's purpose is to drive the program by calling the two subroutines under it.

List to do:

1. Call the subroutine that inputs the sets.
2. Call the subroutine that presents the menu.

Data flow to and from the calling program:

None (MAIN isn't called).

English Fortran:

Not necessary because the main program is a simple, sequential statement structure.

INPUT THE SETS:

List to do:

1. Prompt a user for the name of the file that holds set A.
2. Read the file name for set A from the user.
3. Call the subroutine that inputs a set and get set A.
4. Prompt the user for the name of the file that holds set B.
5. Read the file name for set B.
6. Call the subroutine that inputs a set and get set B.

Data flow to and from the calling program:

The two CHARACTER arrays that represent the sets and the number of rows in the sets are sent back to the main program.

English Fortran:

Not necessary because this subroutine is a simple, sequential statement structure.

INPUTTING A TWO-DIMENSIONAL CHARACTER ARRAY:

The purpose of this subroutine is to read in a two-dimensional CHARACTER array whose dimension is $r \times 75$ where $1 \leq r \leq 20$ from file *file name*.

List to do:

1. OPEN the appropriate file (*file name*) ... The basic OPEN statement was discussed after the design example in Chapter 9.
2. Set a loop index variable named row to 1.
3. Read 75 characters into row row using an implied DO loop.
4. Start a DO WHILE loop whose condition for continuing is that Set (row, 1) is not a "$". (The last line of a set file will contain all "$"s to flag the end of input.)
5. Increment the row counter.
6. Read 75 characters into the next row using an implied DO loop.
7. Return to the top of the DO loop (Step 4).
8. Print the set array.

Data flow to and from the calling program:

The set array and how many rows are in it are sent to the calling program. *File name* comes from the calling program unit.

English Fortran:

```
┌─SUBROUTINE Read_A_Set ( File_Name, Set_Array, Number_Of_Rows )
│      OPEN the set file in file File_Name
│      Row = 1
│      Read the first row
│   ┌─ DO WHILE ( set ( row, 1 ) /= "$" )
│   │      Row = row + 1
│   │      Read the next row
│   └─ END DO
│      Row = row - 1   ! Don't count the "$$$..." row
│      CALL Print the set array
│   RETURN
└─END SUBROUTINE Read_A_Set
```

PRESENT MENU:

This is a supplied black box; the source code is shown below.

```
SUBROUTINE  Present_Menu ( Set_A, Set_B, rows )
      IMPLICIT NONE
!
!------------------------------------------------------------------
!
!   This subroutine will present a menu and take action based on a
!      user's choice.
!
!   Set_A, Set_B    - CHARACTER( LEN = 1 ), DIMENSION ( 1 : 20, 1 : 75 )
!                       INTENT ( IN ) arrays to hold the set data
!
!   rows            - an INTEGER, INTENT ( IN ) variable that holds
!                       the number of rows in each set
!
!   Exit_Menu       - an INTEGER PARAMETER that holds the value of
!                       the "Quit" option
!
!   Choice          - an INTEGER variable that holds the user's menu
!                       selection
!------------------------------------------------------------------
!
      CHARACTER ( LEN = 1 ), DIMENSION ( 1 : 20, 1 : 75 ), &
                                 INTENT ( IN ) :: Set_A, Set_B
      INTEGER, INTENT ( IN )       :: rows
      INTEGER                      :: Choice
      INTEGER, PARAMETER           :: Exit_Menu = 7
!
```

11.5 Program Design Example

```
        Choice = 0   ! Choice must be defined before the loop can test it.
        DO WHILE ( Choice /= Exit_Menu )
            PRINT *, "1. Union."
            PRINT *, "2. Intersection."
            PRINT *, "3. A - B."
            PRINT *, "4. B - A."
            PRINT *, "5. Print Set A."
            PRINT *, "6. Print Set B."
            PRINT *, "7. Quit."
            PRINT *, ""
            PRINT *, "    Please enter your choice."
            READ *, Choice
!
            SELECT CASE ( Choice )
                CASE ( 1 )
                    CALL Union          ( Set_A, Set_B, rows )
                CASE ( 2 )
                    CALL Intersection   ( Set_A, Set_B, rows )
                CASE ( 3 )
                    CALL Subtract_Sets  ( Set_A, Set_B, rows )
                CASE ( 4 )
                    CALL Subtract_Sets  ( Set_B, Set_A, rows )
                CASE ( 5 )
                    CALL Print_Set      ( "Set A:", Set_A, rows )
                CASE ( 6 )
                    CALL Print_Set      ( "Set B:", Set_B, rows )
                CASE ( Exit_Menu )
                    ! Do nothing, the loop will end now.
                CASE DEFAULT
                    PRINT *, ""
                    PRINT *, "*** That was an invalid menu choice..."
                    PRINT *, "    Please try again."
                    PRINT *, ""
            END SELECT
        END DO
!
RETURN
END SUBROUTINE  Present_Menu
```

PRINT RESULTS:

The purpose of this subroutine is to print a set array.

List to do:

1. Print the title.
2. Start a fixed-limit DO loop that goes from 1 to rows.

3. Print a row of the set array with an implied DO loop.
4. Go back to the beginning of the loop.

Data flow to and from the calling program:

The title, set array and the number of rows in the set come from the calling program unit.

English Fortran:

```
SUBROUTINE  Print_Set   ( title, set, rows )
    Print the title
        Start a fixed-limit DO loop that goes from 1 to rows
            Print current row of set with an implied DO loop
        END DO
RETURN
END SUBROUTINE  Print_Results
```

CALCULATE THE UNION:

The purpose of this subroutine is to calculate the union of two conformable CHARACTER arrays.

List to do:

1. Write a WHERE statement whose condition sets the result array to an "*" if there is an "*" in either set, and to a blank otherwise.
2. Print the result of the operation.

Data flow to and from the calling program:

The set arrays and how many rows in a set are received from the calling program.

English Fortran:

```
SUBROUTINE Union_Sets ( Set_A, Set_B, Rows )
        declarations
        WHERE ( Set _A== "*" or Set_B == "*" )
            Result_Array = "*"
        ELSEWHERE
            Result_Array = " "
        END WHERE
        CALL Print Result Array
RETURN
END SUBROUTINE  Union_Sets
```

CALCULATE THE INTERSECTION:

The purpose of this subroutine is to calculate the intersection of two conformable CHARACTER arrays. The design for this subroutine is the same as the one for union except that the WHERE statement is different.

```
SUBROUTINE Intersect_Sets ( Set_A, Set_B, Rows )
    declarations
    WHERE ( Set_A == "*" and Set_B == "*" )
        Result_Array = "*"
    ELSEWHERE
        Result_Array = " "
    END WHERE
    CALL Print Result Array
RETURN
END SUBROUTINE   Intersect_Sets
```

CALCULATE THE DIFFERENCE:

The purpose of this subroutine is to calculate the difference between two conformable CHARACTER arrays. The design for this subroutine is the same as the ones for union and intersection except that the WHERE statement is different.

```
SUBROUTINE Subtract_Sets ( Set_A, Set_B, Rows )
    declarations
    WHERE ( Set_B == "*" )
        Result_Array = " "
    ELSEWHERE
        Result_Array = Set_A
    END WHERE
    CALL Print Result Array
RETURN
END SUBROUTINE   Subtract_Sets
```

The Complete Program

```
      PROGRAM   Set_Operations
                IMPLICIT   NONE
      !
      !-------------------------------------------------------------------
      ! Written by       :   C. Forsythe
      ! Date Written     :   3/25/2004
      !-------------------------------------------------------------------
      ! This program performs various set operations defined by the options
      !    in a menu.
      !
      ! Set_A, Set_B          - two CHARACTER ( LEN = 1 ) arrays that
      !                          represent the sets
      !
      ! rows                  - an INTEGER variable that represents how many
      !                          rows are in a set. Both sets will have the
      !                          same number of rows.
      !-------------------------------------------------------------------
      !
           CHARACTER( LEN = 1 ),DIMENSION ( 1 : 20, 1 : 75) :: Set_A = " " &
                                                             , Set_B = " "
           INTEGER                       ::  rows
      !
           CALL Input_The_Sets   ( Set_A, Set_B, rows )
           CALL Present_Menu     ( Set_A, Set_B, rows )
      !
      END PROGRAM      Set_Operations
!================================================================================
      SUBROUTINE Input_The_Sets ( Set_A, Set_B, rows )
           IMPLICIT   NONE
      !
      !-------------------------------------------------------------------
      ! This subroutine reads in the two sets, Set_A and Set_B.
      !
      ! Set_A, Set_B          - two CHARACTER ( LEN = 1 ), INTENT ( OUT )
      !                          two-dimensional arrays that represent the
      !                          sets
      !
      ! rows                  - an INTEGER, INTENT ( OUT ) argument that
      !                          represents how many rows are in a set
      !
      ! File_Name             - a CHARACTER ( LEN = 80 ) local variable
      !                          that holds the file name of a set array
      !-------------------------------------------------------------------
      !
```

```
              CHARACTER( LEN = 1 ),DIMENSION ( 1 : 20, 1 : 75 ),INTENT ( OUT ) &
                                            :: Set_A, Set_B
              INTEGER, INTENT ( OUT )       :: rows
              CHARACTER ( LEN = 80 )        :: File_Name ! A local variable
       !
              PRINT *, "Please enter the first set's file name."
              READ *, File_Name
              CALL Read_A_Set ( File_Name, Set_A, rows )
              PRINT *, "Please enter the second set's file name."
              READ *, File_Name
              CALL Read_A_Set ( File_Name, Set_B, rows )
              PRINT *, ""
       !
       RETURN
       END SUBROUTINE   Input_The_Sets
!================================================================================
       SUBROUTINE  Read_A_Set ( File_Name, Set, rows )
                   IMPLICIT  NONE
       !
       !-------------------------------------------------------------------------
       !  This subroutine reads in a two-dimensional array from a file whose
       !     name is stored in argument File_Name.
       !
       !  Set                   - a CHARACTER ( LEN = 1 ), INTENT ( OUT )
       !                          array, 20X75
       !
       !  File_Name             - a CHARACTER ( LEN = * ), INTENT ( IN )
       !                          argument that holds the file name for a
       !                          set
       !
       !  End_Of_Set            - a CHARACTER ( LEN = 1 ) PARAMETER that
       !                          represents the end-of-input flag ($)
       !
       !  rows                  - an INTENT ( OUT ) variable whose ultimate
       !                          value will be the number of rows in the
       !                          set
       !
       !  col                   - a local INTEGER loop index variable
       !-------------------------------------------------------------------------
       !
              CHARACTER( LEN = 1 ),INTENT ( OUT ),DIMENSION ( 1 : 20, 1 : 75 ) &
                                            :: Set
              INTEGER, INTENT ( OUT )                :: rows
              CHARACTER ( LEN = 1 ), PARAMETER  :: End_Of_Set = "$"
              CHARACTER ( LEN = * )             :: File_Name
              INTEGER                           :: col
       !
```

```fortran
      OPEN(UNIT = 29, FILE = File_Name, STATUS = "OLD", ACTION = "READ")
      rows = 1
      READ ( UNIT = 29, FMT = 107 ) ( Set ( rows, col ), col = 1, 75 )
  107 FORMAT ( 75A1 )
      DO WHILE ( Set ( rows, 1 ) /= End_Of_Set )
         rows = rows + 1
         READ (UNIT = 29, FMT = 107 ) (Set ( rows, col ), col = 1, 75)
      END DO
      rows = rows - 1 ! Don't count the "$$$$ ..." row.
      CALL  Print_Set ("This is one of the two sets:", Set, rows )
!
RETURN
END SUBROUTINE   Read_A_Set
!===============================================================================
   SUBROUTINE   Print_Set ( Title, Set, rows )
               IMPLICIT  NONE
!
!------------------------------------------------------------------
!  This subroutine prints a title above a two-dimensional array
!     called Set that is rows X 75.
!
!  Title           - a CHARACTER ( LEN = * ), INTENT ( IN ) argument
!                      that identifies what set is being printed
!
!  Set             - a CHARACTER ( LEN = 1 ), INTENT ( IN ) array,
!                      rows X 75
!
!  rows            - an INTENT ( IN ) argument whose value contains the
!                      number of rows in the set
!
!  col, r          - local INTEGER loop index variables
!
!------------------------------------------------------------------
!
      CHARACTER( LEN = 1 ),INTENT ( IN ),DIMENSION ( 1 : 20, 1 : 75 ) &
                                                :: Set
      INTEGER, INTENT ( IN )                    :: rows
      CHARACTER ( LEN = * ), INTENT ( IN )      :: Title
      INTEGER                                   :: col, r
!
```

```
            PRINT *, ""
            PRINT *, Title
            PRINT *, ""
            DO r = 1, rows
                WRITE ( UNIT = *, FMT = 9 ) ( Set ( r, col ), col = 1, 75 )
            END DO
9           FORMAT ( 75A1 )
      !
      RETURN
      END SUBROUTINE  Print_Set
!===============================================================================
      SUBROUTINE   Present_Menu ( Set_A, Set_B, rows )
                   IMPLICIT   NONE
      !
      !-------------------------------------------------------------------------
      !  This subroutine will present a menu and take action based on a
      !     user's choice.
      !
      !  Set_A, Set_B      - CHARACTER( LEN = 1 ),DIMENSION (1 : 20, 1 : 75 )
      !                      INTENT ( IN ) arrays that hold the set data
      !
      !  rows              - an INTEGER, INTENT ( IN ) argument that holds
      !                      the number of rows in the sets
      !
      !  Exit_Menu         - an INTEGER PARAMETER that holds the value of
      !                      the "Quit" option
      !
      !  Choice            - an INTEGER variable that holds the user's
      !                      selection
      !-------------------------------------------------------------------------
      !
            CHARACTER ( LEN = 1 ), DIMENSION ( 1 : 20, 1 : 75 )  &
                                         , INTENT ( IN ) :: Set_A, Set_B
            INTEGER, INTENT ( IN )       :: rows
            INTEGER                      :: Choice
            INTEGER, PARAMETER           :: Exit_Menu = 7
      !
            Choice = 0  ! Choice must be defined before the loop is met.
```

```
            DO WHILE ( Choice /= Exit_Menu )
                PRINT *, "1. Union."
                PRINT *, "2. Intersection."
                PRINT *, "3. A - B."
                PRINT *, "4. B - A."
                PRINT *, "5. Print Set A."
                PRINT *, "6. Print Set B."
                PRINT *, "7. Quit."
                PRINT *, ""
                PRINT *, "   Please enter your choice."
                READ *, Choice
!
                SELECT CASE ( Choice )
                    CASE ( 1 )
                        CALL  Union        ( Set_A, Set_B, rows )
                    CASE ( 2 )
                        CALL  Intersection ( Set_A, Set_B, rows )
                    CASE ( 3 )
                        CALL  Subtract_Sets ( Set_A, Set_B, rows )
                    CASE ( 4 )
                        CALL  Subtract_Sets ( Set_B, Set_A, rows )
                    CASE ( 5 )
                        CALL  Print_Set    ( "Set A:", Set_A, rows )
                    CASE ( 6 )
                        CALL  Print_Set    ( "Set B:", Set_B, rows )
                    CASE ( Exit_Menu )
                        ! Do nothing, the loop will end now.
                    CASE DEFAULT
                        PRINT *, ""
                        PRINT *, "*** That was an invalid menu choice..."
                        PRINT *, "    Please try again."
                        PRINT *, ""
                END SELECT
            END DO
!
        RETURN
        END SUBROUTINE  Present_Menu
!=========================================================================
```

```fortran
      SUBROUTINE  Union ( Set_A, Set_B, rows )
                  IMPLICIT  NONE
!
!------------------------------------------------------------------------
!  This subroutine will find the union of its two arrays and store
!     the result in Result_Array.
!
!  Set_A, Set_B       - two-dimensional CHARACTER ( LEN = 1 ),
!                         INTENT ( IN ) arrays that represent the sets
!
!  Result_Array       - a two-dimensional CHARACTER ( LEN = 1 ),
!                         local array variable that holds the union
!
!  rows               - an INTENT ( IN ) argument whose value
!                         contains the number of rows in the set
!------------------------------------------------------------------------
!
         CHARACTER( LEN = 1 ), DIMENSION ( 1 : 20, 1 : 75 ), INTENT ( IN )&
                                             :: Set_A, Set_B
         INTEGER, INTENT ( IN )              :: rows
         CHARACTER( LEN = 1 ), DIMENSION ( 1 : 20, 1 : 75 ) :: Result_Array
!
         WHERE ( Set_A == "*" .OR. Set_B == "*" )
             Result_Array = "*"
         ELSEWHERE
             Result_Array = " "
         END WHERE
         CALL  Print_Set ( "Union:", Result_Array, rows )
!
      RETURN
      END SUBROUTINE  Union
!==========================================================================
      SUBROUTINE  Intersection ( Set_A, Set_B, rows )
                  IMPLICIT  NONE
!
!------------------------------------------------------------------------
!  This subroutine will find the intersection of its two arrays and
!     store the result in Result_Array.
!
!  Set_A, Set_B       - two-dimensional CHARACTER ( LEN = 1 ),
!                         INTENT ( IN ) arrays that represent the sets
!
!  rows               - an INTEGER, INTENT ( IN ) variable that
!                         represents how many rows are in a set
!
!  Result_Array       - a two-dimensional CHARACTER ( LEN = 1 ),
!                         local array variable that holds the intersection
!------------------------------------------------------------------------
!
```

```
      CHARACTER( LEN = 1 ), DIMENSION ( 1 : 20, 1 : 75 ), INTENT ( IN )&
                                          :: Set_A, Set_B
      INTEGER, INTENT ( IN )              :: rows
      CHARACTER( LEN = 1 ), DIMENSION ( 1 : 20, 1 : 75 ) :: Result_Array
!
      WHERE ( Set_A == "*" .AND. Set_B == "*" )
          Result_Array = "*"
      ELSEWHERE
          Result_Array = " "
      END WHERE
      CALL  Print_Set ( "Intersection:", Result_Array, rows )
!
RETURN
END SUBROUTINE   Intersection
!==========================================================================
      SUBROUTINE   Subtract_Sets ( Set_A, Set_B, rows )
                IMPLICIT  NONE
!
!-------------------------------------------------------------------------
!
!  This subroutine will find the difference of its two arrays and
!     store the result in Result_Array.
!
!  Set_A, Set_B        - two-dimensional CHARACTER ( LEN = 1 ),
!                        INTENT ( IN ) arrays that represent the sets
!
!  rows                - an INTEGER, INTENT ( IN ) variable that
!                        represents how many rows are in a set
!
!  Result_Array        - a two-dimensional CHARACTER ( LEN = 1 ),
!                        local array variable that holds the
!                        difference
!-------------------------------------------------------------------------
!
      CHARACTER( LEN = 1 ), DIMENSION ( 1 : 20, 1 : 75 ), INTENT ( IN )&
                                          :: Set_A, Set_B
      INTEGER, INTENT ( IN )              :: rows
      CHARACTER( LEN = 1 ), DIMENSION ( 1 : 20, 1 : 75 ) :: Result_Array
!
      WHERE ( Set_B == "*" )
          Result_Array = " "
      ELSEWHERE
          Result_Array = Set_A
      END WHERE
      CALL  Print_Set ( "Difference:", Result_Array, rows )
!
RETURN
END SUBROUTINE   Subtract_Sets
```

*************** PROGRAM RESULTS ***************

 Please enter the first set's file name.
pic-1.dat

 This is one of the two sets:

```
**                                                                                        **
 *                                                                                         *
                              *
                            *****
                           ********
                          ************
                         ****************
                        ********************
                        ********************
                         *****************
                           *************
                            *********
                             ******
                              **
```

```
*                                                                                          *
**                                                                                        **
```

 Please enter the second set's file name.
pic-2.dat

 This is one of the two sets:

```
 **                                                                                       **
*                                                                                          *
*                                                                                          *
```

```
                              *
                            *****
                           ********
                          *************
                         ****************
                        ********************
                        *********************
                         *****************
                           *************
                            *********
                             ******
                              **
```

```
*                                                                                          *
*                                                                                          *
 **                                                                                       **
```

```
1. Union.
2. Intersection.
3. A - B.
4. B - A.
5. Print Set A.
6. Print Set B.
7. Quit.

        Please enter your choice.
2

Intersection:

  *                                                                     *
 *                                                                       *
                         ***
                        ******
                       **********
                       **********
                        ******
                         ***
 *                                                                      *
  *                                                                      *
1. Union.
2. Intersection.
3. A - B.
4. B - A.
5. Print Set A.
6. Print Set B.
7. Quit.

        Please enter your choice.
0

*** That was an invalid menu choice ...
    Please try again.

1. Union.
2. Intersection.
3. A - B.
4. B - A.
5. Print Set A.
6. Print Set B.
7. Quit.

        Please enter your choice.
8

   *** That was an invalid menu choice ...
       Please try again.
```

```
1. Union.
2. Intersection.
3. A - B.
4. B - A.
5. Print Set A.
6. Print Set B.
7. Quit.

        Please enter your choice.
1

Union

***                                                                    ***
 *                                                                      *
  *                                                                      *
                          *           *
                        *****       *****
                      ********* *********
                      **********************
                       **************************
                      ******************************
                      ******************************
                       ****************************
                        **********************
                         ********* *********
                          ******   ******
                           **       **
*                                                                          *
 *                                                                          *
***                                                                        ***
```

```
1. Union.
2. Intersection.
3. A - B.
4. B - A.
5. Print Set A.
6. Print Set B.
7. Quit.
            Please enter your choice.
4

Difference:

    *                                                    *
                                                          
 *                                                        *
                          *
                        *****
                      ********
                       *********
                         *********
                           *********
                             *********
                              *********
                             *********
                            *********
                           *********
                          *********
                           ******
                            **


 *                                                        *

    *                                                    *
```

```
1. Union.
2. Intersection.
3. A - B.
4. B - A.
5. Print Set A.
6. Print Set B.
7. Quit.

    Please enter your choice.
3
Difference:

*                                                                    *

                        *
                      *****
                    *********
                   **********
                   **********
                   **********
                   **********
                   **********
                    *********
                     ********
                      ******
                       **
*                                                                    *
1. Union.
2. Intersection.
3. A - B.
4. B - A.
5. Print Set A.
6. Print Set B.
7. Quit.

    Please enter your choice.
7                        <--- the program ends after this option is
                              selected.
```

There are three significant features of this program that deserve special comment:

1. Three subroutines, `Subtract_Sets`, `Read_a_Set` and, particularly, `Print_Set` are reused within the program.

2. When the program was run, it was frustrating to be able to do all the set operations without having the ability to see the original sets on the screen again. This was solved by adding two options to the menu, which were not in the written design. The point is that *running* a program is *part of the design process* and often returns a programmer to the written part of a design to replace, modify or create program units.

3. Fortran's WHERE statement, with its *array* logical expression and *array* assignment statements, make the set operations very easy to accomplish and easy to understand.

Important Concepts Review

- **Array section** references in multidimensional arrays are accomplished by specifying an array section reference for any set of declared bounds.

- **Stride** may be specified in any array section reference.

- See Figure 11.7, 11.8 and 11.9 for examples of array section references with multidimensional arrays.

- **Vector subscripts** can be used for any subscript of a multidimensional array.

- **Nested DO loops** are loops that are contained within other loops. This kind of looping arrangement is very useful for efficiently accessing the contents of multidimensional arrays.

- **Nested implied DO loops** are one-line looping structures that are used in data-transferring statements such as READ, WRITE, PRINT, DATA, etc.
[(]...(list,ix=init,lmt,[incr])[,ix-2=init-2,lmt-2,[incr-2])]...

- When writing information out with a FORMAT, if the FORMAT becomes exhausted because there is more data than the FORMAT can handle, the FORMAT is *reused* and output continues on a *new* line.

- OPEN prepares a file for use and associates it with a UNIT number. The UNIT number is used in READ, WRITE and other I/O statements to identify which file is to be accessed.

EXERCISES

11.1 Use a fixed-limit DO loop and a READ statement with an implied DO to read in values for a REAL two-dimensional array that has three rows and two columns. Using the same types of loops, PRINT out the array.

11.2 Write a complete program that will read in grade data for four students and calculate student averages. The four students' names should be stored in a CHARACTER (LEN = 20) array and the grades should be stored in a 4 × 3 two-dimensional REAL array where each row contains three grades for a student. Imagine the arrays corresponding to each other as seen below:

Name	AS1	AS2	AS3
Linda	3.6	4.0	3.8
Alex	2.9	3.4	3.0
Natasha	3.0	3.1	3.7
Matt	4.0	3.7	3.8

11.3 Calculate the class average on each assignment from Exercise 11.2. Calculate the class average on *all* assignments.

11.4 Read in a three-position one-dimensional array of type REAL that contains weights for each assignment in Exercise 11.2. Do Exercises 11.2 and 11.3 using the weight array in the calculation of the various averages.

11.5 Computer images can be stored as a two-dimensional array of numbers, one number for each picture element (pixel). For example, the letter "H" can be represented in a two-dimensional array of integers, where a 90 represents a black dot (pixel), and a zero represents a white pixel:

90	0	0	0	90
90	0	0	0	90
90	90	90	90	90
90	0	0	0	90
90	0	0	0	90

Usually, either a scanner or a camera connected to a computer is used to get a picture into the computer. The process of getting an image from the world onto a disk file is full of error-producing steps. A camera may be mounted on a robot, for example, and the image may be blurry because the camera was moving at the time the image was captured. One way to deal with these errors (or "noise" in the image) is by *smoothing* it. A simple algorithm for smoothing an image is to replace a pixel's value with the average of its surrounding pixel values:

35	41	30
38	45	42
32	39	31

=>

	37	

Clearly, the original value of 45 for the pixel was too high according to its neighbors, so the 45 is replaced by the average of the 9-pixel field that 45 is in.

$$(35 + 41 + 30 + 38 + 45 + 42 + 32 + 39 + 31)/9 = 37$$

Write a complete program that will read in a two-dimensional array of pixel data and smooth the image. Do not smooth the border pixels. Read the data from a file and write the processed image to another file . . . use the following OPEN and CLOSE statements (these basic file access statements are discussed after the design example in Chapter 9):

```
OPEN ( UNIT=41, FILE="IMAGE.DAT", STATUS="OLD", ACTION="READ")
        Read in the data
CLOSED ( UNIT=41 )
    . . . Smooth the image.
OPEN(UNIT=29,FILE="SMOOTHED.DAT",STATUS="UNKNOWN",ACTION="WRITE")
        Write in the data
CLOSED ( UNIT=29 )
```

11.6 Write a complete program that will read the data created by Exercise 11.5 in file SMOOTHED.DAT and map that data onto a CHARACTER (LEN = 1) two-dimensional array using the following substitution scheme:

	.	,	o	0	8	*
0-14	15-28	29-42	43-57	58-71	72-85	86-100

Print the image.

11.7 Write a complete program that encrypts or decrypts (determined by a menu choice) a message stored in a two-dimensional CHARACTER (LEN = 1) array. To encrypt a message, read each row of it in character by character using a 50A1 FORMAT. A message ends when two tildes (~~) are found on a line as the first two characters. The encryption is performed by switching rows and columns. Each row or column switch will be of the form R or C x, y. A typical encryption sequence will appear as follows:

```
R 2,4
R 3,4
C 1,24
C 32,14
R 1,5
C 26,34
R 3,1
END
```

This encryption sequence continues until the word END is met and should be stored in a file so that it can be retrieved to decrypt the message. To decrypt the message, simply switch rows and columns in reverse order according to the encryption sequence.

11.8 Fill a two-dimensional REAL array with z values that are defined by $f(x, y)$. Declare the array as follows:

REAL, DIMENSION (-9 : 9, -9 : 9) :: f_of_xy

Each position of array f_of_xy should contain the function value of the position's subscripts. For example, f_of_xy (-3, 4) will hold the z value of $f(-3, 4)$. Write out the array using a 19I4 FORMAT. Make up and try various $f(x, y)$s.

11.9 Write the data in Exercise 11.8 to a file and then import it into a spreadsheet and graph it. Increase the size of the array so that there is more data to graph.

11.10 Find the largest value in the array from Exercise 11.8.

11.11 Find the smallest value in the array from Exercise 11.8.

11.12 Write a function that will transpose a square array. An example of an array and its transpose is shown below:

a	b	c	d
e	f	g	h
i	j	k	l
m	n	o	p

a	e	i	m
b	f	j	n
c	g	k	o
d	h	l	p

11.13 Write a function that multiples two square arrays. To accomplish this, each position of the resulting array is filled by the sum of the products of the corresponding row of the first multiplicand and column of the second multiplicand. For example, position 1, 1 in the result array should contain the value of the following expression if the arrays in Exercise 11.12 were the multiplicands:

a*a + b*b + c*c + d*d

Position 3, 2 would be filled with the value of the following expression:

i*e + j*f + k*g + l*h

11.14 Write a complete program that will fill a 3 × 3 two-dimensional array with random integers in which each is between 1 and 10 inclusively and no two integers are the same. Write out any array whose individual columns, rows and diagonals each add up to the same total. Below are some examples of such an array and a black box function that will evaluate to a random integer between Lower_Limit and Upper_Limit each time it is used. Use a fixed-limit DO loop to fill the array 1,000,000 times. The examples below were calculated with integers between 1 and 12, and with the iterations increased to 10 million . . . the program took several minutes to run; the execution time will vary depending on the computer used.

```
****** 1 ******        ****** 4 ******        ****** 7 ******
   4  3  8               11  2  8                4 12  5
   9  5  1                4  7 10                8  7  6
   2  7  6                6 12  3                9  2 10
****** 2 ******        ****** 5 ******        ****** 8 ******
  10  5  6                5  4  9                9  1  8
   3  7 11               10  6  2                5  6  7
   8  9  4                3  8  7                4 11  3
****** 3 ******        ****** 6 ******
   6 12  3                7  1 10
   4  7 10                9  6  3
  11  2  8                2 11  5
```

It is interesting to note that some of the generated squares are not really different but are simply some rotation or reflection of another. For example, squares 3 and 4 above are essentially the same ... the only difference is that rows 1 and 3 have been exchanged.

```
      FUNCTION   Random_Integer ( Lower_Limit, Upper_Limit)
             IMPLICIT   NONE
!
! -----------------------------------------------------------------------
!   The purpose of this function is to generate a random integer
!      between INTENT ( IN ) INTEGER arguments Lower_Limit and
!      Upper_Limit.
!
!      First_Time       - a LOGICAL local variable whose purpose
!                         is to seed the random number generator the
!                         first time the function is invoked
!
!      x                - a REAL local variable that receives
!                         a random decimal from Fortran's intrinsic
!                         RANDOM_NUMBER subroutine where 0 <= x < 1
!
!      Random_Integer - the FUNCTION variable - type INTEGER
! -----------------------------------------------------------------------
!
            INTEGER, INTENT ( IN )      :: Lower_Limit, Upper_Limit
            LOGICAL                     :: First_Time = .TRUE.
            REAL                        :: x
            INTEGER                     :: Random_Integer !  FUNCTION name
!
```

```
        IF ( First_Time ) THEN
            First_Time = .FALSE.   ! This variable will have the same
                                   ! value it was left with when this
                                   ! function was used last time. This
                                   ! is because initializing a variable
                                   ! in a declaration statement in
                                   ! a program unit other than the main
                                   ! program causes the variable to be
                                   ! treated as if it had been listed
                                   ! in a SAVE statement or been given
                                   ! the SAVE attribute.
!
            CALL RANDOM_SEED       ! Lets Fortran pick a starting point
                                   ! for the random numbers.
        END IF
!
        CALL    RANDOM_NUMBER ( x ) ! Result: 0<=x<1
        Random_Integer = INT(x*(Upper_Limit-Lower_Limit+1)+Lower_Limit)
!
END FUNCTION Random_Integer
```

11.15 Solve the following exercises using Fortran intrinsic subroutines or functions rather than originally created algorithms:

    a) Exercise 11.11.
    b) Exercise 11.12.
    c) Exercise 11.13.

Data Files

Introduction

Arrays were introduced because problems involving a lot of data are difficult or impossible to solve with only simple variables. **Files** also solve the following data management problems:

- Files provide a *permanent* means of *storing* data.

- There is *no practical limit* to the *amount* of data that you can *store*.

When a computer is turned off, all the information in RAM is lost. If you've written a program and want to keep the answers it generated, you must direct the **output**, which is information coming *out* of a program, to a more permanent medium such as a disk. Data can be stored indefinitely on disks and a variety of other media.

Another advantage of files is that they can be arbitrarily large. A single, large file can fill many magnetic tapes, each able to hold **gigabytes** of data. A gigabyte is 1024^3 bytes, 1024^2 kilobytes, 1024 megabytes; "**gig**abyte" is pronounced like **gig**-gle. There are also tape devices that can hold **terabytes** (1024 gigabytes) of information on a single cassette. It isn't practical to maintain such large sets of data in a computer's RAM, assuming you had enough, because it would severely limit resources for other users in a shared computing environment.

Fortran offers two styles of files and a rich set of tools to manipulate them:

12.1 I/O . . . What is it?

12.2 List-directed I/O

12.3 Controlled I/O

12.4 Data files

  Important Concepts Review

12.5 Basic file manipulation statements

  Important Concepts Review

12.6 Styles of data files

  Important Concepts Review

  Exercises

- **Sequential files** contain a sequence of **records**, each record usually being one line in the file. Records must be accessed in sequence.

- **Direct-access files** are high-performance files that allow direct access of records by specifying a record number in READ or WRITE statements.

The various controlled input and output statements that are introduced in this chapter can all be customized through the use of **controls**. These controls are numerous and have therefore been gathered together and organized into tables in Appendix E. You will find every commonly used control offered by Fortran in that appendix.

As you pursue your scientific or engineering careers, it is very likely that you will inherit large data files and need to analyze them. Study this very important chapter so you'll be able to control and manipulate data in the limitless domain of data files.

• • •

12.1   I/O . . . What Is It?

I/O means Input/Output. **Input** is useful information entering a program, and **output** is information created by a program. There are many mechanisms that provide input and accept output. They are summarized in Table 12.1.

In Table 12.1, Disk, Touch Screen and Tape under Mechanism are shown in the Both column. This is to identify them as mechanisms that can be used for both input and output; data can be written-to or read-from these devices. Optical Disk is under the Both column but it is predominantly an *input* medium. There are many other apparatuses that can accept data from and/or provide data to programs, such as **scanners**, which input images, and **modems**, which get and send data through networks.

Table 12.1

Mechanism	I/O—Input/Output		
	Output	Input	Both
Screen/Monitor	X		
Keyboard		X	
Mouse		X	
Disk			X
Optical Disks			input
Tape			X
Printer	X		
Microphone (voice recognition)		X	
Speakers (digitized sounds)	X		
Touch Screen			X

12.2 List-Directed I/O

Well-designed, high-level data processing environments endeavor to offer at least two levels of I/O control to users: "Quick-and-dirty" and "Advanced." Fortran is no exception, offering **list-directed I/O** ("Fortran, you figure it out") and **controlled I/O** ("Fortran, I'll figure it out"). Remember that the object of the exercise when designing computer programs is to get results from computers. If some fast, intermediate answers are needed, use a spreadsheet or design a brief program using *list-directed I/O*. If a presentation of some sort such as a detailed menu screen is needed, use *controlled I/O* to get the clearest, most informative appearance possible; use your judgment.

List-Directed Output

List-directed I/O is unformatted input or output that is controlled by an **I/O list**. In the case of output, an I/O list is simply a set of values being written to some device such as a printer, computer monitor or disk. The values being written direct the way the output looks by their types and order of appearance in the output statement:

```
    PRINT *, "Hello", 1., 32.0, -11, "There", +6509231.6112347
    PRINT *, "Looks Ugly", .01, 5d0, 3 ,-90008976.0, -1, -2, "End"
        . . .
Hello   1.0000000   32.0000000  -11 There    6.5092315E+06
Looks Ugly   9.9999998E-03   5.0000000000000000 3  -9.0008976E+07 -1 -2 End
```

It could hardly be said that the above output is under control. Fortran makes all the spacing/how-many-decimal-places/which-numbers-are-in-scientific-notation decisions and does so based on the I/O list's values. Unfortunately, but understandably, Fortran shows maximum accuracy and is unconcerned about alignment . . . list-directed I/O is messy but very easy to use. There are four basic ways to invoke list-directed I/O, which are summarized in Table 12.2 (Table E.1 in Appendix E).

Table 12.2

List-Directed Input	List-Directed Output
READ *, ...	PRINT *, ...
READ (UNIT = *, FMT = *) ...	WRITE (UNIT = *, FMT = *) ...

The **default output device** on a computer is usually its screen/monitor. By implicitly or explicitly specifying the default output device and a default FORMAT, a programmer is saying, "Fortran, you figure it out; I don't want to think about it." Accomplish this with either PRINT or WRITE (Figure 12.1).

In Figure 12.1, the various asterisks tell Fortran to "use its own judgment," about how and where to send output. So what's the difference between PRINT and WRITE? The answer is that PRINT can only be used to send output to the default output device—the computer monitor. WRITE is a full-featured output command that gives a user complete control over output. As with all powerful computing tools, power can imply complexity. The moral of the story is that if simple, straightforward output to the screen is all that's needed, use PRINT. Otherwise, use WRITE.

Figure 12.2 (p. 392) shows an example of list-directed I/O using both PRINT and WRITE. Largely, Fortran does an okay job of formatting output; the main loss of control in list-directed output is in spacing between displayed items and the number of digits/decimal places shown.

List-Directed Input

As seen in Table 12.1, there are many input devices such as disks, touch screens, and microphones ("Beam me up, Scottie"). Fortran, however, views the keyboard as its **default input device**. By implicitly or explicitly specifying the default input device and a default FORMAT, a programmer is saying, "Fortran, you figure out how to read the values; I don't want to think about it." Accomplish this method of inputting information with one of the list-directed READ statements shown in Figure 12.3 (p. 392).

READ's purpose is to get data into a program and *store them in variables*. All READ statements MUST have variables listed so that the incoming data can be preserved for use

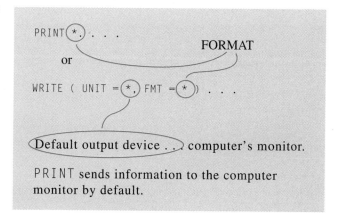

Figure 12.1

Figure 12.2

```
PROGRAM List_Directed_Output
          IMPLICIT NONE
!
!-----------------------------------------------------------------
! Written by     :   C. Forsythe
! Date Written   :   3/25/2004
!-----------------------------------------------------------------
! This program demonstrates how Fortran formats different types of
!    data for output under the conditions of list-directed I/O.
!-----------------------------------------------------------------
!
      PRINT *,                     3.141, -2.718281828d0, 6, -7, 2, -9
      WRITE ( UNIT = *, FMT = * )  3.141, -2.718281828d0, 6, -7, 2, -9
!
      PRINT *, "Hello", "&", "Good-bye", "  ", "<-- Two Spaces."
!
END PROGRAM List_Directed_Output

*************** PROGRAM RESULTS ***************

   3.1410000   -27182818279999998 6 -7 2 -9
   3.1410000   -27182818279999998 6 -7 2 -9
 Hello&Good-bye <-- Two Spaces.
```

Figure 12.3

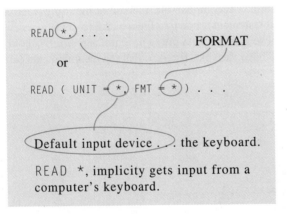

READ *, implicity gets input from a computer's keyboard.

throughout the program. As seen in Figure 12.4 (p. 393), it is also vital that an input statement, list-directed or controlled, receives appropriate types of data for the variables listed in the READ statement. One cannot read a CHARACTER entity into a numeric variable or even a REAL value that contains a decimal point into an INTEGER variable.

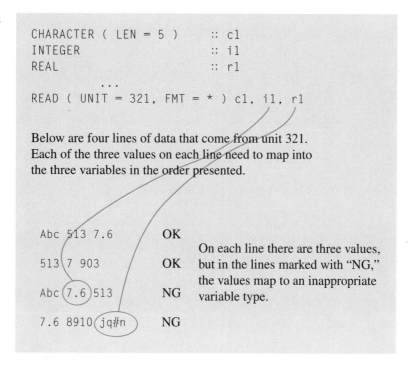

Figure 12.4

Although READ statements must have associated variable lists, WRITE statements can be used without a list of values and/or variables to be printed:

```
        WRITE ( UNIT = *, FMT = 34 )
34      FORMAT ( "Report Header" )
```

The formatted WRITE statement above simply prints the heading Report Header, which is contained in FORMAT 34 as a character string.

When using list-directed input, different items in the incoming data must be delimited by spaces, commas or "Enter"s (pressing the "Enter" key). Data separated by delimiters is called **free-format data** (Figure 12.5).

PROGRAM RESULTS in Figure 12.5 (p. 394) show the various ways data can be entered to a list-directed READ. Notice that the list-directed output is the same in each case even though the list-directed input varies. It is also noteworthy that the CHARACTER variables take only the first two characters of data, ab and xy. This is because c1 and c2 are declared with a *kind* of (LEN = 2); they can only accommodate two characters.

There are additional concerns when using list-directed input with CHARACTER type data. If the information being typed as input needs to include either of the delimiters (space(s) or ","), the input string must be enclosed in apostrophes or quotation marks. Cleverly, the reason for allowing *both* quotation marks or apostrophes is so that either of those symbols can be included in a string as well (Figure 12.6, p. 395).

There is only one minor problem left in transferring CHARACTER data to and from a program: what if one wants to include both an apostrophe *and* a quotation mark within the same string? To accomplish this, double the apostrophe or the quotation

Figure 12.5

```
PROGRAM List_Directed_Input
          IMPLICIT NONE
!
!-------------------------------------------------------------------
! Written by       :   C. Forsythe
! Date Written     :   3/25/2004
!-------------------------------------------------------------------
!  This program demonstrates list-directed input.  Several values of
!     different types are read using list-directed input and then
!     printed using list-directed output.
!-------------------------------------------------------------------
!
      INTEGER                 ::  i1, i2
      REAL                    ::  r1, r2
      CHARACTER ( LEN = 2 ) ::  c1, c2
!
      READ *,  i1, i2, r1, r2, c1, c2
      PRINT *, i1, i2, r1, r2, c1, c2
!
END PROGRAM List_Directed_Input

*************** PROGRAM RESULTS ***************

16758940 -3312 -2.7 99 abcdefghi xyz     <-- Input delimited by spaces.
  16758940 -3312  -2.7000000  99.0000000 abxy

*************** PROGRAM RESULTS ***************

16758940                                  <-- Input delimited by spaces
                                              or "Enter"s.
-3312
-2.7 99 abcdefghi
xyz
  16758940 -3312  -2.7000000  99.0000000 abxy

*************** PROGRAM RESULTS ***************

16758940,-3312,-2.7,99,abcdefghi,xyz     <-- Input delimited by commas.
  16758940 -3312  -2.7000000  99.0000000 abxy
```

mark when either is enclosed by itself. For instance, the following CHARACTER constants contain embedded apostrophes or quotation marks simply by placing two of them immediately next to each other.

```
PRINT *, '"they''re coming!"'        ==> "they're coming!"
PRINT *, "Bear's voice said, ""Boo"""  ==> Bear's voice said, "Boo"
```

Figure 12.6

```
PROGRAM  Character_Input
           IMPLICIT  NONE
!
!-----------------------------------------------------------------
!  Written by     :    C.  Forsythe
!  Date Written   :    3/25/2004
!-----------------------------------------------------------------
!  This is a brief program that demonstrates concerns when inputting
!     CHARACTER data as list-directed input.
!-----------------------------------------------------------------
!
     CHARACTER ( LEN = 10 ) :: c1
!
     PRINT *, "Please enter a string of characters"
     READ *, c1
     PRINT *, "     Result: ", c1
!
END PROGRAM  Character_Input

*************** PROGRAM RESULTS ***************
 Please enter a string of characters
ab,cd,ef,z           <-- The first comma ends the list-directed input.
     Result: ab
*************** PROGRAM RESULTS ***************
 Please enter a string of characters
'ab,cd,ef,z'
     Results: ab,cd,ef,z
*************** PROGRAM RESULTS ***************
 Please enter a string of characters
ab cd ef z           <-- The first space ends the list-directed input.
     Result: ab
*************** PROGRAM RESULTS ***************
 Please enter a string of characters
"ab cd ef zzzzzz"    <-- c1 can't hold all the trailing "z"s.
     Result: ab cd ef z
*************** PROGRAM RESULTS ***************
 Please enter a string of characters
'ab "hi" qr'
     Result: ab "hi" qr
*************** PROGRAM RESULTS ***************
 Please enter a string of characters
" a 'b , d'rrrr"
     Result:  a 'b , d'
*************** PROGRAM RESULTS ***************
 Please enter a string of characters
ab 'c' "d"
     Result: ab
```

12.3 Controlled I/O

Controlled I/O gives program designers complete control over creating, accessing, and presenting data. Formats are essential for control of I/O and they will be used and explained as needed in this chapter. A full explanation of formats is provided in Appendix D.

Controlled I/O is accomplished with seven basic I/O statements (Table 12.3). The very powerful I/O statements listed in Table 12.3 allow a programmer to manipulate files in any way needed. **Controls-list** is a collection of specifiers that define exactly how a given I/O statement is to be used; controls-list customizes I/O statements for specific needs.

Table 12.3

Controlled I/O Statements	
Statement	**Action**
OPEN (controls-list)	Prepares a file for use and describes its status and structure. A UNIT number is also associated with the file for reference in other I/O statements.
CLOSE (controls-list)	Breaks the UNIT number/file connection, writes an *End File* record and releases the file to the *operating system* (the set of computer programs that manage the basic functions in a given computer).
READ (controls-list)	Gets records from an input source according to the specifications in the controls-list.
WRITE (controls-list)	Puts records to an output destination according to the specifications in the controls-list.
INQUIRE (controls-list)	Gathers information about a file according to the controls-list.
BACKSPACE (controls-list)	This is a "file-positioning" statement. It allows a programmer to "back up" so previous records can be accessed.
REWIND (controls-list)	This is a "file-positioning" statement. It positions a file just before the first available record.

12.4 Data Files

A **data file** is a collection of related data. Files are broken down into **records**, each usually representing one physical line of a file. Records can be further broken down into **fields**. The value of files is that they offer a permanent way to keep data without an "amount" limit. Figure 12.7 shows a graphic representation of a file.

Notice in Figure 12.7 that each record is broken down into fields and each field can hold any type of data. In the definition above for files, "A data file is a collection of related data," **related data** means that each record is similar to the others. Values differ from record to record but the types of data in fields usually don't.

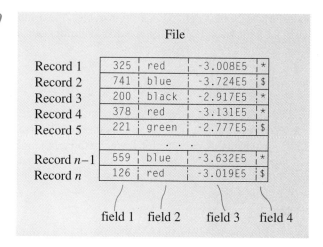

Figure 12.7

Fixed- and Free-Format Data Files

Data files are organized as **fixed-format** or **free-format** collections of data. Fixed-format files' data is divided into specific length fields such as the ones seen in Figure 12.7. Alternatively, the data in a free-format file are separated by delimiters such as commas. The advantage of free-format data files is that they are very easy to construct because it is unimportant where values go in a record as long as the data within each record are in the same order. Figure 12.8 shows the file from Figure 12.7 as a free-format file delimited by commas.

It is noteworthy that the records in a free-format file are often of **variable lengths**. This is not a problem for Fortran; although the record length varies, the number of fields is uniform. "Field" in a free-format file means whatever is between two delimiters. If it is necessary to include a free-format file's delimiter in a CHARACTER type datum, the value must be enclosed in quotation marks or apostrophes. For example, to put a person's last name followed by a comma and her first name in a free-format data file, the following form must be used:

```
value-1,value-2,"last-name, first-name",value-4, . . .
```

Figure 12.8

```
                  Free-Format Files
Record 1     325,red,-3.008E5,*
Record 2     741,blue,-3.724E5,$
Record 3     200,black,-2.917E5,*
Record 4     378,red,-3.131E5,*
Record 5     221,green,-2.777E5,$
                      . . .
Record n-1   559,blue,-3.632E5,*
Record n     126,red,-3.019E5,$
```

By enclosing `last-name, first name` in quotes, the comma separating `last-name` and `first-name` will NOT be "seen" as a delimiter but rather as just another character between quotes in the third datum.

I/O Modes

Movement of data to and from a file depends on the file's **I/O mode**. I/O mode is defined by the `ACTION =` control in an `OPEN` statement's controls-list (see Appendix E for all controls for all I/O statements). Conceptually, there are three I/O modes:

1. Input.
2. Output.
3. Input/Output.

A file whose I/O mode is **input** can be read from: data can move *from* the file *into* a program, but NO attempt must be made to do the reverse. Any attempt to write to an input file will result in a program crash. Files with an **output** mode have exactly the opposite properties: data can be written to them but no attempt may be made to read from them. The final I/O mode is **input/output.** With this mode, data may be written to and read from a file. This I/O mode is useful in applications such as **databases** in which records are often read, modified and written back to a file.

Important Concepts Review

- **I/O** means **I**nput/**O**utput.

- **List-directed I/O** is input or output that is controlled by an I/O list; Fortran decides everything based on the types of variables/values in an I/O statement.

- **I/O list** is a set of variables being read, or values and/or variables being written.

- **Default output** goes to the computer screen/monitor.

- **Default input** comes from the keyboard, touch screen or voice recognition system.

- `READ` gets data from an input device and stores them into variables.

- `WRITE` sends data to an output device.

- **Controlled I/O** gives a program designer complete control over the destination and appearance of data.

- **Data file** is a collection of related data.

- **Record** is a subdivision of files usually consisting of one line of a file.

- **Field** is a subdivision of records and contains data of a certain type and classification. Records often contain several fields.

- **Free-format** data files' data are in fields separated by delimiters.

- **Fixed-format** data files' data are in fields aligned in columns.

- **I/O modes** define the direction data is moved to and from programs: input—data flows into a program; output—data flows out from a program; input/output—data can be transferred in and out of a program.

12.5 Basic File Manipulation Statements

In Table 12.3, seven I/O statements for controlled I/O were shown. With these statements, virtually any kind of file access can be accomplished. There are some limitations regarding the interpretation of incoming data and the appearance of outgoing data, but those problems are solved with formatted I/O as discussed in Appendix D. An abbreviated version of Table 12.3 is shown below for reference (Table 12.4).

Each controlled I/O statement in Table 12.4 is shown with "controls-list" in parentheses. **Controls-list** is a collection of **controls** that customize the way the statements in Table 12.4 work. As each statement is explained, its most basic controls-list options will be illuminated. For a more complete list of I/O controls see Appendix E.

A controls-list is constructed as follows:

```
I/O statement ( control-item-1 = item-1 [ , control-item-2 = item-2] ... )
```

The next sections contain explanations that define basic controls for various I/O statements seen in Table 12.4. Refer to them when doing I/O operations.

Table 12.4

Controlled I/O Statements
OPEN (controls-list)
CLOSE (controls-list)
READ (controls-list)
WRITE (controls-list)
INQUIRE (controls-list)
BACKSPACE (controls-list)
REWIND (controls-list)

Open

OPEN is a connection statement that connects a file to a program so that the data in the file may be accessed. OPEN accomplishes the following:

1. Associates a UNIT number with a file so that the file can be referred to in other I/O statements.
2. Describes features of a file such as its name, whether it already exists or not, etc.
3. Defines I/O mode: input, output, or input/output.
4. Positions a file at the first or last record (assuming the file exists) for sequential files.
5. Makes the opened file available to ALL program units in a program through the associated UNIT number.

In the explanation of arrays in Chapter 9, a figure was shown as part of a crash-course on files for the design of the Forsythe Binary Insertion Sort. Figure 12.9 is a reproduction of that figure and shows the basics of an OPEN statement.

Figure 12.9 shows that an OPEN statement consists of the word OPEN followed by a pair of parentheses that contain controls-list items. Each appropriate control for *basic* OPEN statements is described in Table E.3 in Appendix E. The STATUS = control has several options that need explanation:

1. OLD means the file already exists. If the file does not exist, an error will occur.
2. NEW means the file does not exist. If the file does in fact exist, an error will occur. Files opened with the STATUS = NEW control are created when the OPEN is executed.
3. REPLACE causes the file to be created if it doesn't exist or deletes it and creates a new one if it does.
4. UNKNOWN can vary from compiler to compiler but usually behaves the same as REPLACE. UNKNOWN is the default status.

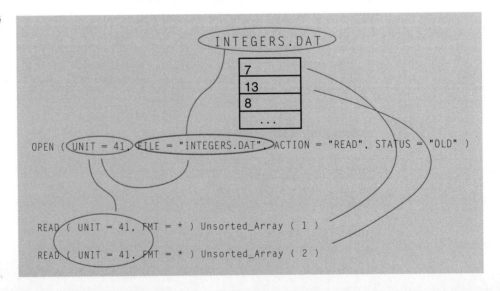

Figure 12.9

5. SCRATCH must not be used with the FILE = control. A status of SCRATCH creates a temporary file that is deleted when it is closed.

Table E.4 in Appendix E shows an additional set of controls that are used in more advanced I/O operations, such as working with **direct-access** files. Several of the controls will not be clear because ideas such as IOSTAT = variable haven't been explained yet.

Close

CLOSE is also classified as a connection statement because it influences connected files. This statement breaks the unit-number/file connection. After executing a CLOSE, its unit and consequently the associated file is no longer accessible unless it is opened again with an OPEN statement. As seen in Table 12.4 earlier in the chapter, the general form of a CLOSE statement is as follows:

CLOSE (controls-list)

CLOSE's controls-list options are listed in Table E.5 in Appendix E.

Read

Controls for READ enable program designers to get input from any appropriate "unit." Data obtained are stored into variables in the READ statement. The first datum is put into the first variable listed in the variable-list, the second datum is put into the second variable in the variable-list, etc.

READ (controls-list) variable-list

Variables that comprise variable-list must correspond to the type of data that are mapped into them from the input source. Controls for the READ statement are itemized in Table E.6.

Write

WRITE statements transfer data to output devices such as a file or a computer screen. This statement is one of the most important in Fortran as it is the means of getting "answers" out of a program. The general form of a WRITE statement is:

WRITE (controls-list) [value-list]

Value-list is surrounded with square brackets indicating that it's optional. How can that be? Why would it ever be necessary to have a WRITE statement that apparently does not write anything? The answer lies in formats. If there is a FMT = label-number in the controls-list of a WRITE statement, it is possible that the format only contains character strings and positional format descriptors and therefore the associated WRITE statement needs no value-list; everything that the WRITE statement is moving to the output device is contained in the format:

```
        WRITE ( UNIT = 45, FMT = 22 )
22      FORMAT ( 33X, "Fatigue Report" )
```

FORMAT 22 contains a positional format descriptor, 33X, whose purpose is to advance thirty-three spaces, followed by the string, "Fatigue Report". Fatigue Report is written to UNIT 45 starting in column 34, so the WRITE statement actually IS writing something.

There is another interesting quality of WRITE statements. They have a value-list rather than, like READ, a variable-list. This is because READ must put the data it gets into variables but WRITE stores nothing ... it simply sends values that may be constants, variables or expressions to some output device. WRITE's controls are listed in Table E.7.

Inquire

INQUIRE is a statement that allows a program to find out information about a file such as is its ACTION "READ", "WRITE" or "READWRITE", or is it open or does it already exist? There are two commonly used forms of the INQUIRE statement.

1. INQUIRE (UNIT = *u*, controls-list)
2. INQUIRE (FILE = *f*, controls-list)

It is possible to make inquiries about a file or a unit; both UNIT = and FILE = may NOT appear in the same INQUIRE statement. This is a very useful statement. For example, if a large project is being coded by several programmers, each might use the INQUIRE statement before an OPEN to make sure he isn't about to use a unit number that is already associated with an open file.

```
    . . .
LOGICAL      ::      Bad_Unit_Number
INTEGER      ::      Unit_Number
    . . .
Bad_Unit_Number = .TRUE.
Unit_Number     = 25
DO WHILE ( Bad_Unit_Number )
    INQUIRE ( UNIT = Unit_Number, OPENED = Bad_Unit_Number )
    IF ( Bad_Unit_Number ) THEN
        Unit_Number = Unit_Number + 1
    ELSE
        OPEN ( UNIT = Unit_Number, FILE = . . .
    END IF
END DO
```

INQUIRE controls other than UNIT = *u* and FILE = *f* set a variable to some value. It is like a reverse assignment statement (Figure 12.10).

INQUIRE's controls are shown in Table E.8. Several controls in that table refer to topics not yet presented.

Rewind and Backspace—File Positioning

Before understanding REWIND and BACKSPACE, the idea of a file-pointer must be grasped. A **file-pointer** is maintained by the operating system and indicates the next available record to read or where the next record will be written by pointing just before it (Figure 12.11).

12.5 Basic File Manipulation Statements

Figure 12.10

```
LOGICAL    ::   Y_N
     . . .
INQUIRE ( FILE = "TEST.DAT", EXIST = Y_N )
IF ( Y_N ) THEN
     OPEN . . .
ELSE
     . . .
END IF
```

EXIST checks to see whether TEST.DAT exists or not. If it does, EXIST sets Y_N to .TRUE. If it doesn't, Y_N is set to .FALSE..

File-pointer is only implicitly influenced by statements such as OPEN, READ and WRITE. To explicitly adjust the file-pointer for a file, use BACKSPACE, REWIND or OPEN with the POSITION = control. The general form of REWIND and BACKSPACE are shown below:

$$\text{REWIND (UNIT} = u, \text{ IOSTAT} = s \text{)}$$
$$\text{BACKSPACE (UNIT} = u, \text{ IOSTAT} = s \text{)}$$

Figure 12.11

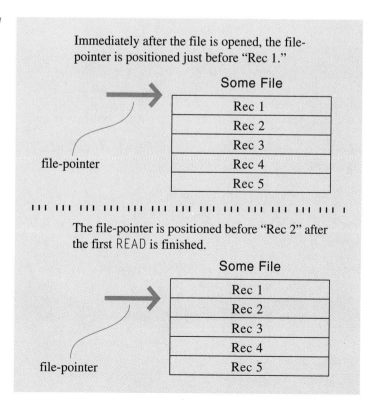

Immediately after the file is opened, the file-pointer is positioned just before "Rec 1."

The file-pointer is positioned before "Rec 2" after the first READ is finished.

REWIND positions file-pointer before the first record of a file as if the file had just been opened. This is useful for rereading files without closing them. BACKSPACE moves the file-pointer to before the previous record unless the ADVANCE = "NO" control is used in the READ. When the ADVANCE = "NO" control is used, BACKSPACE positions the file-pointer just before the current record. If file-pointer is before the first record, neither BACKSPACE nor REWIND have any affect.

Important Concepts Review

- **Controls-list** is a set of controls in an I/O statement.

- **Controls** customize I/O operations.

- OPEN associates a unit number with a file and describes various physical features of and access information for the file.

- CLOSE breaks the unit-number/file connection. After a CLOSE statement, the unit number's file is no longer accessible unless it is opened again with an OPEN statement.

- READ enables program designers to get input from any "unit." Data obtained are stored into variables in the READ statement.

- WRITE transfers data to output devices.

- INQUIRE allows a program to find out information about a file such as whether it is open or already exists.

- **File-pointer** is an entity maintained by operating systems that indicates the next available record to read or where the next record will be written.

- REWIND positions file-pointer before the first record of a file.

- BACKSPACE moves the file-pointer to the previous record of a file when the file-pointer is positioned before the next available record. If file-pointer is at the first record, BACKSPACE has no effect.

12.6 Styles of Data Files

Fortran offers two styles of data files:

1. Sequential.
2. Direct.

Sequential files are a sequential collection of records. All files discussed thus far have been sequential. **Direct-access files** are high-performance files whose records are *directly* accessible by specifying a **record number**.

Sequential Files

A **sequential file** is created by writing records to it one after another, and the file must be similarly read by reading records sequentially. Sequential files are most appropriate when they contain data whose records *all* need to be processed. On the other hand, for operations such as looking up a random record, sequential files are very inefficient. If a sequential file is searched for a given record in a file of n records, it may be necessary to read $n-1$ records before finding the record in question. On average, $n/2$ reads will be required to find a record that is in a random position of a sequential file containing n records. Files requiring activities such as search operations should be designed and implemented as **direct-access files**. Figure 12.12 shows how a sequential file is constructed.

A_FILE's most noteworthy newly presented characteristic is the last record: the **End File record**. This is a specialized record written at the end of every file by the operating system. Any attempt to read the End File record causes an error condition, but Fortran has facilities for using this record to figure out when a file has no more data so input algorithms can be stopped. There are other ways to stop input from a file as well.

Figure 12.12

```
OPEN ( UNIT = 59, FILE = "A_FILE", STATUS = "OLD", ACTION = "READ" )
```

File: A_FILE

Record 1
Record 2
Record 3
Record 4
. . .
Record $n-1$
Record n
* * * End file * * *

Stopping Input from a Sequential file

There are three common ways to stop input algorithms:

1. Record count.
2. Flag.
3. `IOSTAT`: perceiving the End File record.

The **record count** method uses the first record of a file to hold an integer that represents how many additional records are in that file, as shown in Figure 12.13. After the first record is read, a fixed-limit `DO` loop can be used to read the rest of the records.

```
OPEN ( UNIT = 61, . . .
READ ( UNIT = 61, FMT = * ) Record_Count
DO I = 1, Record_Count
        READ ( UNIT = 61, FMT = * ) Real_Values_Array ( I )
END DO
```

The second method of stopping input is accomplished by using a **flag**. Flagging the end of input was introduced in the looping chapter, and the programming example shown there is reproduced here as Figure 12.14.

Although program `Flag_End_Of_Input` is working with input from the standard input device, the principles are the same when reading from a data file. Figure 12.14's input algorithm was dubbed the **standard input algorithm** in the looping chapter. The English Fortran version of the algorithm is shown in Figure 12.15 (p. 408).

Applying the standard input algorithm to files is not difficult. Simply change the input statements so they get data from a file rather than the keyboard and include an `OPEN` statement for the file (Figure 12.16, p. 409).

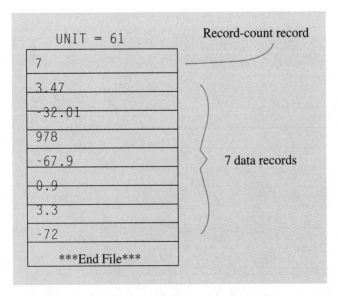

Figure 12.13

Figure 12.14

```
PROGRAM    Flag_End_Of_Input
              IMPLICIT    NONE
!
!-------------------------------------------------------------------
!  The purpose of this program is to demonstrate "flagging" the end
!    of input by "watching" for a flag value (any negative
!    number in this program) by using a conditional loop.
!
!  Length, Width    - REAL variables
!
!  Area             - a REAL variable that holds the value of
!                     Length * Width
!-------------------------------------------------------------------
!
       REAL              ::     Length, Width, Area
!
         PRINT *, "Please enter length and width"
         READ  *, Length, Width
!
         DO WHILE ( Length >= 0.0 .AND. Width >= 0.0 )
            Area = Length * Width
            PRINT *, "Area =", Area
!
            PRINT *, "Please enter length and width"
            READ  *, Length, Width
         END DO
!
END PROGRAM   Flag_End_Of_Input

*************** PROGRAM RESULTS ****************

 Please enter length and width
4.2 3              <----------- Typed by the user.
 Area =   12.5999994
 Please enter length and width
34 34              <----------- Typed by the user.
 Area =    1.1560000E+03
 Please enter length and width
.0001 .00006       <----------- Typed by the user.
 Area =    5.9999996E-09
 Please enter length and width
-1 -1              <----------- The program stopped after
                                the user typed this line.
```

Figure 12.15

```
READ A VALUE
DO WHILE ( THE VALUE ISN'T THE "NO MORE DATA" FLAG )
       PROCESS THE VALUE
          . . .
       READ THE NEXT VALUE
END DO
```

Once again, the PROGRAM RESULTS in Figure 12.16 are U-G-L-Y because of list-directed I/O. This alignment problem is readily remedied with formats, as seen in Figure 12.17 (p. 410). It is always a worthwhile digression to take the time to make output more readable with formats. An EN format descriptor displays numbers in scientific notation such that their exponents are divisible by three and their significands are less than or equal to 999.

As data are read into program Flag_End_Of_Input, a loop is used to "watch" for a flag value. When it is "seen," the input algorithm ends. Often, however, there is no appropriate value for a flag. An example of such a situation is reading from the set of all real numbers: all reals . . . positive, negative . . . all reals—which member of that set can be used as a flag? The answer is that there is no appropriate element. The solution lies in using the system End File record as a flag. Figure 12.17's program can be rewritten to use an End File record for a flag by creating a condition to "watch" for the End File record. This is done with IOSTAT = variable in the READ statement.

IOSTAT = variable will set its INTEGER variable to 0 for all successful READs and -1 when a READ attempts to input an End File record. It is important to understand that IOSTAT not only sets its INTEGER variable to -1, but also *prevents the program from crashing* with a "Read past end of file" error. Program Flag_End_Of_Input is easily modified to use the IOSTAT = variable to perceive the End File record instead of -1.0 flags (Figure 12.18, p. 411).

Normally, if a program attempts to read an End File record the program will crash. The beauty of using IOSTAT is that it allows error conditions such as End File to occur without causing an abnormal end to a program's execution. This is particularly useful in program Flag_End_of_Input of Figure 12.18 where IOSTAT allows the input algorithm to read the End File record and sends a "message" to the program (by setting the status variable, Status_Variable, to -1) that the file's data is exhausted. No further attempts are made to read from file TRINOFLG.DAT because it is out of data.

IOSTAT can be used in any I/O statement to perceive error conditions so that a programmer can take some algorithmic action without having the program terminate. This process is called **error trapping**.

Direct-Access Files

Direct-access files are high-performance files that access records in any order. There is no concept of End File records in direct-access files and they may be sparsely populated with records. Figure 12.19 (p. 412) illustrates a direct-access file.

Figure 12.16

```
PROGRAM  Flag_End_Of_Input
         IMPLICIT  NONE
!
!------------------------------------------------------------------
!  The purpose of this program is to demonstrate "flagging" the end of
!     input by "watching" for a flag value (any negative number in this
!     program) by using a conditional loop.
!
!  Length, Width           - REAL variables
!
!  Area                    - a REAL variable that holds the value of
!                            Length * Width
!------------------------------------------------------------------
!
    REAL              ::    Length, Width, Area
!
    OPEN ( UNIT = 73, FILE = "TRI.DAT", STATUS = "OLD",          &
                                        ACTION = "READ" )
    READ ( UNIT = 73, FMT = * ) Length, Width
!
    DO WHILE ( Length >= 0.0 .AND. Width >= 0.0 )
        Area = Length * Width
        PRINT *, "Length =", Length, " Width =", Width,          &
                                        " Area =", Area
        READ ( UNIT = 73, FMT = * ) Length, Width
    END DO
!
END PROGRAM  Flag_End_Of_Input
```

############### DATA FILE: TRI.DAT

```
6.7,9.2
8.1,7
3.215,11.5309
120934,6341
7e5,8e9
32e-7,607.45e-5
-1,-1
```

*************** PROGRAM RESULTS ***************

```
Length =   6.6999998    Width =    9.1999998    Area =   61.6399956
Length =   8.1000004    Width =    7.0000000    Area =   56.7000046
Length =   3.2149999    Width =   11.5309000    Area =   37.0718422
Length =   1.2093400E+05    Width =   6.3410000E+03    Area =   7.6684250E+08
Length =   7.0000000E+05    Width =   8.0000000E+09    Area =   5.6000001E+15
Length =   3.2000000E-06    Width =   6.0744998E-03    Area =   1.9438399E-08
```

Figure 12.17

```
PROGRAM   Flag_End_Of_Input
          IMPLICIT   NONE
!
!-----------------------------------------------------------------------
!   The purpose of this program is to demonstrate "flagging" the end of
!      input by "watching" for a flag value (any negative number in this
!      program) by using a conditional loop.
!
!   Length, Width          - REAL variables
!
!   Area                   - a REAL variable that holds the value of
!                            Length * Width
!-----------------------------------------------------------------------
!
     REAL              ::       Length, Width, Area
!
     OPEN ( UNIT = 73, FILE = "TRI.DAT", STATUS = "OLD",             &
                                         ACTION = "READ" )
     READ ( UNIT = 73, FMT = * ) Length, Width
!
     DO WHILE ( Length >= 0.0 .AND. Width >= 0.0 )
          Area = Length * Width
          WRITE ( UNIT = *, FMT = 5786 ) Length, Width, Area
          READ ( UNIT = 73, FMT = * ) Length, Width
     END DO
5786 FORMAT ( 1X, "Length = ", EN12.4, 3X, "Width = ", EN12.4,       &
                                       3X, "Area - ", EN12.4 )
!
END PROGRAM   Flag_End_Of_Input

############### DATA FILE:   TRI.DAT

6.7,9.2
8.1,7
3.215,11.5309
120934,6341
7e5,8e9
32e-7,607.45e-5
-1,-1


*************** PROGRAM RESULTS ***************

 Length =    6.7000E+00   Width =     9.2000E+00   Area =    61.6400E+00
 Length =    8.1000E+00   Width =     7.0000E+00   Area =    56.7000E+00
 Length =    3.2150E+00   Width =    11.5309E+00   Area =    37.0718E+00
 Length =  120.9340E+03   Width =     6.3410E+03   Area =   766.8425E+06
 Length =  700.0000E+03   Width =     8.0000E+09   Area =     5.6000E+15
 Length =    3.2000E-06   Width =     6.0745E-03   Area =    19.4384E-09
```

Figure 12.18

```fortran
PROGRAM Flag_End_Of_Input
           IMPLICIT   NONE
!
!----------------------------------------------------------------
!   The purpose of this program is to demonstrate "flagging" the end of
!     input by "watching" for the End File record. A different data
!     file, TRINOFLG.DAT, is used that doesn't contain the -1.0 numeric
!     flags.
!
!   Length, Width            - REAL variables
!
!   Area                     - a REAL variable that holds the value of
!                              Length * Width
!----------------------------------------------------------------
!
      REAL                ::   Length, Width, Area
      INTEGER, PARAMETER  ::   EOF_Condition = -1
      INTEGER             ::   Status_Variable
!
      OPEN ( UNIT = 73, FILE = "TRINOFLG.DAT", STATUS = "OLD",        &
                                               ACTION = "READ" )
      READ ( UNIT = 73, FMT = *, IOSTAT = Status_Variable ) Length, Width
!
      DO WHILE ( Status_Variable /= EOF_Condition )
          Area = Length * Width
          WRITE ( UNIT = *, FMT = 5786 ) Length, Width, Area
          READ ( Unit = 73, FMT = *, IOSTAT = Status_Variable )       &
                                               Length, Width
      END DO
      5786 FORMAT ( 1X, "Length = ", EN12.4, 3X, "Width = ", EN12.4, 3X,&
                                               "Area = ", EN12.4 )
!
END PROGRAM   Flag_End_Of_Input
```

############### DATA FILE: TRINOFLG.DAT

```
6.7,9.2
8.1,7
3.215,11.5309
120934,6341
7e5,8e9
32e-7,607.45e-5
```
            <-- Flag values are gone; the End File record is the flag.

*************** PROGRAM RESULTS ***************

```
Length =     6.7000E+00   Width =     9.2000E+00   Area =    61.6400E+00
Length =     8.1000E+00   Width =     7.0000E+00   Area =    56.7000E+00
Length =     3.2150E+00   Width =    11.5309E+00   Area =    37.0718E+00
Length =   120.9340E+03   Width =     6.3410E+03   Area =   766.8425E+06
Length =   700.0000E+03   Width =     8.0000E+09   Area =     5.6000E-09
Length =     3.2000E-06   Width =     6.0745E-03   Area =    19.4384E-09
```

Figure 12.19

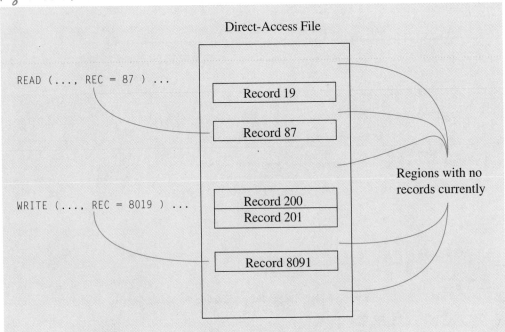

Rather than an End File concept, direct-access files have a **No Record condition**. If a READ statement successfully reads a record, IOSTAT = variable sets its INTEGER status variable to 0. If no record is found because no record has ever been written to the particular record position attempting to be read, a No Record condition occurs and the status variable is set to some other value that varies from compiler to compiler. Again, a program can take advantage of IOSTAT's status variable's value to change the way a program deals with a No Record condition and prevent a program crash.

Direct-access files achieve their ability to access records "directly" by insisting that all records be the same length. Conceptually, if the 4,017th record is needed, the *fixed record length* is multiplied by 4,016 and the operating system advances the file-pointer that number of characters so it is positioned just before the correct record.

There are several considerations in OPEN, READ, and WRITE when using direct-access files. Examples are shown below.

```
OPEN ( UNIT = 69, FILE = "TESTDIRC.DAT",    &
       ACCESS = "DIRECT",                    &
       RECL = 63, FORM = "FORMATTED"         &
       any other controls . . . )
READ  ( UNIT = 91, FMT = *, IOSTAT = Stat, REC = 96 ) . . .
       modify the record
WRITE ( UNIT = 91, FMT = *, IOSTAT = Stat, REC = 96 ) . . .
```

OPEN needs three special controls to open a direct-access file: ACCESS = "DIRECT" to let Fortran know it's dealing with a direct-access file, RECL = integer-expression so

Fortran knows the size of each record, and `FORM = "FORMATTED"` because direct-access files default to `"UNFORMATTED"`. When a file's `FORM =` control is set to `"FORMATTED"`, the contents of the file are readable. Choosing a `FORM =` control of `"UNFORMATTED"` allows a given compiler to format information any way that it chooses. Unless there is some compelling reason (such as creating massive files in limited storage), it is best to use `FORM = "FORMATTED"` for clarity. Sequential files default to `FORM = "FORMATTED"`.

`READ` and `WRITE` only need one additional control: `REC =` record-number. By specifying a record-number, Fortran knows where to get the record. `REC =` is a lot like a subscript for a one-dimensional array. A subscript specifies an element of an array, and a `REC =` record-number control specifies a record of a direct-access file.

Important Concepts Review

- **Sequential files** are a sequential collection of records. A sequential file is created by writing records to it one after another, and the file must be read the same way.

- **Direct-access files** are high-performance files whose records are *directly* accessible by specifying a **record number**.

- There are three common ways to stop input algorithms:
    1. Record count.
    2. Flag.
    3. `IOSTAT`. perceiving the End File record.

- **Flags** in data files are special values that signal the end of data.

- **End File record** is a special record written at the end of every file by the operating system.

- `IOSTAT =` can be used to perceive when an attempt is made to read an End File record. `IOSTAT` is also used in any other I/O statement to perceive an error condition. One of `IOSTAT`'s great strengths is that it can report an error without the program crashing.

- **Error trapping** is the act of creating algorithms that deal with error situations without the program stopping as a result of the error.

- A **No Record condition** happens when an attempt is made to read a nonexistent record in a direct-access file.

- `ACCESS = "DIRECT"` is a control required in the `OPEN` statement for a direct-access file; the default is `ACCESS = "SEQUENTIAL"`.
- `RECL = integer-expression` is a control required in the `OPEN` statement for direct-access files that describes the length of each record.
- `REC = integer-expression` is a control required in `READ` and `WRITE` statements to specify which record to read or write in a direct-access file.

EXERCISES

12.1 Describe and compare list-directed I/O and controlled I/O, including when each is appropriate.

12.2 Write an `OPEN` statement that opens a sequential file so that it can receive output by adding records to the end of the file.

12.3 Write a complete Fortran program that uses list-directed input to read in the name/description of a spring and its k constant. Write each record to a file called `SPRING.DAT` and put one spring description and its k constant per line.

12.4 Modify the program from Exercise 12.3: `REWIND` the file before it is closed and, using a fixed-limit `DO`, print out all the records in `SPRING.DAT`.

12.5 Write a selection structure that determines whether a file named `ALLOY.DAT` is open or not. If it is open, figure out whether it is open for `READ`, `WRITE` or `READWRITE` and what its unit number is.

12.6 Open a file, `STAR.DAT`, that has an unknown number of records each containing a 12-digit `CHARACTER` string and a four-digit `REAL` value. Read all records and keep a count of how many were read. `REWIND STAR.DAT` and open a `SCRATCH` file. Write the record count value to the `SCRATCH` file followed by all of the records from `STAR.DAT`. Close `STAR.DAT` and open it again for output. Copy all records from the `SCRATCH` file to `STAR.DAT`. Be sure to use subroutines wisely and archive any that are of general use.

12.7 Exercise 10.5 demonstrated how to merge two sorted one-dimensional arrays into one larger sorted array. Use the exact same principle and merge two sorted files. A merge works as seen in the illustration below:

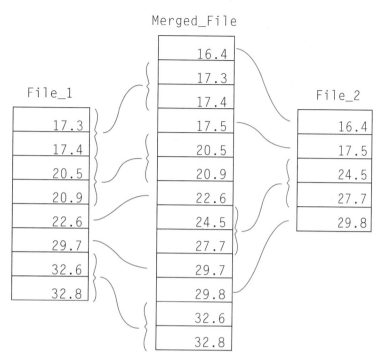

12.8 Write a complete Fortran program that creates a database of names and phone numbers. Use the style, XXX-XXX-XXXX, for phone numbers and allow 20 characters for a name. Use a direct-access file for this exercise and use the following menu as a guide for what kind of functions to include for the database.

    1. Add a record.
    2. Print the data base.
    3. Quit.

12.9 Expand Exercise 12.8 to include the following functionality:

    3. Find a record by name.
    4. Find a record by phone number.
    5. Modify a person's information.
    6. Delete a record.
    7. Quit.

When finding a record by name, be sure to display all records with the same names. For instance, there may be two John Smiths. When option 5 is chosen, it will be necessary to present a list of records having the same names so that the user can choose the correct one to modify. For option 6, simply fill the record with blanks and write it back to the file.

12.10 Write a complete program whose purpose is to append one file to another. Get the file names from a user as CHARACTER input including the name of the result file. Special handling may be required if the "result file" is the same as either of the two source files.

12.11 Write a complete program that accumulates data about a patient's blood pressure. Organize the program so that each time a reading is taken it will be appended to the existing file. Blood pressure is specified by two numbers: systolic and diastolic (*i.e.*, 120/75 is a blood pressure reading with a systolic of 120 and a diastolic of 75). The systolic pressure is a measure of how much pressure the heart generates when it pumps. Diastolic pressure readings are a measure of how much cardiovascular pressure is present when the heart is at rest between pumps.

Store two integers per line using a 2I4 format to represent the blood pressure. Also store the date and time that the readings were taken: *mm/dd/yy, hh:mm A* or *P*. When ten blood pressure readings have been stored, print a message to the user that it is time to produce a graph of the data.

12.12 Write a subroutine to improve the program in Exercise 12.11. The subroutine should create a histogram of the data in the daily blood pressure file. Only call this subroutine if the user, when presented with the "Time to graph the data," message, chooses to do so. Whether a graph is displayed or not, append the current data file of ten readings to a year-to-date version of the blood pressure file, and prepare to collect the next ten readings. Graph the data as seen below and scale the systolic and diastolic values by dividing each by 5.

```
02/03/04 07:00 P,  145 . . . sssssssssssssssssssssssssssss
                    80 . . . ddddddddddddddddd
02/03/04 10:00 P,  135 . . . sssssssssssssssssssssssssss
                    78 . . . ddddddddddddddddd
02/04/04 08:00 A,  130 . . . ssssssssssssssssssssssssss
                    60 . . . dddddddddddd
02/04/04 07:30 P,  155 . . . sssssssssssssssssssssssssssss
                    95 . . . dddddddddddddddddddd
                       . . .
```

12.13 Import the year-to-date file from Exercise 12.12 into a spreadsheet and graph it. Try a variety of graph styles and print the one that best expresses the blood pressure data.

12.14 Read a file that contains data on the growth rates of amphibians. The file will contain three fields . . . the name of the amphibian and its growth rate in two forms. Growth rates will be expressed by both of the following: length/unit-of-time and weight/unit-of-time. Print the amphibian who has the slowest growth rate based on length, and the amphibian who has the slowest growth rate based on body mass.

Advanced Functions and Modules

13

Introduction

In Chapter 5, MODULEs were introduced as program units that are used to package data definitions—especially derived data type declarations. Modules are much more powerful than just some means of housing data. Now that you understand other powerful program design features such as functions and dot operators, you are ready to learn how to couple modules with functions and/or subroutines to create *functional objects* and *overloaded interfaces*. These terms may sound somewhat frightening now, but they are very comprehensible.

In the following sections, modules and the ways that they can be combined with functions and subroutines are explained. You will learn to accomplish three new and useful programming techniques:

1. **Create custom programmer-defined dot operators.** An example of an existing dot operator is .AND..

2. **Overload existing operators** such as + to give them meaning for specific derived data types.

3. **Overload function/subroutine interfaces**, which means to use one generic interface to invoke different specific program units based on the type of actual arguments.

13.1 Creating custom dot operators

13.2 Overloading existing operators

13.3 Overloading function/subroutine interfaces

Important Concepts Review

Exercises

13.1 Creating Custom Dot Operators

The reason that operators such as .AND. are called **dot operators** is because of the periods (dots) on each side of the operator. Fortran offers several dot operators such as .OR. and .NOT. but, as always, basic general tools don't always do the job. Understanding how to create custom dot operators is a very powerful skill.

A **dot operator** can be thought of as either of the following:

1. A **binary** function with *two* INTENT (IN) arguments that produces some ultimate result. .AND., for example, is a binary dot operator that requires *two operands* and produces the logical "and" of them.

2. A **unary** function with *one* INTENT (IN) argument that produces some ultimate result. .NOT. is an example of a unary dot operator that requires only *one operand* and produces the logical "not" of it.

The word *binary* in definition 1 above is not being used in the sense of binary numbers, but in the sense that there are *two* operands involved in the operation. The following logical expression is an example of a binary operation; .OR. is called a **binary dot operator** because it is operating on *a* and *b*: *two* operands.

$$a \text{ .OR. } b$$

On the other hand, **unary dot operators** operate on only one operand and are used as seen below:

$$\text{.NOT. } a$$

Fortran DOES offer some very useful dot operators such as the ones listed below:

.AND.
.OR.
.NOT.
.EQV. and .NEQV.

However, these operators are only for *logical* expressions and don't even include such useful logical operators as .XOR. (exclusive or), .NOR. (not or), .NAND. (not and), etc. By creating custom dot operators with **modules** and functions, dot operators are not limited to relational and logical operators.

The easiest way to understand how to create a dot operator is by seeing an example. A dot-operator example follows that solves the problem of how to correctly compare values of type REAL for equality.

What problem? $a == b$ should work, right?

No. The problem is this: the == relational operator compares two values to see if they're *identical*. But computers often represent the least-significant digits of floating point (REAL) numbers inaccurately. This can happen for a variety of boring reasons, but it

Figure 13.1

```
PROGRAM Test_REAL_Accuracy
        IMPLICIT NONE
    REAL                :: x = 0.0
    REAL, PARAMETER     :: One_Eleventh = 1.0 / 11.0
    INTEGER             :: Loop_Counter
!
    DO Loop_Counter = 1, 11
        x = x + One_Eleventh
    END DO
    PRINT *, "x should be 1.0 but is:", x
END PROGRAM Test_REAL_Accuracy

************** PROGRAM RESULTS ***************

x should be 1.0 but is:    1.0000001
```

suffices to say that after a series of calculations, two expressions that are algebraically identical may differ in the way they are stored in a computer.

If there is ANY difference between two values being tested for equality, they WILL NOT compare as equal. An example of this problem is seen in the program of Figure 13.1.

This program accumulates eleven elevenths into variable *x*. Variable *x* should contain `1.0` but doesn't because of inaccuracies in the way the computer does arithmetic and stores decimal numbers. If the value stored in variable *x* is compared for equality to `1.0`, the comparison *should* be true because it's algebraically true that eleven elevenths equals one. Unfortunately, Fortran's equality relational operator, ==, would find `1.0` and *x* NOT equal because `1.0000000` isn't identical to *x*'s value of `1.0000001`.

To solve the problem of comparing two REAL values for equality, one must test whether the *absolute value* of the difference between them is fairly small, say, less than 10^{-5}. If DOUBLE PRECISION values are tested for equality, one must require the difference to be less than 10^{-12}.

In the example of Figure 13.1, the absolute value of the difference between `1.0` and *x*'s value of `1.0000001` is: `0.0000001` ($1 \cdot 10^{-7}$), which is less than $1 \cdot 10^{-5}$. Therefore, the two values, `1.0000000` and `1.0000001`, can be considered equal within the accuracy of the way computers store REAL numbers. Wouldn't it be useful to have a custom dot operator to take care of such messy comparisons? Notice how much easier the dot operator version is to understand in the selection-structure fragments below.

```
          IF ( ABS ( X - 1.0 ) < 1.0E-05 ) THEN ...
```
or
```
          IF ( x .REALLYEQUAL. 1.0 ) THEN ...
```

Figure 13.2

```
MODULE module-name
      INTERFACE OPERATOR ( .dot-operator-name. )
          MODULE PROCEDURE dot-operator-function-name
      END INTERFACE
CONTAINS
      FUNCTION dot-operator-function-name ( argument_1, argument_2 )
          function-statements
      END FUNCTION dot-operator-function-name
END MODULE  module-name
```

As mentioned before, custom dot operators are created with modules. The general form of a module used to create a *binary* dot operator is shown in Figure 13.2.

There are several key elements in the general form seen in Figure 13.2:

1. Module heading.
2. `INTERFACE OPERATOR ( .name. )` statement.
3. `MODULE PROCEDURE procedure-name` statement.
4. `CONTAINS` statement.
5. The function that creates the dot operator's action.

Module heading is used to tell the compiler that a `MODULE` is being created. It includes the word `MODULE` and a mnemonic name that will be used by other program units to refer to the module. Modules end with an `END MODULE` statement.

`INTERFACE OPERATOR ( .name. )` begins the operator's definition. This **interface block** is used to make a connection between the dot operator, `.name.`, and the function that defines how the operator works; the dot operator *will invoke* its corresponding function. `name` is surrounded by "dots" (periods) and *can only contain letters*—no digits or underscores are allowed as in other Fortran identifiers. Interface blocks end with an `END INTERFACE` statement.

`MODULE PROCEDURE function-name` specifies the name of the function *in* the module that will make the dot operator work. Finally, the function that makes the operator do whatever it is supposed to do follows the word `CONTAINS`. Figure 13.3 shows a dot operator designed to test `REAL` values for equality. The `.REALLYEQUAL.` dot operator created in Figure 13.3 can be used as shown in Figure 13.4 (p. 422).

Notice the `USE` statement in Figure 13.4. As explained when modules were introduced in Chapter 5, `USE` is employed to gain access to module definitions. Include the following statement as the *first* statement after a program unit header to give that unit access to module `Dot_Operators`:

```
USE Dot_Operators
```

Figure 13.3

```
MODULE Dot_Operators
          IMPLICIT  NONE
!------------------------------------------------------------------
!  This module is for custom dot operators. More operators will be
!     added to this module as they are developed.
!
!     .REALLYEQUAL.  -   a relational operator that determines
!                        whether two REAL values are equal within a
!                        computer's accuracy
!------------------------------------------------------------------
!
     INTERFACE OPERATOR ( .REALLYEQUAL. )
         MODULE PROCEDURE  Real_Equal
     END INTERFACE
!
CONTAINS
!
     FUNCTION Real_Equal ( Real_Value_1, Real_Value_2 )
              IMPLICIT  NONE
     !------------------------------------------------------------------
     ! The purpose of this LOGICAL function is to compare Real_Value_1
     !    and Real_Value_2 (both INTENT ( IN ) REAL arguments) to
     !    determine whether they are equal or not. This is done by
     !    figuring out whether the absolute value of the difference between
     !    them is less than 10**(-5).
     !
     !  Real_Equal    -  a LOGICAL variable: the function name
     !------------------------------------------------------------------
     !
         LOGICAL                     ::     Real_Equal    ! Function name
         REAL,    INTENT ( IN )  ::     Real_Value_1, Real_Value_2
     !
         IF ( ABS ( Real_Value_1 - Real_Value_2 ) < 1.0E-5 ) THEN
              Real_Equal = .TRUE.
         ELSE
              Real_Equal = .FALSE.
         END IF
     END FUNCTION  Real_Equal
END MODULE   Dot_Operators
```

Figure 13.4

```
PROGRAM Test_REAL_Accuracy
        USE Dot_Operators   ! This USE statement gives this
                            ! program access to the dot operators
                            ! defined in MODULE Dot_Operators.
        IMPLICIT NONE
        REAL              :: x = 0.0
        REAL, PARAMETER   :: One_Eleventh = 1.0 / 11.0
        INTEGER           :: Loop_Counter
!
        DO Loop_Counter = 1, 11
            x = x + One_Eleventh
        END DO
        PRINT *, 'x should be 1.0 but, unfortunately, is:', x
        CALL Print_Blank_Lines ( 1 )   ! BLACK BOX
        IF ( x .REALLYEQUAL. 1.0 ) THEN
            PRINT *, "However, x == 1.0 when using .REALLYEQUAL."
        ELSE
            PRINT *, "Oops, x is still /= 1.0"
        END IF
!
END PROGRAM   Test_REAL_Accuracy

*************** PROGRAM RESULTS ***************

x should be 1.0 but, unfortunately, is: 1.0000001

However, x == 1.0 when using .REALLYEQUAL.
```
The above line was printed because .REALLYEQUAL. found x and 1.0 to be within 10^{-5} of each other.

This gives the program unit in Figure 13.4 (in this case, a main program) access to the dot operator, .REALLYEQUAL., as if it were an intrinsic operator such as .AND. or .OR..

13.2 Overloading Existing Operators

Dot operators can also be defined to manipulate **derived data types** by using the methods described in the previous section. Another more elegant way to create customized operators is by extending the function of operators that already exist so they can operate on derived data types. This is called **operator overloading**. When an operator is *overloaded*, it behaves differently based on the *types* of the operands involved in the expression.

In Fortran, operator overloading is already implicitly in effect for the intrinsic arithmetic operators: **, *, /, + and -. For example, the multiplication operator, *, takes different actions when it multiplies two complex values (numbers with imaginary parts) than it does when multiplying two integers. Fortran knows which multiplication actions to take because of the types of the operands that are in a given expression.

Fortran provides the facility for **overloading operators** to define specialized meanings for intrinsic operators when used with derived data types. To see how this works, a derived data type must be created first. Figure 13.5 illustrates a derived data type for simple vectors on a Cartesian plane. The vector is represented by a line segment between two points. Once type X_Y_Vector is defined, the + operator will be overloaded so that vectors can be added.

Before moving on to the actual operator overloading, examine the simple program in Figure 13.6 (p. 424) to see how variables of type X_Y_Vector are declared and assigned values. The PROGRAM RESULTS in Figure 13.6 are really U-G-L-Y because of

Figure 13.5

```
MODULE    Simple_Vector
              IMPLICIT   NONE
!
!------------------------------------------------------------------
!  The initial purpose of this module is to define a derived data
!     type for a simple vector represented by two points on a Cartesian
!     plane.
!
!  Operations such as adding vectors and calculating their magnitude and
!     direction (angle to the x axis) can be added later.
!
!  Tail_X, Tail_Y - type REAL; represents the vector's "tail"
!
!  Head_X, Head_Y - type REAL; represents the vector's "head"
!------------------------------------------------------------------
!
      TYPE X_Y_Vector
           REAL              ::    Tail_X, Tail_Y
           REAL              ::    Head_X, Head_Y
      END TYPE X_Y_Vector
END MODULE    Simple_Vector
```

Figure 13.6

```
PROGRAM  Play_With_Type_Vector
         USE  Simple_Vector     ! This USE statement gives
                                ! this program unit
                                ! access to the definitions
                                ! in MODULE Simple_Vector.
         IMPLICIT  NONE
!
!-----------------------------------------------------------------
! Two variables of type X_Y_Vector, Vec_1 and Vec_2, are given
!    initial values with structure constructors when they are
!    declared, and are then printed out.
!-----------------------------------------------------------------
!
     TYPE ( X_Y_Vector )    :: Vec_1 = X_Y_Vector ( 0.0, 0.0, 4.0, 9.0 )
                                        !     X1, Y1, X2, Y2
                                        !      tail - head
     TYPE ( X_Y_Vector )    :: Vec_2 = X_Y_Vector ( 4.0, 9.0, 11.0, 11.0 )
                                        !     X1, Y1, X2, Y2
                                        !      tail - head

     PRINT *, "Vec_1 =", Vec_1
     PRINT *, "Vec_2 =", Vec_2
!
END PROGRAM  Play_With_Type_Vector

*************** PROGRAM RESULTS ***************

Vec_1 =   0.0000000E+00    0.0000000E+00    4.0000000    9.0000000
Vec_2 =   4.0000000    9.0000000    11.0000000    11.0000000
```

list-directed I/O. Figure 13.7 uses a format to pretty them up a bit. This is always a worthwhile digression; formats really improve the appearance and clarity of answers.

Figure 13.7 uses a derived data type, X_Y_Vector, to create two vectors, Vec_1 and Vec_2. The vectors from Figure 9.14 and their sum are graphically displayed in Figure 13.8 (p. 426).

The following few figures show how to extend the addition operator so that it can "add" variables of type X_Y_Vector. Fortran's + operator is overloaded so that when its operands are of type X_Y_Vector, vector addition is performed. + *still performs normally* when adding INTEGERs, REALs, etc. Figure 13.9 (p. 426) shows the module in Figure 13.5 after it is expanded to include the vector addition function.

Figure 13.7

```
PROGRAM   Play_With_Type_Vector
          USE   Simple_Vector   ! This USE statement gives
                                ! this program unit
                                ! access to the definitions
                                ! in MODULE Simple_Vector.
          IMPLICIT   NONE
!
!---------------------------------------------------------------------
! Two variables of type X_Y_Vector, Vec_1 and Vec_2, are given
!    initial values with structure constructors when they are
!    declared, and are then printed out.
!---------------------------------------------------------------------
!
    TYPE ( X_Y_Vector )    :: Vec_1 = X_Y_Vector ( 0.0, 0.0, 4.0, 9.0 )
                                          !    X1,  Y1,  X2,  Y2
                                          !    tail   -   head
    TYPE ( X_Y_Vector )    :: Vec_2 = X_Y_Vector ( 4.0, 9.0, 11.0, 11.0 )
                                          !    X1,  Y1,  X2,  Y2
                                          !    tail   -   head
    WRITE ( UNIT = *, FMT = 43 ) "Vec_1 =", Vec_1
    WRITE ( UNIT = *, FMT = 43 ) "Vec_2 =", Vec_2
 43 FORMAT ( 1X, A7, 4F7.2 )
          !
          ! The 1X moves one column to the right (skips 1 space),
          !   A7 allows the Vec_x = to be printed and
          !   4F7.2 specifies four seven-digit fields each
          !   having two places to the right of the decimal.
          !
END PROGRAM   Play_With_Type_Vector

*************** PROGRAM RESULTS ***************

Vec_1 =   0.00   0.00   4.00   9.00
Vec_2 =   4.00   9.00  11.00  11.00
```

Notice that the convention of the head of Vec_1 physically corresponding to the tail of Vec_2 is used to simplify the overall program. Figure 13.10 (p. 427) shows the overloaded + operator being used to add two vectors of type X_Y_Vector.

If +'s operands are INTEGERs, integer addition is performed. If +'s operands are TYPE (X_Y_Vector), vector addition is performed. That's what operator overloading is all about.

Figure 13.8

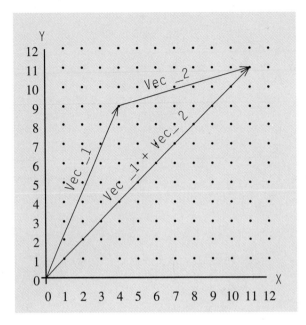

Figure 13.9

```
MODULE Simple_Vector
          IMPLICIT   NONE
!
!-------------------------------------------------------------------
! The purpose of this module is to define a derived data type for a
!     simple vector represented by two points on a Cartesian
!     plane. Then the + operator is overloaded so that vector
!     addition can be performed by using a +.
!
! Tail_X, Tail_Y - type REAL; represents the vector's tail
!
! Head_X, Head_Y - type REAL; represents the vector's head
!-------------------------------------------------------------------
!
      TYPE X_Y_Vector
          REAL                      ::      Tail_X, Tail_Y
          REAL                      ::      Head_X, Head_Y
      END TYPE   X_Y_Vector
!
      INTERFACE OPERATOR ( + )
          MODULE PROCEDURE Add_Vectors
      END INTERFACE
!
CONTAINS
!
```

Figure 13.9 (continued)

```
      FUNCTION Add_Vectors ( V_1, V_2 )
              IMPLICIT    NONE
      !
      !-----------------------------------------------------------------
      !  This function will overload the + operator so that it adds
      !     variables of type X_Y_Vector. This function assumes that
      !     the coordinates of the head of V_1 are the same as the
      !     coordinates of the tail of V_2 for simplicity.
      !
      !  Add_Vectors - the function name: a variable of type X_Y_Vector
      !
      !  V_1, V_2  -  two INTENT ( IN ) arguments of type X_Y_Vector
      !               that represent the two vectors to be added.
      !-----------------------------------------------------------------
      !
         TYPE ( X_Y_Vector )                           :: Add_Vectors
         TYPE ( X_Y_Vector ), INTENT ( IN )            :: V_1, V_2
      !
         Add_Vectors % Tail_X = V_1 % Tail_X
         Add_Vectors % Tail_Y = V_1 % Tail_Y
      !
         Add_Vectors % Head_X = V_2 % Head_X
         Add_Vectors % Head_Y = V_2 % Head_Y
      !
      END FUNCTION Add_Vectors
      !
END MODULE  Simple_Vectors
```

Figure 13.10

```
PROGRAM   Play_With_Type_Vector
          USE Simple_Vector    ! This USE statement gives
                               ! this program unit
                               ! access to the definitions
                               ! in MODULE Simple_Vector.
          IMPLICIT   NONE
!
!---------------------------------------------------------------------
!  Two variables of type X_Y_Vector, Vec_1 and Vec_2, are given
!     initial values with structure constructors when they are
!     declared. They are then added into Resulting_Vector, and all are
!     printed out.
!
!  Vec_1, Vec_2      - variables of type X_Y_Vector
!
!  Resulting_Vector - the vector resulting from adding Vec_1 and Vec_2
!                     also of type X_Y_Vector
!---------------------------------------------------------------------
```

continues

Figure 13.10 (continued)

```
      !   BLACK BOX:   SUBROUTINE Print_Blank_Lines ( n )
      !
      !   The purpose of this subroutine is to print n blank lines.
      !
      !   n           - an INTENT ( IN ) INTEGER argument that defines how
      !                    many blank lines to print
      !--------------------------------------------------------------------
!
          TYPE ( X_Y_Vector )      :: Vec_1 = X_Y_Vector ( 0.0, 0.0, 4.0, 9.0 )
                                   !             X1,  Y1,  X2,  Y2
                                   !             tail  -  head
          TYPE ( X_Y_Vector )      :: Vec_2 = X_Y_Vector ( 4.0, 9.0, 11.0, 11.0 )
                                   !             X1,  Y1,  X2,  Y2
                                   !             tail  -  head
          TYPE ( X_Y_Vector )      :: Resulting_Vector
!
          Resulting_Vector = Vec_1 + Vec_2   ! Since the operands are of
                                             ! type X_Y_Vector, the +
                                             ! operator will perform
                                             ! addition as defined in
                                             ! MODULE Simple_Vector.
                                             !
          WRITE ( UNIT = *, FMT = 43 ) "Vec_1 =", Vec_1
          WRITE ( UNIT = *, FMT = 43 ) "Vec_2 =", Vec_2
             CALL Print_Blank_Lines ( 1 ) ! BLACK BOX
          WRITE ( UNIT = *, FMT = 43 ) "Result:", Resulting_Vector
             43     FORMAT ( 1X, A7, 4F7.2 )
!
      END PROGRAM  Play_With_Type_Vector

      *************** PROGRAM RESULTS ***************

      Vec_1 =   0.00   0.00   4.00   9.00
      Vec_2 =   4.00   9.00  11.00  11.00

      Result:   0.00   0.00  11.00  11.00
```

13.3 Overloading Function/Subroutine Interfaces

In several example programs in this text, the following subroutine is used:

> Get_Generic_INTEGER_Input (Prompt, Response)

It's a useful black box that prints the character value in Prompt and returns an integer value entered from the keyboard into variable Response. It would also be reasonable to create a subroutine called Get_Generic_REAL_Input (Prompt, Response) and one called `Get_Generic_DOUBLE_Input ( Prompt, Response )`, etc. In each of these additional subroutines, the *type* of argument `Response` would differ depending on the type of input being sought and would therefore need to be a completely different subroutine. If a programmer was using three or four of these functionally similar but data dissimilar subroutines, the documentation would become cumbersome and the likelihood of accidentally using the wrong subroutine would increase.

To overcome this problem and make programs easier to understand and less error prone, **interface overloading** should be used. **Interface overloading** is when **one** generic **interface** works for *multiple program units* whose interfaces are similar as described in the previous paragraph. The single generic interface selects which program unit to invoke based on the *actual argument* types. Examine the module in Figure 13.11.

Figure 13.11

```
MODULE  Generic_Input
            IMPLICIT  NONE
        !
        !---------------------------------------------------------------
        !   This module creates a generic interface for three Get_ ... _Input
        !       subroutines.
        !---------------------------------------------------------------
        !
            INTERFACE Get_Generic_Numeric_Input
                MODULE PROCEDURE   Get_Generic_INTEGER_Input,  &
                                   Get_Generic_REAL_Input,     &
                                   Get_Generic_DOUBLE_Input
            END INTERFACE
!
CONTAINS
!
        SUBROUTINE Get_Generic_INTEGER_Input ( Prompt, Response )
                IMPLICIT NONE
            !
            !---------------------------------------------------------------
            !   This subroutine presents the prompt contained in CHARACTER
            !       INTENT ( IN ) argument Prompt and then reads a value from the
            !       keyboard into INTEGER INTENT ( OUT ) argument Response.
            !---------------------------------------------------------------
            !
                CHARACTER ( LEN = * ),     INTENT ( IN  )   ::   Prompt
                INTEGER,                   INTENT ( OUT )   ::   Response
            !
                PRINT *, Prompt
                READ  *, Response
            !
            RETURN
        END SUBROUTINE  Get_Generic_INTEGER_Input
!===============================================================
```

continues

Figure 13.11 (continued)

```
      SUBROUTINE Get_Generic_REAL_Input ( Prompt, Response )
              IMPLICIT   NONE
!
!-------------------------------------------------------------------
!   This subroutine presents the prompt contained in CHARACTER
!       INTENT ( IN ) argument Prompt and then reads a value from the
!       keyboard into REAL INTENT ( OUT ) argument Response.
!-------------------------------------------------------------------
!
      CHARACTER ( LEN = * ),     INTENT ( IN )   ::    Prompt
      REAL,                      INTENT ( OUT )  ::    Response
!
          PRINT *, Prompt
          READ  *, Response
!
      RETURN
      END SUBROUTINE  Get_Generic_REAL_Input
!===================================================================
      SUBROUTINE Get_Generic_DOUBLE_Input ( Prompt, Response )
              IMPLICIT   NONE
!
!-------------------------------------------------------------------
!   This subroutine presents the prompt contained in CHARACTER
!       INTENT ( IN ) argument Prompt and then reads a value from the
!       keyboard into DOUBLE PRECISION INTENT ( OUT ) argument Response.
!-------------------------------------------------------------------
!
      CHARACTER ( LEN = * ),     INTENT ( IN )   ::    Prompt
      DOUBLE PRECISION,          INTENT ( OUT )  ::    Response
!
          PRINT *, Prompt
          READ  *, Response
!
      RETURN
      END SUBROUTINE  Get_Generic_DOUBLE_Input
!
END MODULE  Generic_Input
```

The module interface Get_Generic_Numeric_Input, seen near the beginning of Figure 13.11, contains three MODULE PROCEDURES:

1. Get_Generic_INTEGER_Input.
2. Get_Generic_REAL_Input.
3. Get_Generic_DOUBLE_Input.

Get_Generic_Numeric_Input is overloaded with the three specific subroutines. Now, any program that requires INTEGER, REAL or DOUBLE PRECISION input

Figure 13.12

```
PROGRAM  Test_Overloaded_SUBROUTINEs
         USE   Generic_Input
         IMPLICIT  NONE
!
!-------------------------------------------------------------------------
! This program tests the generic interface Get_Generic_Numeric_Input.
!
! Int_1           - an INTEGER input variable
!
! Real_1          - a REAL input variable
!
! Dbl_1           - a DOUBLE PRECISION input variable
!-------------------------------------------------------------------------
    ! BLACK BOX:  Overloaded subroutines:
    !
    !        Get_Generic_Numeric_Input ( Prompt, Response)
    !
    ! Prompt       - a CHARACTER INTENT ( IN ) argument that holds the
    !                 prompt
    !
    ! Response     - varies...it's overloaded..it can be INTEGER,
    !                 REAL, or DOUBLE PRECISION
    !-------------------------------------------------------------------------
    !
        INTEGER                   ::    Int_1
        REAL                      ::    Real_1
        DOUBLE PRECISION          ::    Dbl_1
    !
        CALL  Get_Generic_Numeric_Input ( "Enter an Integer:", Int_1 )
        CALL  Get_Generic_Numeric_Input ( "Enter a  Real:",    Real_1 )
        CALL  Get_Generic_Numeric_Input ( "Enter a  Dbl:",     Dbl_1 )
    !
        PRINT *, "Integer:",    Int_1,              &
                 " Real:",      Real_1,             &
                 " Dbl Prec:",  Dbl_1
    !
END PROGRAM  Test_Overloaded_SUBROUTINEs

*************** PROGRAM RESULTS ***************

 Enter an Integer:
-9168022
 Enter a Real:
67.98009309331299564
 Enter a  Dbl:
6.02d23
 Integer: -9168022    Real:  67.9800949 Dbl Prec:    6.0200000000000000E+23
```

can simply use the generic overloaded interface, `Get_Generic_Numeric_Input`. Most of Fortran's intrinsic library is overloaded. `TAN ( var )`, for example, will use a different algorithm if a `DOUBLE PRECISION var` is used rather than a single precision `REAL var`. Figure 13.12 demonstrates module `Generic_Input` in action.

See how convenient this makes numeric input in Figure 13.12? ONE overloaded interface, `Get_Generic_Numeric_Input`, is used to read in values for three variables of different types: `Int_1`, `Real_1`, and `Dbl_1`. The way that `Get_Generic_Numeric_Input` "knows" which of the three module procedures to use is by examining arguments. If a type `REAL` variable is used as the `Response` actual argument, `Get_Generic_Numeric_Input` selects the `Get_Generic_REAL_Input` subroutine to execute. One generic interface invokes the appropriate subroutine just by examining the actual argument types in the `CALL` and choosing the routine that has corresponding formal arguments.

Important Concepts Review

- **Custom dot operators** are created with modules and functions.

- **Overloading intrinsic operators** such as "+" gives the operator extended meaning when its operands are of a derived type.

- **Overloading function/subroutine interfaces** uses one generic interface to invoke one of several program units. The particular program unit invoked is determined by the type(s) of the actual argument(s) used in the generic interface.

- A **binary operator** is an operator with two operands.

- A **unary operator** is an operator with only one operand.

- An **interface operator block** is used to make a connection between the dot operator, `.dot-operator-name.`, and the function that defines how the operator works; the dot operator will invoke its corresponding function.

- `USE module-name` gives a program unit access to the definitions in `module-name`.

EXERCISES

13.1 Create a unary custom dot operator that will "round" positive real numbers. A number should be rounded up if its decimal part is 0.5 or greater and rounded down otherwise.

13.2 Create the following logical dot operators:

    a) .XOR. (exclusive or)
    b) .NAND. (not and)
    c) .NOR. (not or)

13.3 Create a derived data type called Plane_Landing that represents a plane's altitude and distance from a landing strip. Put this definition in a module. Add a unary dot operator to the module that calculates the angle that the plane is from the ground.

Read in values of type Plane_Landing and do the following: If the angle is less than 10°, print, "Take'er up, Captain!". If the angle is greater than 40°, print, "The engines can't take it, Captain!" otherwise print, "Take'er in, Captain." Use a CASE statement for the selection structure.

13.4 Create a unary dot operator called .VECTORANGLE. for TYPE (X_Y_Vector) (Figure 13.5) that will calculate the vector's direction (angle from the *x* axis). Assume the vector's tail is at the origin.

13.5 Create a unary dot operator called .MAGNITUDE. for TYPE (X_Y_Vector) (Figure 13.9) that will calculate the vector's magnitude (length). (*Hint: steal the function that calculates the distance between points on a Cartesian plane that was developed in Chapter 8, Figure 8.4.*)

13.6 Overload the .REALLYEQUAL. operator from Figure 13.3 so that it works for comparing DOUBLE PRECISION values, too.

13.7 Add two more subroutines to module Generic_Input (and rename identifiers as appropriate) so that one can use the generic interface to get CHARACTER and COMPLEX input as well.

13.8 Overload Fortran's intrinsic subtraction operator, "-", so that it subtracts vectors of TYPE (X_Y_Vector). Remove the restriction that the head of Vec_1 must coincide with the tail of Vec_2.

Appendix A
ASCII Collating Sequence

Decimal	Character	Binary	Byte	Octal	Hex
0	NUL	0000 0000		000	00
1	SOH	0000 0001		001	01
2	STX	0000 0010		002	02
3	ETX	0000 0011		003	03
4	EOT	0000 0100		004	04
5	ENQ	0000 0101		005	05
6	ACK	0000 0110		006	06
7	BEL	0000 0111		007	07
8	BS	0000 1000		010	08
9	HT	0000 1001		011	09
10	LF	0000 1010		012	0A
11	VT	0000 1011		013	0B

A ASCII Collating Sequence

Decimal	Character	Binary	Byte	Octal	Hex
12	FF	0000 1100		014	0C
13	CR	0000 1101		015	0D
14	SO	0000 1110		016	0E
15	SI	0000 1111		017	0F
16	DLE	0001 0000		020	10
17	DC1	0001 0001		021	11
18	DC2	0001 0010		022	12
19	DC3	0001 0011		023	13
20	DC4	0001 0100		024	14
21	NAK	0001 0101		025	15
22	SYN	0001 0110		026	16
23	ETB	0001 0111		027	17
24	CAN	0001 1000		030	18
25	EM	0001 1001		031	19
26	SUB	0001 1010		032	1A

A ASCII Collating Sequence

Decimal	Character	Binary	Byte	Octal	Hex
27	ESC	0001 1011		033	1B
28	FS	0001 1100		034	1C
29	GS	0001 1101		035	1D
30	RS	0001 1110		036	1E
31	US	0001 1111		037	1F
32	space	0010 0000		040	20
33	!	0010 0001		041	21
34	" quote	0010 0010		042	22
35	#	0010 0011		043	23
36	$	0010 0100		044	24
37	%	0010 0101		045	25
38	&	0010 0110		046	26
39	' apostrophe	0010 0111		047	27
40	(	0010 1000		050	28
41	)	0010 1001		051	29

A ASCII Collating Sequence

Decimal	Character	Binary	Byte	Octal	Hex
42	*	0010 1010		052	2A
43	+	0010 1011		053	2B
44	, comma	0010 1100		054	2C
45	– minus sign	0010 1101		055	2D
46	. period	0010 1110		056	2E
47	/	0010 1111		057	2F
48	0	0011 0000		060	30
49	1	0011 0001		061	31
50	2	0011 0010		062	32
51	3	0011 0011		063	33
52	4	0011 0100		064	34
53	5	0011 0101		065	35
54	6	0011 0110		066	36
55	7	0011 0111		067	37
56	8	0011 1000		070	38

A ASCII Collating Sequence

Decimal	Character	Binary	Byte	Octal	Hex
57	9	0011 1001		071	39
58	:	0011 1010		072	3A
59	;	0011 1011		073	3B
60	<	0011 1100		074	3C
61	=	0011 1101		075	3D
62	>	0011 1110		076	3E
63	?	0011 1111		077	3F
64	@	0100 0000		100	40
65	A	0100 0001		101	41
66	B	0100 0010		102	42
67	C	0100 0011		103	43
68	D	0100 0100		104	44
69	E	0100 0101		105	45
70	F	0100 0110		106	46
71	G	0100 0111		107	47

A ASCII Collating Sequence

Decimal	Character	Binary	Byte	Octal	Hex
72	H	0100 1000		110	48
73	I	0100 1001		111	49
74	J	0100 1010		112	4A
75	K	0100 1011		113	4B
76	L	0100 1100		114	4C
77	M	0100 1101		115	4D
78	N	0100 1110		116	4E
79	O	0100 1111		117	4F
80	P	0101 0000		120	50
81	Q	0101 0001		121	51
82	R	0101 0010		122	52
83	S	0101 0011		123	53
84	T	0101 0100		124	54
85	U	0101 0101		125	55
86	V	0101 0110		126	56

A ASCII Collating Sequence

Decimal	Character	Binary	Byte	Octal	Hex
87	W	0101 0111		127	57
88	X	0101 1000		130	58
89	Y	0101 1001		131	59
90	Z	0101 1010		132	5A
91	[	0101 1011		133	5B
92	\	0101 1100		134	5C
93	]	0101 1101		135	5D
94	^ circumflex	0101 1110		136	5E
95	_ underscore	0101 1111		137	5F
96	` left apostrophe	0110 0000		140	60
97	a	0110 0001		141	61
98	b	0110 0010		142	62
99	c	0110 0011		143	63
100	d	0110 0100		144	64
101	e	0110 0101		145	65

A ASCII Collating Sequence

Decimal	Character	Binary	Byte	Octal	Hex
102	f	0110 0110		146	66
103	g	0110 0111		147	67
104	h	0110 1000		150	68
105	i	0110 1001		151	69
106	j	0110 1010		152	6A
107	k	0110 1011		153	6B
108	l	0110 1100		154	6C
109	m	0110 1101		155	6D
110	n	0110 1110		156	6E
111	o	0110 1111		157	6F
112	p	0111 0000		160	70
113	q	0111 0001		161	71
114	r	0111 0010		162	72
115	s	0111 0011		163	73
116	t	0111 0100		164	74

A ASCII Collating Sequence

Decimal	Character	Binary	Byte	Octal	Hex
117	u	0111 0101		165	75
118	v	0111 0110		166	76
119	w	0111 0111		167	77
120	x	0111 1000		170	78
121	y	0111 1001		171	79
122	z	0111 1010		172	7A
123	{	0111 1011		173	7B
124	| vertical bar	0111 1100		174	7C
125	}	0111 1101		175	7D
126	~ tilde	0111 1110		176	7E
127	DEL	0111 1111		177	7F

Appendix B
Generic Intrinsic Functions

For all of the intrinsic procedures, the arguments shown are the names that must be used for keywords when using the keyword form for actual arguments. For example, a reference to CMPLX may be written in the form CMPLX (A,B,M) or in the form CMPLX (Y=B,KIND=M,X=A). Many of the argument keywords have names that are indicative of their usage. For example:

KIND	Describes the KIND of the result.
STRING, STRING_A	An arbitrary character string.
BACK	Indicates a string scan is to be from right to left (backward).
MASK	A mask that may be applied to the arguments.
DIM	A selected dimension of an array argument.
A	Type REAL.
A, Ax	Where x is some integer and arguments are of type REAL.
Z	Type COMPLEX.
X and Y	Usually REAL but can be dependant on each other.

Syntax Definitions

[ ]	Optional.
...	Repeat as many times as necessary.
|	Either/Or.

Example: TEST_EXAMPLE_1(A1,A2 [,A3]...)

A1 and A2 are not optional, A3 and up are optional but there can be as many Ax's as needed.

Example: TEST_EXAMPLE_2([PUT|GET])

This function could have either argument, PUT or GET, but not both.

Argument Presence Inquiry Function

PRESENT (A)   Argument presence.

Numeric Functions

ABS (A)	Absolute value.
AIMAG (Z)	Imaginary part of a COMPLEX number.
AINT (A[,KIND])	Truncation to a whole number.
ANINT (A[,KIND])	Nearest whole number.
CEILING (A)	Least integer greater than or equal to number.
CMPLX (X[,Y][,KIND]	Conversion to COMPLEX type. X must be of type INTEGER, REAL or COMPLEX. Y must be of type INTEGER or REAL and it must not be present if X is of type COMPLEX.
CONJG (Z)	Conjugate of a COMPLEX number.
DBLE (A)	Conversion to DOUBLE PRECISION real type.
DIM (X,Y)	Returns X-Y if the difference is positive, zero otherwise. X must be of type INTEGER or REAL. Y must be of the same type as X.
DPROD (X,Y)	DOUBLE PRECISION real product.
FLOOR (A)	Greatest integer less than or equal to number.
INT (A[,KIND])	Conversion to INTEGER type.
MAX (A1,A2[,A3]...)	Maximum value.
MIN (A1,A2[,A3]...)	Minimum value.
MOD (A,P)	Remainder function. A must be of type INTEGER or REAL. P must be of the same type as A.
MODULO (A,P)	Modulo function. A must be of type INTEGER or REAL. P must be of the same type as A.
NINT (A[,KIND])	Nearest integer.
REAL (A[,KIND])	Conversion to REAL type.
SIGN (A,B)	Transfer of sign. A must be of type INTEGER or REAL. B must be of the same type as A.

Mathematical Functions

ACOS (X)	Arccosine.
ASIN (X)	Arcsine.
ATAN (X)	Arctangent.
ATAN2(Y,X)	Arctangent. Y must be of type REAL. X must be of the same type as Y. If Y has the value zero, X must not be zero.

COS (X)	Cosine.
COSH (X)	Hyperbolic cosine.
EXP (X)	e^x.
LOG (X)	Natural logarithm.
LOG10 (X)	Common logarithm (base 10).
SIN (X)	Sine.
SINH (X)	Hyperbolic sine.
SQRT(X)	Square root.
TAN (X)	Tangent.
TANH (X)	Hyperbolic tangent.

Character Functions

ACHAR(I)	Character in a given position in the ASCII collating sequence.
ADJUSTL (STRING)	Adjust left.
ADJUSTR (STRING)	Adjust right.
CHAR (I[,KIND])	Character in a given position in the processor collating sequence.
IACHAR (C)	Position of a character in the ASCII collating sequence.
ICHAR (C)	Position of a character in the processor collating sequence.
INDEX (STRING,SUBSTRING [,BACK])	Starting position of a substring.
LEN-TRIM (STRING)	Length without trailing blank characters.
LGE (STRING_A,STRING_B)	Lexically greater than or equal.
LGT (STRING_A,STRING_B)	Lexically greater than.
LLE (STRING_A,STRING_B)	Lexically less than or equal.
LLT (STRING_A,STRING_B)	Lexically less than.
REPEAT (STRING,NCOPIES)	Repeated concatenation.
SCAN (STRING,SET[,BACK])	Scan a string for a character in a set.
TRIM (STRING)	Remove trailing blank characters.

`VERIFY (STRING,SET[,BACK])` — Verify that `CHARACTER` entity `SET` contains all the characters in `CHARACTER` entity `STRING`. Returns a zero if so, and the position in `STRING` of the first nonmember otherwise.

Character Inquiry Function

`LEN (STRING)` — Length of a character entity.

Kind Functions

`KIND (X)` — Kind type parameter value.

`SELECTED_INT_KIND (R)` — `INTEGER` kind type parameter value, given range.

`SELECTED_REAL_KIND ([P][,R])` — `REAL` kind type parameter value, given precision and range. Returns a value of the kind type parameter of a `REAL` data type with decimal precision of as least P digits and a decimal exponent range of at least R. At least one argument must be present. P and R must be type `INTEGER`.

Numeric Inquiry Functions

`DIGITS (X)` — Number of significant binary digits in the model.

`EPSILON (A)` — Number that is almost negligible compared to one. A must be of type `REAL`.

`HUGE (X)` — Largest number in the model.

`MAXEXPONENT (A)` — Maximum exponent in the model. A must be of type `REAL`.

`MINEXPONENT (A)` — Minimum exponent in the model. A must be of type `REAL`.

`PRECISION (X)` — Decimal precision. X must be `REAL` or `COMPLEX`.

`RADIX (X)` — Base of the model.

`RANGE (A)` — Decimal exponent range.

`TINY (A)` — Smallest positive number in the model. A must be of type `REAL`.

Bit Inquiry Function

    `BIT_SIZE (I)`   Number of bits in the model.

Bit Manipulation Functions

`BTEST (I,POS)`	Bit testing.
`IAND (I,J)`	Logical AND.
`IBCLR (I,POS)`	Clear bit.
`IBITS (I,POS,LEN)`	Bit extraction.
`IBSET (I,POS)`	Set bit.
`IEOR (I,J)`	Exclusive OR.
`IOR (I,J)`	Inclusive OR.
`ISHFT (I,SHIFT)`	Logical shift.
`ISHFTC (I,SHIFT[,SIZE])`	Circular shift.
`NOT (I)`	Logical complement.

Transfer Function

    `TRANSFER (SOURCE,MOLD[,SIZE])`   Treat first argument as type of second argument.

Floating Point Manipulation Functions

`EXPONENT (A)`	Exponent part of a number.
`FRACTION (A)`	Fractional part of a number.
`NEAREST (A,S)`	Nearest different processor representable number in given direction. S must be REAL and nonzero.
`RRSPACING (A)`	Reciprocal of the relative spacing of model numbers near a given number.
`SCALE (A,I)`	Multiply a REAL by its base to an INTEGER power.
`SET_EXPONET (A,I)`	Set exponent part of a number.
`SPACING (A)`	Absolute spacing of model numbers near given number.

Vector and Matrix Multiply Functions

`DOT_PRODUCT (VECTOR_A,VECTOR_B)`	Dot product of two one-dimensional arrays.
`MATMUL (MATRIX_A,MATRIX_B)`	Matrix multiplication.

Array Reduction Functions

`ALL (MASK[,DIM])`	True if all values are true.
`ANY (MASK[,DIM])`	True if any value is true.
`COUNT (MASK[,DIM])`	Number of true elements in an array.
`MAXVAL (ARRAY[,DIM][,MASK])`	Maximum value in an array.
`MINVAL (ARRAY[,DIM][,MASK])`	Minimum value in an array.
`PRODUCT (ARRAY[,DIM][,MASK])`	Product of array elements.
`SUM (ARRAY[,DIM][,MASK])`	Sum of array elements.

Array Inquiry Functions

`ALLOCATED (ARRAY)`	Array allocation status.
`LBOUND (ARRAY[,DIM])`	Lower dimension bounds of an array.
`SHAPE (SOURCE)`	Shape of an array or scalar.
`SIZE (ARRAY[,DIM])`	Total number of elements in an array.
`UBOUND (ARRAY[,DIM])`	Upper dimension bounds of an array.

Array Construction Functions

`MERGE (TSOURCE,FSOURCE,MASK)`	Merge under mask.
`PACK (ARRAY,MASK[,VECTOR])`	Pack an array into a one-dimensional array under a mask.
`SPREAD (SOURCE,DIM,NCOPIES)`	Replicates an array by adding a dimension.
`UNPACK (VECTOR,MASK,FIELD)`	Unpack a one-dimensional array into an array under a mask.

Array Reshape Function

`RESHAPE (SOURCE,SHAPE[,PAD][,ORDER])`	Reshape an array.

Array Manipulation Functions

`CSHIFT (ARRAY,SHIFT[,DIM])`	Circular shift.
`EOSHIFT (ARRAY,SHIFT[,BOUNDRY][,DIM]`	End-off shift.
`TRANSPOSE (MATRIX)`	Transpose a two-dimensional array.

Array Location Functions

MAXLOC (ARRAY[,MASK])   Location of a maximum value in an array.

MINLOC (ARRAY[,MASK])   Location of a minimum value in an array.

Intrinsic Subroutines

DATE_AND_TIME ([DATE][,TIME][,ZONE][,VALUES])   Gets date and time.

MVBITS (FROM,FROMPOS,LEN,TO,TOPOS)   Copies bits from one integer to another.

RANDOM_NUMBER (HARVEST)   Returns pseudorandom number.

RANDOM_SEED ([SIZE|PUT|GET])   Initializes or restarts the pseudo-random number generator.

SYSTEM_CLOCK ([COUNT][,COUNT_RATE][,COUNT_MAX])   Obtains data from the system clock.

Appendix C
The Art of Debugging—Error Removal

Introduction

Programs almost never run correctly the first time (surely you've noticed). Have you been getting a sea of indecipherable error messages when you try to compile and run your programs? . . . Errors such as these? . . .

```
Malformed statement at line 85 [Fatal 107
- Compiler ABEND]

SYMBOLIC STACK DUMP AT 21A4DD7E
000AEEF1        MAIN
AR=0000AED2, BC=9999FFFF . . .
```

Heartwarmingly friendly and informative, aren't they? These error messages present no problem . . . all you need to understand is what a symbolic stack is, interpret hexadecimal addresses and register dumps and figure out just how line 85 is "malformed." Arrrrgh.

Every compiler has its own set of hieroglyphic error codes, each vying to set new industry standards for the incomprehensible while giving fresh meaning to words such as unintelligible, meaningless, confusing, perplexing, baffling, puzzling, and impenetrable.

You're going to love this appendix! It gives step-by-step advice for tracking down program errors despite lousy error messages. Errors (**bugs**) will be classified and the appropriate insecticides supplied.

To debug a program, you must become a detective. Bugs are identified with a systematic procedure of computer-aided deduc-

- **C.1** Debugging
- **C.2** Antibugging
- **C.3** Syntax errors
- **C.4** Runtime errors

    Important Concepts Review

- **C.5** Logic errors
- **C.6** Testing programs
- **C.7** Debugging in different computing environments
- **C.8** Debuggers and tracers

    Important Concepts Review

450

tion that leads to the errors' eradication. As an interesting historical aside, the term "bug" was popularized when Grace Hopper found a moth in a relay of a Mark I computer causing it to malfunction. She later joked that she had found the first real computer bug. The terms "bug" and "debug" are still in continual use in computer science to describe error situations of all kinds.

Obviously, it would be best to avoid letting bugs get into programs in the first place. A section in this appendix is devoted to this process of **antibugging** to help you reduce inadvertently building errors into your program designs. Many skills that you learned in the first three chapters contain inherent antibugging techniques.

Testing and debugging a program after it is entered into a computer is part of the design process. Design flaws are often revealed during testing. Depending on the severity of a flaw, you may need to return to the "drawing board" to redesign parts of a program.

As you do more program designs, testing the program will be less fraught with problems because your written designs will be more error free. Computer science is an athletic discipline; with effort and practice you will become a skilled, strong designer whose working programs take the minimum time to create.

● ● ●

C.1 Debugging

Debugging is the act of removing errors (bugs) from a program. Programs rarely work the first time they are run. The process of debugging is a fundamental detective skill and is similar to other error-identification-and-removal skills used throughout engineering and science.

Table C.1

Error Classification	Symptom
Syntax/Compile time	The computer doesn't understand what is typed in the program and/or how the program statements are organized.
Runtime	The program is attempting the impossible, such as taking the square root of -1, and crashes without finishing.
Logic	The program reaches a normal conclusion without error messages but produces wrong answers.

In the realm of program design there are three categories of errors (Table C.1). **Syntax errors**, also known as **compile time** errors, only occur when a program is compiled. **Runtime errors** happen while a program is running, causing it to crash. The accursed **logic errors** are the most insidious of all; they are typified by normal completion of a program but the results are wrong.

Error messages are displayed when a program is run and it has bugs other than logic errors, which give no error messages. The messages are supposed to help programmers understand how to correct the errors. Unfortunately, the messages are composed by those whose first spoken language is a computer language. Techniques must be learned to isolate and eradicate program bugs in an empirical way through observation and inference. An example of this method is shown in Figure C.1.

Figure C.1

```
      1      PROGRAM Error_Messages
      2                IMPLICIT NONE
      3      !-----------------------------------------------------------------
Error: Unrecognized statement, line 3

Error: Implicit type for THIS, line 3

Error: Syntax error at line 3
****Malformed statement

      4      This program is designed to frustrate the compiler.
      5      !-----------------------------------------------------------------
      6      DOUBLE PRECISION :: n, g = 1.0, Tolerance = .0001D0, g_old = 0.0
      7      INTEGER, PARAMETER :: Counter_1 = 7.5D3
      8      !
      9      PRINT * "Please enter the number you want the square root of:"
Error: Syntax error, line 9
       ****Malformed statement
      . . .
```

Figure C.1 shows a **listing file**. Listing files are files generated by compilers and show numbered program lines that the compiler tried to compile. Listing files also contain any error messages that the compiler generated.

Looking at the error messages for line 3 in Figure C.1, it can soon be figured out that line 4 is missing an exclamation point to make it a comment line. The cause of this error can be deduced by careful observation of the lines near the error message, although the error messages are mostly useless and refer to the wrong line number. Interestingly, `IMPLICIT NONE` offers the best guidance for figuring out the line 3 error. `IMPLICIT NONE` causes `THIS` to be pointed out as an undeclared item. Since `THIS` is being "seen" by the compiler, it's not in a comment—the exclamation point is missing.

Now the error message associated with line 9 must be analyzed:

```
Error: Syntax error, line 9
      ****Malformed statement
```

This message is even less helpful but, again, examination of line 9 will lead to the solution. Don't be afraid to compare a program's statements to examples in the text; this will often point out improper statement construction (syntax errors). Comparing line 9's `PRINT` statement to any `PRINT` statement in this text reveals that a comma is missing after `PRINT *`. This kind of look-at-the-line empirical analysis will solve many problems when error messages don't help.

C.2 Antibugging

Antibugging means to use design and coding techniques that prevent bugs from ever entering a program. Table C.2 (p. 454) lists a set of antibugging methods.

The idea of top-down designs is the most important antibugging technique. In a top-down design, a problem is broken down into its component subproblems and organized into a structure chart that describes the complete structure of the programming solution. Well-designed structure charts will have blocks that are easy to code, each representing only one small well-defined task.

Beginning program designers often type entire programs into a computer and consequently are besieged by screen after screen of error messages. Don't be a victim of this depressing mistake. As explained in the chapter on subroutines, develop programs on a computer using **program stubs**.

To use the program stub method, completely enter the *main program* first. Next, enter all subroutines called by the main program with each subroutine containing only a simple `PRINT` statement to display the name of its corresponding program unit. This method lays in the entire structure of the program and the program WILL RUN even though all its functionality isn't included. Each subroutine is then developed in any sensible order that the programmer chooses. At any time during this process there is only one small piece of code being entered and debugged. Always keep things as simple as possible.

The rest of the antibugging approaches in Table C.2 are self-explanatory and were discussed in previous chapters.

Table C.2

	Antibugging
Technique	**Explanation**
Top-Down Design	Break a problem down into small, manageable pieces. These pieces are inherently less error prone because of their small size and the fact they are designed to accomplish only one straightforward task. Subroutines also help prevent accidental modification of variables' values because the only way to get data in and out of subroutines is through arguments.
Library Routines	When an existing working routine is used, it is error free and won't require debugging or introduce any errors.
Program Stubs	**Program stubs: the main program is completely coded and all subordinate subroutines initially contain only a `PRINT` statement to display the name of the program unit. This lays in the backbone of the program. Subroutines are then developed and debugged one at a time, minimizing errors.**
`IMPLICIT NONE`	Catches mistyped identifiers.
Mnemonic Identifiers	Prevent confusion about the purpose of variables, named constants, subroutines, etc., by making their names imply their use.
Named Constants (`PARAMETER`s)	Constants are only typed once, which reduces the chance of mistyping them when they are used several times in a program unit. Named constants together with `IMPLICIT NONE` (which ensures that you don't mistype a `PARAMETER`'s name) will reduce errors involving constants to nearly zero. Named constants can't be changed in a program.
Good Programming Style	Good programming style makes programs easier to understand, less confusing and therefore, less susceptible to errors. Indentations in block statements help ensure all parts of the block statement are present and correctly organized, etc.

C.3 Syntax Errors

Syntax errors (also known as **compile time errors**) are the results of violating the grammar of a computer language. To help understand how strict a computer language's syntax is, imagine that someone said, "I ain't got none bananas" (perhaps spoken by one of the people who composes compiler error messages). He would be understood as being short of bananas by other humans even though he used poor grammar.

People can infer meaning from grammatically incorrect utterances but such inferences are currently beyond computers because, as noted before, they are stupid. One must "speak" with perfect Fortran grammar (syntax) to get a program to compile successfully.

Syntax errors are the first error messages a programmer sees and usually appear in a *program listing file* (line-numbered Fortran source code with embedded error messages). A fragment of a compiler-generated program listing is shown in Figure C.2.

Listing files include line numbers for convenience. **Be careful NOT to correct program errors in the listing file**; make all corrections in the actual program source code. The listing file is only a tool to help point out where errors were spotted by the compiler.

Common Causes

Syntax errors arise from two basic causes: either the person entering the program into the computer makes typing errors or he doesn't use correct Fortran syntax. Table C.3 (p. 456) lists some common causes of syntax errors. Table C.3 will help identify common syntax errors, and the next section explores the plan for getting rid of the bugs (**insecticide**).

Insecticide for Syntax Bugs

The steps for eliminating syntax errors are listed in Table C.4 (p. 456). When confronted with many syntax errors the first time a program is compiled (as we ALL are), DON'T DESPAIR! Follow the steps outlined in Table C.4. Look at the first error and fix it; *ignore the other errors*—focus on one problem at a time.

Often, when a true error occurs, many additional misleading error messages will be displayed because the compiler becomes confused. These error messages are called **generated errors**. An example of this situation is illustrated in Figure C.3 (p. 457).

Figure C.2

```
     1     PROGRAM _Syntax_Errors
Error: Syntax error at line 1
       ***Improper identifier

     2                   IMPLICET NONE
Error: Unrecognized statement at line 2
***Unrecognized structure

     3     !--------------------------------------------------------------------
Error: Unrecognized statement at line 3
***Malformed statement

     4     ! This program is designed to show syntax errors.
     5     !--------------------------------------------------------------------
     6     DOUBLE PRECISION : n,g = 1.0, Tolerance = .0001D0,g_old = 0.0D0
     7     INTERGER, PARAMETER :: Counter_1 = 7.5D3
Error: Unrecognized statement at line 7
***Misspelled type
```

Table C.3

Common Causes for Syntax Errors
Typos—`IMPLICIT NONE` is a great help.
Not understanding correct Fortran syntax.
Not declaring variables.
Unmatched parentheses.
Unmatched quotes.
Actual arguments don't correspond to formal arguments.
Missing parts of block statements, e.g., `END IF`, `END DO`, etc.
Overlapping block statements—no straddling allowed.
Using `0` instead of `O` (zero instead of the letter "O") . . . vice versa.
Using `l` (lowercase "L") instead of `1` . . . vice versa.
Missing commas in `READ`, `WRITE`, or `PRINT` statements.
Not enough/too many constants in a `DATA` statement.
Omitting exclamation points at the beginning of comments.

Table C.4

Syntax Error Insecticide	
Step	Rationale
1. Read the error message.	. . . it just might help.
2. Examine the line's syntax.	Syntax may have been violated by a typing error or accidentally using incorrect Fortran syntax. Look at some examples in the text to see appropriate Fortran syntax.
3. Check the lines *above* and *below* the error line for syntax mistakes.	Often compiler error messages incorrectly identify which line is in error. If the identified line looks okay, the error may be in an adjacent line.
4. Go through the "Common Causes" list in Table C.3	A common error may have been committed.

Notice the syntax error in line 6. The colon should have been double colons, and consequently the compiler couldn't interpret the declaration statement. As a result, all the associated variables (n, g, Tolerance and g_old) ARE NOT declared. The error messages for line 14 include the following:

```
Error: Implicit type for G_OLD at, line 14
Error: Implicit type for G at line 14
```

These are generated errors caused by the error on line 6. Correcting the double colon problem will eliminate the line 14 error messages. Take one error at a time and solve it. Generated error messages will disappear as true errors are corrected.

On the other side of the peso, errors can hide other true errors. Program code is often ignored by the compiler when there is a syntax error. Typically, the statement that follows an errored statement will be ignored. Sometimes the compiler will just give up if very

Figure C.3

```
    6          DOUBLE PRECISION : n, g = 1.0, Tolerance = .0001D0, g_old = 0.0D
Error: Syntax error at line 6
       ***Invalid item in declaration

    7    !     INTEGER, PARAMETER :: Counter_1 = 7.5D3
    8    !
    9          PRINT *, "Please enter the number you want the square root of:"
   10          READ *, n
Error: Implicit type for N at line 10
    Error: Unrecognized statement, line 12

   11          CALL Print_Blank_Lines ( 1 )
   12    !
Error: Implicit type for DO at line 12

Error: Syntax error, line 12
****Malformed statement

   13          DO WHILE ( ABS ( g - g_old ) > Tolerance )
   14              g_old = g = 5
Error: Implicit type for G_OLD at line 14
    Error: Implicit type for G at line 14
    Error: Syntax error, line 14
    ****Malformed statement
```

severe errors are found or too many errors are encountered. When one error is corrected, new errors may be "seen" by the compiler that were previously skipped because of compiler confusion. The new errors blossom to thrill and delight the programmer with the prospect of getting to do more challenging debugging.

C.4 Runtime Errors

Runtime errors occur after all the syntax errors have been killed with the insecticides prescribed in the previous section. If a program compiles and links without any error messages but crashes when it is run, the problem is a runtime error. Runtime errors are the result of asking the computer to "Dream the impossible dream." Computers are quite obtuse and definitely can't work magic; they can't calculate the tangent of $\pi/2$ or take the logarithm of -3. When a program is asked to do this kind of operation, the program has a kind of mental breakdown wanting to follow its instructions but being unable to comply. The program crashes.

Runtime error messages don't come in bunches as syntax errors often do. A single message is displayed describing what caused the error and where it occurred in the program (Figure C.4).

The error in Figure C.4 was caused by an attempt to take the square root of -9.0 in subroutine `Square_Root_No_Good`. If the program hadn't explicitly assigned -9.0 to variable N to force the runtime error, the analysis of what caused the error might have gone something like this:

> Although the hexadecimal numbers in the error messages aren't very helpful, the problem IS identified as an illegal use of a function AND the subroutine name, `Square_Root_No_Good`, is mentioned. The only function used in `Square_Root_No_Good` is `SQRT()`, so it should be suspected that `SQRT(N)` might be the problem. If these deductions are correct, the value contained in variable N must be suspected of being negative. Then the programmer needs to ask how N got its value. Work back to where the subroutine was called and try to find out where N got its value.

Figure C.4

```
 1      PROGRAM Run_Time_Error
 2                      IMPLICIT NONE
 3          REAL                        ::      N = -9.0
 4      !
 5          CALL Square_Root_No_Good ( N )
 6          Print *, "THE END"
 7      !
 8      END PROGRAM Run_Time_Error
 9      !\\\\\\\\\\\\\\\\\\\\\\\\\\\\\\\\\\\\\\\\\\\\\\\\
10      SUBROUTINE Square_Root_No_Good ( N )
11                      IMPLICIT NONE
12          REAL, INTENT( IN )          ::      N
13          REAL                        ::      Square_Root_Of_N
14      !
15          Square_Root_Of_N = SQRT ( N )
16          Print *, "N =", N, "SQRT ( N ) =", Square_Root_Of_N
17      !
18      RETURN
19      END SUBROUTINE Square_Root_No_Good

*************** PROGRAM RESULTS ***************

Illegal function use at User - 0000012E
In routine SQUARE_ROOT_NO_GOOD_ at 1D
In routine main at 1A
```

The second two lines of the error messages in Figure C.4 are called a **traceback**. Tracebacks identify which program unit caused the program crash and the sequence of calls back to the main program that lead to the crash.

Common Causes

Some recurring themes for program crashes due to runtime errors are itemized in Table C.5. Some common causes for runtime errors listed in Table C.5 may be unfamiliar because the related programming skills haven't been studied yet. These causes of runtime errors will become more meaningful when the associated skills are learned.

Insecticide for Runtime Bugs

Table C.6 lists the steps for eliminating runtime errors. "Read the error message . . ." is a recurring theme. Although error messages are often difficult or impossible to understand,

Table C.5

Common Causes for Runtime Errors
Wrong variables/values passed as actual arguments.
Not returning necessary values to a calling program unit.
Dividing by zero.
Integer overflow—creating an `INTEGER` value that is too large.
Illegal function usage, e.g., `SQRT(-9.0)`, `LOG(0.0)`. etc.
Input value doesn't match data type, e.g., `REAL` value with a decimal point is entered when an `INTEGER` was requested.
Using a variable whose value hasn't been defined.
Read past end-of file; attempting to read nonexistent data from a data file.
Subscript out of range—array variables.
Invalid substring reference of `CHARACTER` data.
Attempting to modify the value of a named constant.

Table C.6

Runtime Error Insectide	
Step	Rationale
1. Read the error message.	Again, it just might help.
2. Look at the traceback to figure out which program unit caused the problem.	The error message together with the traceback will help locate where the error happened.
3. Use `PRINT` statements to display involved variables' and expressions' values.	An incorrect value may be spotted. For example, in a quotient, the variable or expression in the denominator may have a value of zero causing a division-by-zero runtime error. The programmer must then backtrack to figure out how the errant value was generated. Perhaps a wrong value was passed as an actual argument or an incorrect value was read, etc.

they may help. The traceback from runtime errors will define which program unit had flow of control (was executing) when the program crashed. The problem can then be isolated further by inserting PRINT statements in the program to display relevant values involved in questionable expressions and running the program again. Place these PRINT statements *before* the suspected problem or the program will crash without printing the debugging information. When the program crashes again, the final values from the PRINT statements will be on the screen just before the runtime error message and traceback. Assuming incorrect values are found, the programmer can systematically work back to find out how the erroneous values were created, and the problem is solved.

Good programming style dictates that debugging PRINT statements be placed at the extreme left of all other program statements so that they are readily visible. When the problem is resolved, don't delete the debugging PRINT statement; **comment it out** by placing an "!" in front of it. It is surprising how often one has to use those PRINT statements again.

Important Concepts Review

- **Bugs** are errors in a program.

- **Antibugging** means to adopt design techniques that prevent bugs from being introduced into a program.

- **Debugging** is the act of removing errors from a program.

- **Syntax/compile time errors** are errors that prevent the program from successfully compiling.

- **Runtime errors** occur after all syntax errors are fixed. While the program is running, it tries to do something impossible and the program crashes.

- A program has **logic errors** when it successfully runs to completion but produces wrong or no answers.

- **Error messages** are designed to guide programmers to a program's problems.

- Develop programs on a computer using **program stubs.** Completely enter the main program and then enter all subroutines called by the main program. Each subroutine should only contain a simple PRINT statement that displays the name of its corresponding program unit. Subroutines are then developed in any sensible order that the programmer chooses.

- A **program listing,** also called a **listing file,** is line-numbered source code with any error messages from the compiler.

- **Insecticide** is the set of steps a programmer follows to remove (kill) errors (bugs).

- **Generated errors** are error messages that point out errors that aren't really errors. The messages are displayed because the compiler becomes confused.

- A **traceback** is a runtime error message that tells where an error happened. It also lists the sequence of `CALL`s that lead to the crash from the main program down to the errant program unit.

- **Debugging `PRINT` statements** are `PRINT` statements used to display variables' and expressions' values to help isolate runtime and logic errors. Good programming style dictates that these `PRINT` statements be placed at the extreme left of all other program statements.

- **Comment out** debugging `PRINT` statements when the related problem is fixed; don't delete them.

C.5 Logic Errors

Logic errors are the worst . . . a program compiles without any error messages, runs without crashing and then produces wrong answers or no answers at all. To isolate this type of error, it is necessary to do one of the following:

1. Start in the program unit where the answers are printed and systematically work backward (from a flow point of view) printing relevant values to pin down where wrong values were generated.

or

2. Make an educated guess about where the problem might be and, again, systematically work back to the error by using `PRINT` statements to find out which wrong values were created or obtained.

Common Causes

Unfortunately, logic errors can take such a variety of forms it is not easy to offer a very comprehensive set of common causes. A simplistic list is shown in Table C.7.

Table C.7

Common Causes for Logic Errors
Incorrect flow of control.
Sending wrong values as actual arguments.
Faulty or missing error checking algorithms.
Initializing variables incorrectly.
Using /= when == should be used and vice versa.
Using .AND. when .OR. should be used and vice versa.
Infinite loops.
Arithmetic expressions constructed incorrectly.
Using wrong formulas.

Insecticide for Logic Bugs

Follow the steps that are itemized in Table C.8 to isolate logic errors.

Table C.8

Logic Error Insecticide	
Step	Rationale
1. Put a PRINT statement in each program unit . . . the original *program stub PRINTs* will do.	The first step is to make sure that the program units are being called in the correct order—that flow of control is correct. Often the problem is immediately obvious because the program should have called some subroutine but didn't or vice versa. The question then becomes: why? Is some condition not being met? Is a CALL omitted or in the wrong place, etc.?
2. Use PRINT statements before and after CALLs to verify the arguments' values.	Argument passing is often at the root of logic errors. For instance, switching the order of two arguments that are of the same type will lead to a logic eror. Printing arguments before and after a CALL will verify the integrity of data flowing in and out of program units through arguments.
3. Use PRINT statements to display variables that are involved in conditional expressions.	An incorrect value may be spotted that is causing a condition that calls the wrong subroutine or executes the wrong block of an IF statement or executes a loop too many or too few times, etc.
4. Check formulas with a calculator.	Do a couple of specific cases of each program formula "by hand." This will ensure that formulas were constructed correctly and that arithmetic order of precedence, mixed-mode expressions and integer division errors were not made. Be careful to perform appropriate integer division as integer division on the calculator.

Logic errors require the greatest detective skills. Program flow should be analyzed first to see if the program is calling its subroutines in the correct order. If the order is wrong, why? Is there a selection structure that is selectively calling subroutines? If so, use `PRINT` statements to examine the values of the variables involved in the logical expression. If an errant value is found, how did that variable get the incorrect value? Backtrack and, again, use `PRINT` statements to display earlier stages of the variable's contents. Keep working back until the problem is found.

When a logic error is finally located and corrected, DON'T DELETE the **debugging `PRINT` statements**. Sadly, they are often needed again. Put an exclamation point in front of each debugging `PRINT` statement to turn it into a comment. It is also recommended that debugging `PRINT` statements be placed at the extreme left of all program statements. This programming style makes the debugging statements easy to identify and they cause minimum interference with the clarity of the actual program code (Figure C.5, p. 464).

Figure C.5's program seems to be calculating powers of 5. These results lead to the conclusion that the assignment statement is in error:

```
Fact = Fact * N
```

The assignment statement should be the one below because `i` is the variable that is incrementing:

```
Fact = Fact * i
```

Once the program is corrected, the debugging `PRINT` statements are **commented out** (Figure C.6, p. 465). They may ultimately be removed but it is only advisable to do this after the entire program is working completely correctly.

Breakpoints: `STOP`

The `STOP` **statement** abruptly stops program execution. *When debugging*, create breakpoints with the `STOP` statement. A **breakpoint** is a place in a program where a programmer wants to "break" (stop) the execution of the program to examine values. This can be very useful when diagnosing logic errors. `STOP` may be used by itself or may have the following form:

```
STOP "some message"
```

Breakpoints are particularly helpful when large programs are being debugged, and it is not desirable to let the program continue past a suspected error and associated debugging `PRINT` statements. Important debugging information will often scroll off the screen if execution is allowed to finish.

DON'T use the `STOP` statement as the "normal" end of a program. Programs should be organized so that they end when the main program's `END PROGRAM` statement is met. **Use `STOP` *only* for creating breakpoints while debugging.**

Figure C.5

```
        PROGRAM Logic_Errors
                IMPLICIT NONE
        !------------------------------------------------------------
        !   This program should calculate the factorial of N.
        !
        !   N     - an INTEGER variable
        !
        !   Fact  - a DOUBLE PRECISION variable to hold the
        !           factorial
        !
        !   i     - a loop counter type INTEGER
        !------------------------------------------------------------
        !
            INTEGER          ::    N, i = 2
            DOUBLE PRECISION ::    Fact = 1
    !
                PRINT *, "Please enter N"
                READ *, N
                DO WHILE ( i <= N )
!
PRINT *, "N =    ", N
PRINT *, "Fact = ", Fact
PRINT *, "i =    ", i
!
                    Fact = Fact * N
                    i = i + 1
                END DO
                PRINT *, "N!=", Fact
        END PROGRAM   Logic_Errors

*************** PROGRAM RESULTS ***************

 Please enter N
5
 N =    5
 Fact =    1.0000000000000000
 i =    2
 N =    5
 Fact =    5.0000000000000000
 i =    3
 N =    5
 Fact =    25.000000000000000
 i =    4
 N =    5
 Fact =    1.2500000000000000E+02
 i =    5
 N!=    6.2500000000000000E+02
```

Figure C.6

```
        PROGRAM Logic_Errors
                 IMPLICIT  NONE
        !----------------------------------------------------------
        !  This program should calculate the factorial of N.
        !
        !  N      - an INTEGER variable
        !
        !  Fact   - a DOUBLE PRECISION variable to hold the
        !             factorial
        !
        !  i      - a loop counter type INTEGER
        !----------------------------------------------------------
        !
              INTEGER          ::      N, i = 2
              DOUBLE PRECISION ::      Fact = 1
        !
                    PRINT *, "Please enter N"
                    READ *, N
                    DO WHILE ( i <= N )
        !
        !PRINT *, "N =     ", N
        !PRINT *, "Fact = ", Fact
        !PRINT *, "i =     ", i
        !
                          Fact = Fact * i
                          i = i + 1
                    END DO
                    PRINT *, "N!=", Fact
        END PROGRAM Logic_Errors

        *************** PROGRAM RESULTS ***************

        Please enter N
        5
        N!= 1.2000000000000000E+02
```

C.6 Testing Programs

To ensure that a program is working correctly, it must be tested. Several steps for proper program testing are summarized in Table C.9 (p. 466).

It is obvious that programs must be tested. If they were put into use without being tested they would be tested anyway—by the user. Attempt to "be a user" and route out as many errors as possible. Use incorrect data to make sure error-checking algorithms prevent

Table C.9

	Testing Programs
Step	Rationale
1. Try to make the program crash.	If the programmer can crash the program, the user will. Improve and refine the program with error checking, prompting messages—anything that was overlooked in the Thinking Phase of the design.
2. Use varied data, especially incorrect data.	This will test error-checking algorithms.
3. When a problem is found and repaired, test the program again with the SAME data.	Focus on one problem at a time. If a piece of data causes a problem, redesign the code and make sure the correction does work; try the same data that originally caused the problem.
4. Test all branches of a program.	Take the time to test all features of a program. This shouldn't be too time consuming; the program stub method of program development inherently minimizes problems because each program unit is coded and tested one at a time. If the program units are thoroughly tested, the only problems left should be the overall way the program units interact with each other.
5. Have another person run the program and observe him.	It is amazing to see how others operate programs. Notice when they don't understand prompts and when they type something that crashes the program. DON'T EXPLAIN MISTAKES THEY'RE MAKING WHILE THEY'RE RUNNING THE PROGRAM! Learn from their confusion. When programmers test their own programs they tend to follow the same steps repeatedly. It is easy to miss problems when the same testing pattern is followed every time.

inappropriate data from entering a program. Test every feature of a program. If a program has four parts, use data that will exercise all four. Try to crash the program. Hammer programs as hard as possible and then let others run them. Letting others use a program usually points out a whole new set of errors that the programmer's testing didn't reveal. The software industry illustrates a macro view of these principles.

When a software company completes a software package, the software is sent to various sites for testing before it is released to the public. The first round of testing after the program designers' personal testing is called **alpha testing**. After the reported bugs from the alpha testers are fixed, the software is **beta tested** on a larger scale at more sites and is now nearly ready for release. The beta test bugs are fixed, the software is released and is **user tested** (not by the user's choice, however). User testing reveals yet another set of bugs that are fixed because users complain and comment to the software company. Throughout this process, **enhancements** (**refinements**) are recommended by testers and changes may be made. This evolutionary set of steps that software goes through release after release is another example of the **software life cycle**. Through evolving, being tested, and growing to meet new needs, software has a life of its own.

C.7 Debugging in Different Computing Environments

Computing environments differ depending on where one ultimately does science or engineering. Debugging techniques vary based on the type of computing environment being used. The three types of systems to consider are:

1. Networked workstations/PCs—client/server schemes.
2. Stand-alone resources—PCs or workstations.
3. Mainframes (mainframe-a-saurus).

In a **networked** environment, many powerful computers called workstations are connected together. Computing resources such as printers and Fortran compilers are shared among the workstations. When a programmer at a workstation makes a request for a Fortran compiler, for example, he waits while the necessary software is transferred over the network to his workstation (the **client**) from another computer that has the necessary files (the **server**). If the network is busy with many users, there can be lots of waiting for responses from the network. But usually, after the necessary files have been transferred to a workstation, programming activities are much the same as for standalone PCs. The debugging methods in a client/server networked environment are the same as for PCs discussed below.

On **standalone** (non-networked) workstations and PCs, a programmer usually isn't constrained by limited CPU resources, as with mainframes. Debug programs as follows: look at the first error, correct it, recompile and attack the next error; focus on one error at a time.

Mainframe computers are big, hulking monstrosities that often fill large rooms. All users are tethered to the same machine. In this sort of environment, resources can often be at a premium. Many engineers and scientists compiling their programs repeatedly to debug them while other users are running a variety of applications can tend to overload the system.

To remove syntax errors in a situation where computing resources are limited, it is best to compile the program and get a **printout**. A printout is a printed version of the program listing file; it is also known as a **hardcopy**. Sit down with pen-in-hand and attempt to identify ALL errors. After a best effort is made to spot all errors in the program listing, make the corrections and recompile the program. The main Achilles' heel of this method of debugging is recognizing **generated errors**. Programmers shouldn't agonize over errors that have no apparent origin; they may be a result of compiler confusion. As far as eliminating runtime and logic errors in an environment where resources are tight, be more liberal with debugging `PRINT` statements. Don't use `STOP` statements and get a program listing. These steps will give a programmer the most information per compilation/run of the program.

C.8 Debuggers and Tracers

Many programming environments supply **debuggers** or **tracers** as program development tools. These software devices allow the programmer to execute a program line by line while watching flow of control and monitoring variables' values. This is a tremendous advantage, as it helps reduce the necessity for designing and coding debugging PRINT statements to find errors in a program.

Important Concepts Review

- A breakpoint is a place in a program where the programmer abruptly stops program execution to examine values.

- STOP "some message" is a Fortran statement that stops program execution and displays its optional message. Create breakpoints with the STOP statement; DON'T ever use STOP as the "natural" end of a program.

- Testing programs is the process where a program is run as many times as necessary with varied data to flush out any bugs.

- Alpha testing is the first testing done outside a software development group.

- Beta testing is the second round of off-site software testing. It is done on a larger scale than alpha testing and the code should be more error free.

- User testing is done after software is released and reveals bugs not uncovered by alpha and beta testing. Users are victims not volunteers in this testing.

- Enhancements/refinements are improvements in program designs made when the program is tested.

- Software life cycle is the evolutionary phases software goes through from testing and adding enhancements over time.

- Limited-resource debugging means to find and correct as many errors as possible between runs by using a printout of the program listing file.

- Stand-alone debugging means to find an error and correct it. Find the next error; fix it, and so on. Focus on one problem at a time and don't worry about recompiling the program frequently.

- Tracers/debuggers are available in some environments allowing programmers to step through their programs, create breakpoints and track the values of variables.

Appendix D
Formats

Introduction

Throughout this text, formats have been introduced by examples to help make answers more readable. One of our main goals is to make all our programming efforts clear and understandable. And as the whole purpose of computer programs is to get answers, any answers that are intended for human viewing should be easy to read and understand. Your most powerful tools for presenting results in an organized, comprehendible way are **formats**.

There is another side to formats: use with input. During your careers as engineers or scientists, you will probably create or inherit data files that need analysis. Often, only some of the data from a file is required for a specific investigation. The way to deal with that situation is to use a format to selectively extract the necessary data (Figure D.1).

In the "Files" Chapter, **controlled I/O** was introduced and how to move data to and from I/O devices was explained. Adding FORMATs to the **controlled I/O** picture completes it, empowering you with a powerful set of data transferring and editing tools. Data can be moved in and out of programs fluidly and the data may edited in the process.

This appendix is full of tables and brief examples that show how FORMATs work in a broad variety of situations. Use this chapter as a handbook to look up exactly what you need for a

D.1 What are formats?

D.2 Format descriptors

D.3 Using formats to control I/O

Important Concepts Review

D.4 Repeating format descriptors

D.5 Stopping the output of unwanted strings

D.6 Output considerations for descriptors

D.7 Carriage control

D.8 Input considerations for descriptors

Important Concepts Review

Figure D.1

```
              DATA FILE
003AXT*1.78902DIMINISHED4E-6
001AXS 2.00314DIMINISHED4E-6
004AXT*1.90901NORMAL    5E-6
002AXS*2.10000DIMINISHED3E-5
 ‿‿ ‿ | ‿‿    ‿‿‿‿‿‿‿‿ ‿‿
 f1 f2   f4      f5      f6
      f3
       fx means field x
```

To read only fields 2 and 4, the following FORMAT could be used:

```
107    FORMAT ( 3X. A3. 1X. F7.5 )
```

given formatting problem to seize control over the appearance of your answers and get data from files.

• • •

D.1 What Are Formats?

In **list-directed I/O** (I/O controlled by an **I/O list**), Fortran formats values as it sees fit, but it DOES format them. **Formatting** is the process of taking data from one form and changing it to another. This is most easily understood by considering how data is maintained in a program. A number such as 2033182 is *not* stored as seven digits; that is unnecessary and inefficient. Instead, 2033182 is stored as a **binary number** organized in **hexadecimal bytes** that can be stored more compactly and is ready to be involved in arithmetic operations. Unfortunately, that form isn't very readable (Figure D.2, p. 471).

Printing the binary or hexadecimal version of an INTEGER as stored in a computer is not very useful to most users. Formatting transforms the internal representation of a value to a form that is easier to understand.

FORMATs are used by referencing a **format label** in READ, WRITE or PRINT statements:

```
                       . . .
             READ (. . . FMT = 65 ) ( Ary ( I ), I = 1, 100 )
        65   FORMAT ( 10X, 3I5 )
             WRITE (. . . FMT = 701 ) ( Ary ( I ), I = 1, 100 )
       701   FORMAT ( 1X, 10F7.2 )
             PRINT 65, ( Ary ( I ), I = 1, 100 )
                       . . .
```

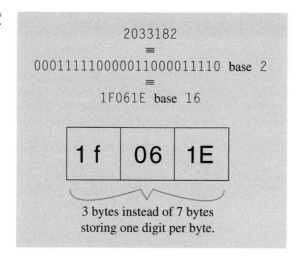

Figure D.2

FMT = *n* and the 65 in the PRINT statement control the I/O statements above and tell Fortran to transfer data according to the formats specified by the label numbers. The general form of a FORMAT statement is shown below:

label FORMAT (format-descriptors)

Label is a literal integer constant (not a named constant) such that $0 < \text{label} < 100{,}000$ and **format descriptors** describe how data is interpreted and edited. Following sections explain Fortran format descriptors.

D.2 Format Descriptors

There are format descriptors to deal with each of Fortran's data types. Format descriptors are broken down into four classifications:

1. Numeric.
2. Nonnumeric.
3. Positioning.
4. Special.

Numeric descriptors are used to edit and interpret numeric data. General descriptions of numeric and other descriptors are listed in Tables D.1, D.2, D.3 and D.4. Details of how all descriptors behave differently for input and output will be explained in later sections.

Table D.1

	Numeric Format Descriptors	
Descriptor	Form	Explanation
I	Iw, I$w.m$	I is for INTEGER values. w is width of the integer field; e.g., I5 can accommodate up to a five-digit positive number or a four-digit negative number ("-" occupies one of the five places). m is the *m*inimum digits displayed when used for output.
B, O, Z	Bw, B$w.m$ Ow, O$w.m$ Zw, Z$w.m$	Same as the I descriptor except that values read or written are binary, octal or hexadecimal, respectively.
F	F$w.d$	F is for REAL values of fairly small magnitude. w is the overall *w*idth of the number and d is the number of places to the right of the *d*ecimal.
E	E$w.d$ E$w.d$Ee	E is for REAL values in scientific notation. w is the overall *w*idth of the number and d is the number of places to the right of the *d*ecimal. e specifies how many digits are in the *e*xponent; this must be used when displaying numbers whose exponents' magnitudes are greater than 999.
EN	EN$w.d$ EN$w.d$Ee	Same as E except values are edited so that the exponent is divisible by 3 and $1 \leq$ significand ≤ 999.
ES	ES$w.d$ ES$w.d$Ee	Same as E except values are edited so that the most-significant digit of the mantissa is to the left of the decimal; $1 \leq$ significand ≤ 9.
SP	SP	Write a plus sign in front of positive values.
SS	SS	Suppress leading plus signs.
P	xP	This descriptor has various effects on numeric descriptors. Exactly how it impacts each is explained in subsequent sections.

Table D.2

	Nonnumeric Format Descriptors	
Descriptor	Form	Explanation
A	A, Aw	A is for CHARACTER values. Use A with no field width to handle unknown length CHARACTER values.
L	Lw	L is for LOGICAL values. w is width of the field.

Table D.3

Descriptor	Form	Explanation
X	nX	X is for relative spacing. n specifies how many positions to move to the right from the current position in an input or output record (screen, file, etc.).
T	Tc	T accomplishes absolute spacing. c is the column where I/O takes place next. Think of T as doing a TAB operation.
TR	TRn	Relative spacing; TR works the same as nX.
TL	TLn	TL works the same as TR except TL moves to the left n positions. If n is specified to position before the beginning of a record, position is set AT the beginning.
/	$n/$	/ moves n records forward before doing the I/O operation.

Positional Format Descriptors

Table D.4

Special Format Descriptors

Descriptors	Form	Explanation
G	*any*	G is a general descriptor that can be used to read or write any data. Its main strength is that it will switch between E and F dynamically, depending on the magnitude of the value being displayed.
:	:	: is used to prevent printing inappropriate final CHARACTER literals in a format.
BN	BN	BN is used to interpret blanks as "nulls" on input.
BZ	BZ	BZ is used to interpret blanks as zeros on input.

D.3 Using Formats to Control I/O

When values are output to the screen or a file or any output device, think of them as being written to an **output record**. As each character or digit is written, the **record pointer** is updated to always point at the next position in the output record that will receive data. Similarly, as each character or digit is transferred from an **input record**, the record pointer always points at the next available data.

Record pointer helps explain the details of I/O from input and to output records by indicating a "current" reference point. The record pointer is analogous to a **file-pointer**, which points at the current record during an I/O operation and just before the next record after the operation for sequential I/O devices (Figure D.3).

When data is transferred according to a format, there must be a correspondence between the data being read or written and the format descriptors. Some descriptors are not used to edit data and are not considered in the correspondence such as the X descriptor. An example of how data is mapped through data editing descriptors is shown in Figure D.4.

In the OUTPUT half of Figure D.4 (p. 475), output data are mapped through descriptors that are appropriate for the data's type; they are edited and transferred to the output record. Other entities such as positional descriptors (e.g., 2X) and literal CHARACTER constants (e.g., "T =" and "J =") may be intermixed with the data editing descriptors freely.

The INPUT half of Figure D.4 shows how positional descriptors can be used to skip over some data and selectively extract relevant data. Data editing descriptors are used to specify how data from the input record are mapped into the READ's variable list. Editing descriptors must be the same type as the corresponding values in the input record and must

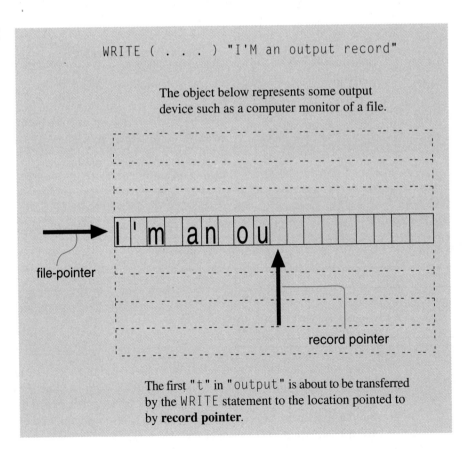

Figure D.3

Figure D.4

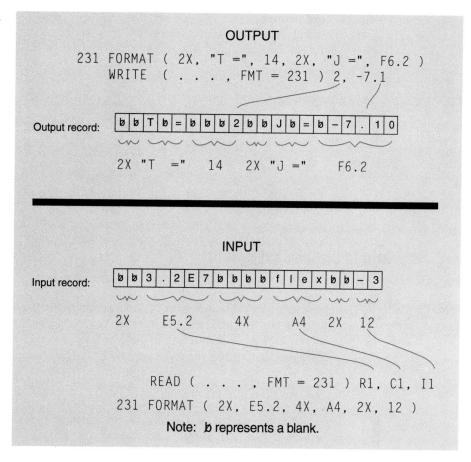

be transferred to variables of appropriate types (assume R1 is REAL, C1 is CHARACTER and I1 is INTEGER).

Important Concepts Review

- **Formats** are used to edit internal representations of data for output and interpret data in input records.

- **Format label** is a literal integer constant such that 0 < label < 100,000. Labels must appear on FORMAT statements so they can be referred to in READ, WRITE, and PRINT statements.

- FMT = is an I/O control that specifies a format label number.
- **Format descriptors** are used to edit data and change the record pointer.
- **Record pointer** points to the next position in an I/O record. For input, the record pointer points to the next available data; for output, the record pointer indicates where the next output will go.

D.4 Repeating Format Descriptors

All the format descriptors may have an optional **repeat factor** except T, TR, TL, SS, SP, P and :. Repeat factors are placed in front of descriptors and, in effect, a descriptor with a repeat factor of n causes n occurrences of the descriptor (note: n must be a positive *literal* INTEGER constant). Consider the following example of two equivalent formats:

```
77      FORMAT ( 1X, 3I5, 4F7.2 )
88      FORMAT ( 1X, I5, I5, I5, F7.2, F7.2, F7.2, F7.2 )
```

Groups of descriptors may also be enclosed in parentheses with an optional repeat factor. Again, examine a pair of equivalent formats:

```
61      FORMAT ( 1X, 2( 6X, A4, E12.6 ) )
82      FORMAT ( 1X, 6X, A4, E12.6, 6X, A4, E12.6 )
```

It is reasonable to nest repeated items as needed:

```
107     FORMAT ( 1X, 2( 2I3, 2( A4, 1X ) ) )
4001    FORMAT ( 1X, I3, I3, A4, 1X, A4, 1X, I3, I3, A4, 1X, A4, 1X )
```

If an output list has more items than a format has data editing descriptors, a new output record is started and the associated format is <u>reused</u>:

```
        WRITE ( . . . . FMT = 60312 ) 5, 6, -7, 8, -9, 10, 2
60312   FORMAT ( 3I5 ),
         . . .
        5    6   -7    <-- first output record
        8   -9   10    <-- second output record
        2              <-- third output record
```

Similarly, if the number of variables in a READ statement exceeds the number of data editing descriptors, a new record is obtained from the input source and the format is reused.

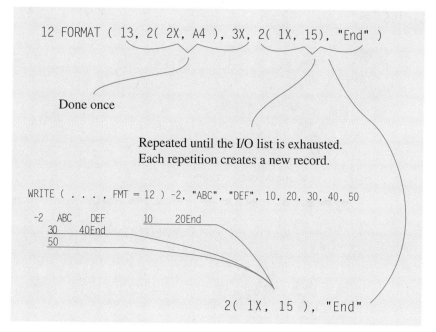

Figure D.5

When using groups of descriptors in parentheses, Fortran enforces a feature called **reversion**. Reversion is the situation where more values or variables are in an I/O list than a format can deal with so the last group of repeated descriptors is used over. Each time the group is repeated, a new I/O record is created (Figure D.5).

A noteworthy feature of the example in Figure D.5 is that the last value, 50, does not use up all the data editing format descriptors. There is still one repetition of (1X, I5) and the last "End" left. When a format is not used up, the unused data editing descriptors are simply ignored except in the case discussed in the next section.

D.5 Stopping the Output of Unwanted Strings

A problem often crops up in output formats that use character strings among data editing descriptors. The problem is that when descriptive strings *precede* data editing descriptors that edit no data, those strings are left twisting in the wind and get written to the output record when they shouldn't. This is solved with "The Terminator"—the colon descriptor (:). Figure D.6 shows how this works.

Figure D.6

```
                        PROBLEM

    WRITE (. . . , FMT = 63 ) 25
63 FORMAT ( 1X, "Pounds =", I3, 4X,"Ounces = ", I3 )
     . . .
Pounds = 25  Ounces =
```

Yuk! This particular WRITE should only display Pounds. Ounces should not be written.

```
                        SOLUTION

    WRITE (. . . , FMT = 63 ) 25
63 FORMAT ( 1X, "Pounds =", I3, :, 4X,"Ounces = ", I3 )
     . . .
Pounds = 25
```

The colon stops Ounces from being written to the output record.

D.6 Output Considerations for Descriptors

Many descriptors behave differently in output statements than they do in input statements. The next sections will show each data editing descriptor and some examples of how they work.

I, B, O, Z—Integer

I, B, O and Z are all INTEGER descriptors; they are used to edit integers into decimal, binary, octal and hexadecimal numbers, respectively. Each of the INTEGER descriptors have two forms:

$$I w \quad \text{and} \quad I w.m$$

(B, O and Z are the same)

I, B, O, Z Examples

Value	Descriptor	Output ("Δ" represents a blank)
7031	B15	ΔΔ1101101110111
7031	O8	ΔΔΔ115567
7031	Z5	Δ1B77
7031	I7	ΔΔΔ7031
1	I5.3	ΔΔ001
67012	I4	****
-902	I10	ΔΔΔΔΔΔ-902

F—Reals

F is for REAL values of fairly small magnitude. *w* is the overall *w*idth of the number and *d* is the number of places to the right of the *d*ecimal.

$$F w.d$$

F Examples

Value	Descriptor	Output ("Δ" represents a blank)
-3.2	F3.0	-3.
6849.763	F9.5	*********
7910.555	F7.2	7910.56 . . . (note the rounding)
-0.007845	F15.3	ΔΔΔΔΔΔΔΔ-0.008

Least-significant decimal places are rounded in an output record if the digit to the right of the least-significant decimal place in the output field is greater than or equal to 5. If a number is too small to fill the field width of any numeric descriptor, it is right justified and will contain leading blanks. Values too large for any numeric descriptor's field are displayed as asterisks.

E, EN, ES—Reals in Scientific Notation

Use E, EN and ES to display REAL values in scientific notation. In the second of the two forms for the E descriptor below, *e* specifies how many digits are in the *e*xponent; this must be used when displaying numbers whose exponents are greater than 999. *w* specifies the overall *w*idth of the number and *d* is the number of places to the right of the *d*ecimal.

$$E w.d \quad \text{and} \quad E w.d E e$$

EN (engineering version of E) works exactly the same as E except that EN normalizes displayed numbers so that their exponents are divisible by three and their significands are

$1 \leq$ significand ≤ 999. ES (scientific version of E) works exactly the same as E except that ES normalizes displayed numbers so that $1 \leq$ significand ≤ 9.

E, EN and ES **Examples**		
Value	Descriptor	Output ("Δ" represents a blank)
0.000001	E9.3	Δ0.100E-5
0.000001	EN9.3	Δ1.000E-6
-21.76	E15.4	ΔΔΔΔ-0.2176E+02
-21.76	ES15.4	ΔΔΔΔ-2.1760E+01
-21.76	EN15.4	ΔΔΔ-21.7600E+00
-0.71	EN15.4	ΔΔ-710.0000E-03

SS, SP—Sign Display

SS and SP are used to control whether a plus sign is shown in front of positive numbers or not. SP (sign plus) turns on the plus sign display, which remains in effect during an I/O operation unless an SS (suppress sign) descriptor is used to turn off SP for any remaining numeric descriptors.

SS, SP **Examples**		
Values	Descriptor	Output ("Δ" represents a blank)
7, -9, 0	SP, 3I3	Δ+7Δ-9ΔΔ0
8, 5, 2, 6	I4, SP, I5, SS, I2, SP, I7	ΔΔΔ8ΔΔΔ+5Δ2ΔΔΔΔΔΔ+6

P—Scale Factor

P is used to scale REAL values as they are written to an output record. When used with E, the value of the output is the same; only the appearance is changed. The decimal is shifted n places if the scale factor is n (n MUST be an INTEGER **literal constant**—no named constants). If n is positive, the decimal is shifted to the right; the decimal is shifted to the left if n is negative. P works like SS and SP insofar as when a scale factor is turned on, it remains in effect during an I/O operation unless a different scale factor is met.

The P descriptor has no effect on EN and ES but DOES affect numbers displayed with the F descriptor. The *magnitude* of a number being displayed with an F descriptor IS CHANGED in the output record—beware!

*n*P Examples

Value	Descriptor	Output ("Δ" represents a blank)
-13.45	3P, F9.2	-13450.00
2.0, 3.1, -5.8	2P, F4.0, 0P, 2F4.1	200.Δ3.1-5.8
-14.3E5	6P, ES12.4	Δ-1.4300E+06
.00908	-5P, EN12.4	ΔΔ9.0800E-03
510000.	2P, E12.5	51.00000E+04
-430967.	-3P, E17.4	ΔΔΔΔΔ-0.0004E+09

A—Character

There are two forms for the A descriptor:

$$A \quad \text{and} \quad Aw$$

A alone facilitates outputting character strings of unknown length. When A is used with a field width, *w*, character values that are too big for the field are sent to the output record from character 1 through as many characters as the field can hold. Character values smaller than *w* characters are right justified in the field.

A Examples

Value	Descriptor	Output ("Δ" represents a blank)
"ΔABΔΔ"	A7	ΔΔΔABΔΔ
"ABCDEFG"	A4	ABCD
"A"	A4	ΔΔΔA
"Hello"	A	Hello
"SeeΔyouΔlater"	A	SeeΔyouΔlater

L—Logical

L*w* allows LOGICAL data to be printed. Either a T or an F is right justified in the *w* field.

D.7 Carriage Control

When Fortran sends information to certain output devices such as printers, it uses the first character of a record to control vertical spacing. This is particularly useful, for example, to double-space reports or move to the top of a new page to present different information on a printout. *This only works with formatted I/O*; list-directed I/O is single-spaced by definition. The important control characters are listed below.

Character	Vertical Spacing before Printing
Blank	One line
0 (Zero)	Two lines
1 (One)	First line of next page
+ (Plus symbol)	No advance

One of the problems that can arise is not realizing when one of these control characters is accidentally used. Below is an example of just such a problem.

```
      WRITE ( . . . FMT = 12 ) 123, 109, 188
12    FORMAT ( I3 )
              . . .
```

The problem here is that the first digit printed in each of the three numbers is 1. Output would be as follows:

```
                         23
                        . . .
                          9
                        . . .
                         88
```

Each of the numbers above has had its leading 1 used up for carriage control. In addition, each of the numbers above would be at the top of separate pages! The common way to deal with this situation, assuming that single-spaced output is the general goal, is to start every output format to a printer-like device with 1X. This effectively offers a blank for the carriage control character. FORMAT 12 above can be fixed as seen below.

```
12    FORMAT ( 1X, I3 )
```

D.8 Input Considerations for Descriptors

The following sections will discuss the way data editing descriptors interpret data in an input record. As with output, data being worked on by a given data editing descriptor must correspond in type to the data descriptor. And the data editing descriptors must map their data onto variables of appropriate types.

E, EN, ES, F—Reals in Scientific Notation

All the data descriptors for REALs interpret data from an input record the same way. Interpretations depend on the data, which comes in four forms (Table D.5).

Table D.5

REALs in an Input Record	
Description	Example
Scientific notation *without* a decimal in the mantissa.	123456E-07 or 123456-07 or 123456D-07
Scientific notation *with* a decimal in the mantissa.	1234.56E-07 or 1234.56-07 or 1234.56D-07
Nonscientific notation *without* a decimal.	-7301238
Nonscientific notation *with* a decimal.	-7301.238

Field *w*idth (e.g., E*w*.d) in an input format descriptor designates how many adjacent positions of an input record hold a given datum. An F12.6 descriptor, for example, will take data from 12 contiguous positions of an input record.

When there is no decimal in a datum being processed by a REAL input descriptor, the ".*d*" part (e.g., F*w*.*d*, E*w*.*d*, EN*w*.*d* or ES*w*.*d*) of the descriptor determines where the decimal is placed. For example, if -7301238 is read from an input record under an F8.3 descriptor, the datum would be interpreted as -7301.238; the rightmost ".*d*" digits in the field become the fractional part. If a decimal exists in a datum, the ".*d*" part of the descriptor is overridden. For instance, if -73.01238 is read with an E9.1 descriptor, the ".*d*" (.1) is ignored and the number is interpreted as -73.01238 and transferred as such to the input variable. Exponent parts such as E-09, D-09 and -09 are all interpreted the same way. Each of the example forms represents 10^{-9}, which is used to give the mantissa its magnitude.

I, B, O, Z—Integer

I, B, O and Z are all INTEGER descriptors that are used to interpret INTEGER data from an input record. I, B, O and Z are for decimal, binary, octal and hexadecimal numbers, respectively. Each of the INTEGER descriptors has one input form:

I*w*, B*w*, O*w*, Z*w*

As always, *w* refers to the *w*idth of the input field. When using B, O or Z, the corresponding values in the input record must be valid binary, octal or hexadecimal numbers, respectively.

P—Scale Factor

P is used in an input format to scale REAL values as they are read from an input record. If the datum in an input record is in scientific notation form, the P descriptor has no effect. Normal REAL data (not in scientific notation) are effectively multiplied by 10^n where *n* is the literal `INTEGER` constant scale factor (nP). If a scale factor is put into effect in a format, it remains in effect until either the I/O operation is complete or another scale factor is met.

A—Character

There are two forms for the A descriptor:

$$\text{A} \quad \text{and} \quad \text{A}w$$

A alone facilitates inputting up to an entire record depending on the declared size of the corresponding variable. Characters are transferred from the left of the input record to the left of the variable one by one until the variable is full. If there aren't enough characters to fill the variable, it's blank filled on the right.

Use the Aw form to get data from a specific field of characters. If a variable's declared size is smaller than w, the rightmost characters in the input field are used to fill the variable. If the variable's declared size is larger than the field $w$idth, the data is transferred into the left of the variable and it's blank filled.

L—Logical

Lw allows `LOGICAL` data to be transferred into a program from an input record. For L1, the only valid input data is T or F. For wider fields, the T or F may be preceded by optional blanks and then an optional period. Characters after the T or F are irrelevant.

Important Concepts Review

- **Repeat factors** repeat descriptors or groups of descriptors. Repeat factors must be positive, literal `INTEGER` constants.

- **Reversion** is where more values or variables are in an I/O list than a format's descriptors can deal with, so the last group of repeated descriptors is used repeatedly until the I/O list is exhausted.

- : is used to prevent displaying unwanted strings in a format.

- **Character** constants may be used freely in formats. The strings must be literal `CHARACTER` constants—no named constants.

Appendix E
I/O Controls

List-Directed I/O

Table E.1

List-Directed Input	List-Directed Output
`READ *, . . .` `READ ( UNIT = *, FMT = * ) . . .`	`PRINT *, . . .`

Controlled I/O

Table E.2

Controlled I/O Statements	
Statement	Action
`OPEN (controls-list)`	Prepares a file for use and describes its status and structure. A `UNIT` number is also associated with the file for reference in other I/O statements.
`CLOSE (controls-list)`	Breaks the `UNIT` number connection, writes an End File record and releases the file to the operating system.
`READ (controls-list)`	Gets records from an input source according to the specifications in the controls-list.
`WRITE (controls-list)`	Puts records to an output device according to the specifications in the controls-list.
`INQUIRE (controls-list)`	Gathers information about a file according to the controls-list.
`BACKSPACE (controls-list)`	This is a "file-positioning" statement. It allows a programmer to "back up" so previous records can be accessed.
`REWIND (controls-list)`	This is a "file-positioning" statement. It positions a file just before the first available record.

Table E.3

Basic Controls for OPEN (defaults, if any, are underlined)	
Control	Explanation
UNIT = u	u is an integer expression such as that $0 \leq u$. Unit numbers represent a connection to the file named in the FILE = control. For example, if a file is opened with unit number 37, other I/O statements can access that file by specifying the control UNIT = 37.
FILE = f	f is a CHARACTER expression that specifies the name of the file being opened.
ACTION = a	a is a CHARACTER expression that specifies the way the file will be used. Acceptable values are, "READ", "WRITE" and "READWRITE". These options correspond to I/O modes: input, output and input/output. The default is processor dependent.
STATUS = s	s is a CHARACTER expression that specifies the file's status. Acceptable values are "OLD", "NEW", "REPLACE", "<u>UNKNOWN</u>" and "SCRATCH".
POSITION = p	p is a CHARACTER expression that specifies whether the file should be opened at the beginning or end. Useful values are "APPEND" and "REWIND". Use "APPEND" to add more records to the end of an existing file. If this control is omitted, the file is positioned at the beginning.

Table E.4

Advanced Controls for OPEN (defaults, if any, are underlined)	
Control	Explanation
ACCESS = a	a is a CHARACTER expression that specifies the kind of file. Acceptable values are "<u>SEQUENTIAL</u>" and "DIRECT".
FORM = f	f is a CHARACTER expression that specifies whether a file is formatted or unformatted. Values include "FORMATTED" or "UNFORMATTED". For sequential files the default is "FORMATTED". For direct access files the default is "UNFORMATTED".
RECL = r	r is an INTEGER expression that defines the maximum length of records.
IOSTAT = v	v is an INTEGER variable that is given a value by this control. The value is a code that describes how I/O operation worked . . . Successful? . . . Unsuccessful? . . . What went wrong?
BLANK = b	b is a CHARACTER expression that defines how blanks are interpreted in a data file. Fortran can consider them as zeros or "nothings" called *nulls* that have no effect on input. Acceptable values are "<u>NULL</u>" and "ZERO".
DELIM = d	d is a CHARACTER expression that defines how CHARACTER constants are delimited in a data file. Values include "APOSTROPHE", "QUOTE" or "<u>NONE</u>".
PAD = p	p has a CHARACTER value of either "<u>YES</u>" or "NO".

Table E.5

Control	Controls for CLOSE (defaults, if any, are underlined)
	Explanation
UNIT = u	u is an INTEGER expression that specifies which file to close.
STATUS = s	s is a CHARACTER expression that specifies whether the file is saved or deleted. Values include "KEEP" or "DELETE".
IOSTAT = v	v is an INTEGER variable that is given a value by this control. The value is a code that describes how the I/O operation worked.

Table E.6

Control	Controls for READ (defaults, if any, are underlined)
	Explanation
UNIT = u	u is an INTEGER expression that specifies which file to read from.
FMT = f	f is a *label* number: $0 < f < 1000000$ or "*". If f is an asterisk, list-directed I/O is used; otherwise, f refers to a FORMAT that defines how incoming data is interpreted.
IOSTAT = v	v is an INTEGER variable that is given a value by this control. The value is an error code that describes how the I/O operation worked. Was it successful or not? If not, why?
REC = n	n is an INTEGER expression that identifies which record of a direct-access file is to be read.
ADVANCE = a	a is a CHARACTER expression that is either "YES" or "NO". This is useful for reading a record piece by piece. This control can be used with formatted input.

Table E.7 Controls for WRITE (defaults, if any, are underlined)

Control	Explanation
UNIT = u	u is an INTEGER expression that specifies which file will receive the output.
FMT = f	f is a *label* number: $0 < f < 1000000$ or "*". If f is an asterisk, list-directed I/O is used; otherwise, f refers to a FORMAT that defines how outgoing data is edited.
IOSTAT = v	v is an INTEGER variable that is given a value by this control. The value is a code that describes how the I/O operation worked.
REC = n	n is an INTEGER expression that identifies which record of a direct-access file is to be written.
ADVANCE = a	a is a CHARACTER expression that is either "<u>YES</u>" or "NO". This is useful for writing to the same record more than once. This control can only be used with formatted output.

Table E.8 Controls for INQUIRE

Control	Explanation
IOSTAT = v	INTEGER, same as defined previously.
EXIST = e	LOGICAL, .TRUE. if the file exists.
OPENED = o	LOGICAL, .TRUE. if the file is open.
NUMBER = n	INTEGER, unit number of the file or -1 if not open.
NAME = n	CHARACTER, name of the file associated with unit n.
SEQUENTIAL = s	CHARACTER, "YES", "NO" or "UNKNOWN".
DIRECT = d	CHARACTER, "YES", "NO" or "UNKNOWN".
ACCESS = a	CHARACTER, "SEQUENTIAL", "DIRECT" or "UNDEFINED" if the file is not connected.
FORM = f	CHARACTER, "FORMATTED", "UNFORMATTED" or "UNDEFINED".
FORMATTED = f	CHARACTER, "YES", "NO" or "UNKNOWN".
UNFORMATTED = u	CHARACTER, "YES", "NO" or "UNKNOWN".
RECL = r	INTEGER, maximum record length.
NEXTREC = n	INTEGER, last record read or written + 1.
BLANK = b	CHARACTER, "NULL", "ZERO" or "UNDEFINED" if the file is not connected or is unformatted.
POSITION = p	CHARACTER, "REWIND", "APPEND" or "UNDEFINED" if the file is not connected or is direct access.
ACTION = a	CHARACTER, "READ", "WRITE", "READWRITE" or "UNDEFINED" if the file is not connected.
READ = r	CHARACTER, "YES", "NO" or "UNKNOWN".
WRITE = w	CHARACTER, "YES", "NO" or "UNKNOWN".
READWRITE = r	CHARACTER, "YES", "NO" or "UNKNOWN".
DELIM = d	CHARACTER, "QUOTE", "APOSTROPHE", "NONE" or "UNDEFINED" if the file is not connected.
PAD = p	CHARACTER, "YES" or "NO".

Quick Reference Syntax Guide

PROGRAM UNITS

PROGRAM program-name
    [USE module-name]
    [IMPLICIT NONE]
    [declarations]
    [main-program-statements]
END [PROGRAM [program-name]]

SUBROUTINES subroutine-name [([argument-list])]
    [USE module-name]
    [IMPLICIT NONE]
    [declarations]
    [subroutine-program-statements]
RETURN
END [SUBROUTINE [subroutine-name]]

[RECURSIVE] FUNCTION function-name [([argument-list])]
[RESULT(result-name)]
    [USE module-name]
    [IMPLICIT NONE]
    [declarations]
    [function-program-statements]
END [FUNCTION [program-name]]

MODULE module-name
    [module-specification-stmts]
[CONTAINS module-subprograms]
    [module-subprograms-specifications]
END [MODULE [module-name]]

IDENTIFIERS

letter [identifier-character] . . .
Identifier-character is defined as a letter, digit or "_" (underscore). Identifiers may be up to 31 characters long.

CALL

CALL subroutine-name[([argument-list])]

VARIABLE DECLARATIONS
```
type [ (kind) ] [ ,attribute ] . . . :: variable-list
```

`Type` is one of the following: `INTEGER, REAL, DOUBLE PRECISION, COMPLEX, CHARACTER, LOGICAL, TYPE (type-name)`.
`Attribute` may be: `PARAMETER, INTENT ( IN ), INTENT ( OUT ), INTENT ( INOUT ), DIMENSION (bounds-specifications), PRIVATE, PUBLIC, SAVE, ALLOCATABLE, POINTER, TARGET`.
`Variable-list` is a list of valid Fortran identifiers that may be optionally assigned values. If the type definition contains the attribute, `PARAMETER`, each of the identifiers in the variable list is a *named constant* and *must* be assigned a value.

DERIVED DATA TYPES
```
TYPE type-name
    type-declarations
END TYPE type-name
```

SELECTION STRUCTURES
Block `IF`:
```
IF ( condition ) THEN
    block-1
[ ELSE
    block-2 ]
ENDIF
```

Multialternative `IF`:
```
IF ( logical-expression-1 ) THEN
    block-1
[ ELSE IF ( logical-expression-2 ) THEN
    block-2 ] . . .
[ ELSE
    block-if-all-other-conditions-fail ]
END IF
```

Select Case:
```
SELECT CASE ( test-expression )
   [ CASE ( set-expression-1 )
       block-1 ] . . .
   [ CASE DEFAULT
       block-n ]
END SELECT
```

Where:
```
WHERE ( array-logical-expression )
    array-assignments
[ ELSEWHERE
    array-assignments ]
END WHERE
```

LOOPING STRUCTURES

Conditional Loop:
```
DO WHILE ( condition )
    block-of-statements
END DO
```

Fixed-Limit Loop:
```
DO index-variable = initial-value, limit [ , increment ]
    block-of-statements
END DO
```

Implied `DO`:
```
[ ( ] . . . ( list,ix = init, lmt, [ incr ] ) [ ,ix-2 = init-2, lmt-2, [ incr-2 ] ) ] . . .
```
(Note: that `ix-2` *means the second "*`ix`*" variable, not "*`ix` *variable minus 2"*)

Optional parentheses on the left above, which are indicated by, "`[ ( ] . . .`", are used to balance any parentheses that are introduced on the right by adding additional "nests" to the implied DO.

DATA TRANSFER
`DATA variable-list / data-items /`

Where `variable-list` is a set of already declared variables and individual `data-items` each take one of the following forms:

`constant`

`named-constant`

`repeat-factor*constant`

`repeat-factor*name constant`

`OPEN ( controls-list )`

Where `controls-list` may be any appropriate combination of the following:

UNIT $=u$, FILE $=f$, ACTION $=a$, STATUS $=s$, POSITION $=p$, ACCESS $=a$, FORM $=f$, RECL $=r$, IOSTAT $=v$, BLANK $=b$, DELIM $=d$, PAD $=p$

`CLOSE ( controls list )`

Where `controls-list` may be any appropriate combination of the following:

UNIT $=u$, STATUS $=s$, IOSTAT $=v$

INQUIRE (controls list)
   Where `controls-list` may be any appropriate combination of the following:

$$\text{IOSTAT} = v, \text{EXIST} = e, \text{OPENED} = o, \text{NUMBER} = n, \text{NAME} = n, \text{SEQUENTIAL} = s, \text{DIRECT} = d,$$
$$\text{ACCESS} = a, \text{FORM} = f, \text{FORMATTED} = f, \text{UNFORMATTED} = u, \text{RECL} = r, \text{NEXTREC} = n, \text{BLANK} = b,$$
$$\text{READ} = r, \text{WRITE} = w, \text{READWRITE} = r, \text{DELIM} = d, \text{PAD} = p$$

READ (controls list) variable-list
   Where `controls-list` is an appropriate combination of the following and `variable-list` is a set of declared variables that will receive the input values.

$$\text{UNIT} = e, \text{FMT} = f, \text{IOSTAT} = v, \text{REC} = n, \text{ADVANCE} = a.$$

WRITE (controls list) [output-list]
   Where an `output-list` is a list of values to be printed and `controls-list` may be an appropriate combination of the following:

$$\text{UNIT} = e, \text{FMT} = f, \text{IOSTAT} = v, \text{REC} = n, \text{ADVANCE} = a$$

BACKSPACE (controls list)
   Where `controls-list` is an appropriate combination of the following:

$$\text{UNIT} = u, \text{IOSTAT} = s$$

REWIND (controls list)
   Where `controls-list` is an appropriate combination of the following:

$$\text{UNIT} = u, \text{IOSTAT} = s$$

FORMAT
   label-number FORMAT (format-descriptors)
     Where `format-descriptors` are any of the following:

$$\text{I, B, O, Z, F, E, EN, ES, SP, SS, P, A, L, X, T, TR, TL, /, G, :, BN, BZ}$$

And `label-number` is a positive integer whose value is: $0 < \text{label-number} < 10000$.

Glossary

!	starts a comment line.
% operator	allows a programmer to access individual elements of a derived data type.
&	enables a line to be continued on the next line.
:	used to prevent displaying unwanted strings in a format.
::	separates a type description from variables that will conform to that description.
Absolute cell replication	cell references that have a dollar sign preceding the row or column designator such as $J$67 don't change when copied—they refer to an "absolute" location on a spreadsheet.
Active cell	where the next entry will be stored in a spreadsheet.
Actual arguments	arguments in a `CALL` statement.
Algorithm	a method of solution . . . a step-by-step solution to a problem.
Allocatable arrays	arrays whose size and dimensionality are defined and redefined as a program runs.
Alpha testing	the first testing done outside a software development group.
Anchor point	used to create or modify spreadsheet series.
Antibugging	to adopt design techniques that prevent bugs from being introduced into a program.
Arguments	the way data flows in and out of subroutines.
Arithmetic order of precedence	the order that arithmetic operators are performed. Refer to Table 4.3.
Array constructors	allow arrays with specific values to be constructed.
Array expressions	meaningful combinations of conformable arrays, simple variables, constants, and operators.
Array logical expressions	formed with relational operators, logical operators and conformable operands.

Array section — a collection of array elements forming a subset of an array.

Array section reference — accomplished in multidimensional arrays by specifying an array section reference for any set of declared bounds.

```
( lower : upper [ : stride ] )
( upper : lower [ : -stride] )  ! negative stride
```

Arrays — allot multiple storage locations under one variable name.

Assignment statement — the expression on the right side of the equal sign is evaluated and that value is then stored into the variable on the left side of the equal sign.

Assumed-shape arrays — *formal arguments* whose bounds may change each time their program unit is used; assumed-shape arrays adopt the bounds of their actual arguments.

Attribute — a list of Fortran **attributes** that define any additional special characteristics for *type*.

Automatic arrays — arrays created to be conformable with other arrays. An automatic array is never a formal argument, but a conformable clone of another array.

`BACKSPACE` — moves the file-pointer to the previous record of a file when the file-pointer is positioned before the next available record. If file-pointer is at the first record, `BACKSPACE` has no effect.

Beta testing — the second round of off-site testing for software.

Binary operator — an operator with two operands.

Binary search — a very efficient way to search *sorted* data; data are found by repeatedly reducing the sorted list by half.

Bit — a binary digit that is either **on** (one) or **off** (zero).

Black box — a library program unit whose purpose and interface are all that needs to be understood to use the program unit. It is unimportant for a programmer to understand **how** a black box works.

Blank lines and indentations — make structures more obvious and the program easier to understand.

Block	a collection of any combination of the four basic flow-control structures: (1) `CALL`, (2) Sequential, (3) Selection, (4) Looping.
Block `IF`	conditionally executes blocks of statements.
Block statements	operate on blocks (block `IF`, `DO WHILE`, etc.)
Bounds	the limits of a dimension of an array.
BOZ constants	may be used in `DATA` statements or read with BOZ formats to assign binary, octal or hexadecimal values to integer variables. BOZ values can also be written using BOZ formats.
Breakpoint	a place in a program where a programmer abruptly stops program execution to examine values.
Bugs	errors in a program.
Byte	composed of eight bits.
Called program unit	a program unit that is called.
Calling program unit	the program unit that called the currently active program unit.
Cell	a spreadsheet container that can hold a number, label or formula. Cells are identified by the intersection of a column (designated by a letter) and a row (designated by a number).
Cell attributes	used to enhance the appearance of cell contents.
Character truncation	more characters are assigned to a variable than it can hold; the rightmost characters that don't fit will be lost.
`CLOSE`	breaks the unit-number/file connection. After a `CLOSE` statement, the unit number's file is no longer accessible unless it is opened again with an `OPEN` statement.
Collating sequence	a system where common symbols, such as those on a keyboard, are associated with numbers.
Columns	vertical stacks of adjacent cells.
Combinations	$_nC_k = \dfrac{n!}{(n-k)!k!}$
Comment blocks	occur at the beginning of the main program and every other program unit. A comment block provides a brief description of the purpose of the program unit and descriptions of all variables. For main programs, they include the author's name, date written (or modified), etc.

Comments	notes within a program that explain and clarify a program. A comment starts with an exclamation point.
Compilers	translate programs written in high-level languages into machine language.
Compiling	the process where source code (Fortran program statements) is converted to machine language.
Concatenation	attaching two strings together. The concatenation operator is //.
Condition	an expression that has a value of either true or false and may be tested by a selection structure or conditional looping structure.
Conditional loops	continue to iterate as long as their condition is satisfied.
Conformable arrays	have the same shape.
Constant	an unchanging instance of a Fortran type; it remains constant. For example, 2.3 is an instance of Fortran's REAL type; it is a constant.
Controlled I/O	gives a program designer complete control over the destination and appearance of data.
Controls	customize I/O operations.
Controls-list	a set of controls in an I/O statement.
Creating a series	create a linear extrapolation of two or more cell entries.
Custom dot operators	created with modules and functions.
Data	information that is required by a program, generated by it, or both.
Data file	a collection of related data.
DATA statements	allow variables to be initialized at compile time.
Database managers (DBMs)	store large amounts of data and allow easy flexible searching of it.
Deallocate	to recover allocated memory from a program and return that memory to the computer's memory pool.
Debugging	the act of removing errors from a program.
Debugging PRINT statements	PRINT statements used to display variables' and expressions' values to help isolate runtime and logic errors. Good programming style dictates that these PRINT statements be placed at the extreme left of all other program statements.

Declarations	create and define program variables and named constants, and allow data initializations.
Default input	comes from the keyboard, touch screen or voice recognition system.
Default output	goes to the computer screen/monitor.
Delimiter	a special character used to separate values. The most common data file delimiters are **space**, **comma** and **TAB**.
Derived data type arrays	arrays whose elements are of some derived type.
Derived data types	types that are defined by the programmer and are composed of simple Fortran data types or simple Fortran data types and/or other derived data types.
`DIMENSION`	attributes that specify array bounds.
Dimensionality	the number of declared sets of bounds of an array. `INTEGER, DIMENSION ( 1 : 4, 6 : 9 )` ... is a two-dimensional array.
Direct-access files	high-performance files whose records are *directly* accessible by specifying a **record number**.
`DO WHILE`	a looping structure that enables a programmer to repeat a block of statements while a condition is true.
Dynamic allocation	memory being allocated to a program's variables as it runs.
`ELSE`	optional in block `IF`s and multialternative `IF`s. If `ELSE` is included, its block is executed if all other conditions in the selection structure fail.
E-mail	an efficient way to communicate with other computer users where computers act as the postmaster and mail carriers.
`END DO`	the physical bottom of a loop, which returns control to the top of the loop.
End File record	a special record written at the end of every file by the operating system.
Enhancements/ refinements	improvements in program designs made when the program is tested.
Error checking	the act of preventing erroneous data from entering a program.
Error messages	compiler messages designed to guide programmers to a program's problems (ha! ha!).

Error trapping	the act of creating algorithms that deal with error situations without the program crashing as a result of the error.
Field	a subdivision of records containing data of a certain type and classification. Records often contain several fields.
File-pointer	an entity maintained by operating systems that indicates the next available record to read or where the next record will be written.
File transfer programs	make it easy to transfer information from one person to another through computer networks.
Five-Paragraph Essay Approach to Program Design	a formulized, methodical approach to breaking down a problem into its component parts so a modular solution can be written.
Fixed format	data files' data are in fields aligned in columns.
Fixed-limit DO loops	eliminate the necessity of manually incrementing a loop counter variable and having to figure out a logical expression to terminate the loop.

```
DO index-variable = initial-value, limit [, increment]
    Fortran statements
END DO
```

Flag	a specialized value that can be "watched" for by a conditional loop or a selection structure.
Flags	in data files, special values that signal the end of data.
Flow of data	the movement of data in and out of program units.
Formal arguments	in the subroutine header, formal arguments *represent* actual arguments. Formal arguments that are arrays must be declared with the DIMENSION attribute.
Format descriptors	used to edit data.
Format label	a literal integer constant such that $0 <$ label $< 100,000$. Labels must appear on FORMAT statements so they can be referred to in READ, WRITE and PRINT statements.
Formats	used to edit internal representations of data for output and interpret data in input records.
Formula area	create/edit spreadsheet formulas or values.
Formulas	spreadsheet expressions composed of **operands** and **operators**.

Free-format data file	usually has variable-length lines with values separated by some special delimiter.
Friendly computer programs	use techniques such as prompting messages to minimize confusion for the user.
Function	a specialized variable whose value is derived from the function's algorithm operating on any accompanying argument(s). For example, `LOG10 (X)` is a function that calculates $\log_{10}(x)$. `LOG10`, the function name, is a variable and is set to the value of $\log_{10}(x)$.
Function evaluation	done before any other operations in an expression.
Generated errors	error messages that point out errors that aren't really errors. The messages are displayed because the compiler becomes confused.
Hexadecimal	base-16 numbering system that uses the digits 0, 1, 2, 3, 4, 5, 6, 7, 8, 9, A, B, C, D, E and F. A byte is composed of two hexadecimal digits.
High-level language	an understandable (written in somewhat English terms) computer language that follows strict rules and definitions and requires translation into machine language.
I/O	means **I**nput/**O**utput.
I/O list	a set of variables being read, or values, variables, and/or expressions being written.
I/O modes	define the direction data is moved to and from programs: input—data flows into a program; output—data flows out from a program; input/output—data can be transferred in and out of a program.
Identifier	a programmer-supplied name that should be well chosen to imply its use. An identifier must start with a letter and may be followed by any combination of letters, digits or underscores and be up to 31 characters long.
IF (logical-expression, value-if-true, value-if-false)	a spreadsheet function that displays different values in a cell depending on a logical expression.
`IF (question)  THEN . . .  ELSE . . . ENDIF`	one of Fortran's selection structures.
`IMPLICIT NONE`	turns implicit typing off (Yea!).

Implicit typing	a language feature (a B-A-D feature) that gives undeclared variables a default type. Variables starting with `I`, `J`, `K`, `L`, `M` or `N` are `INTEGER` and all others are `REAL`. We hate this.
Implied DO loops	abbreviated one-line looping structures used in data-transferring statements such as `PRINT` and `READ`.

`( value, index = initial-value, limit[, increment] )`

Importing data	bringing an external data set into a spreadsheet for analysis.
Inactive cells	cells other than the active cell.
`Increment`	the amount added to `index-variable` each time a fixed-limit `DO` loop iterates.
Incremental assignment statement	adds a value to a numeric variable: $j = j + 2$
`Index-variable`	changes by `increment` each time a fixed-limit loop iterates. DON'T change `index-variable` within a fixed-limit `DO` loop!
Infinite loop	a loop whose condition never becomes false.
Informational region	which cell is active.
`Initial-value`	`index-variable`'s first value in a fixed-limit `DO` loop.
Input	the data coming into a program.
`INQUIRE`	allows a program to find out information about a file such as whether it is open or even exists.
Insecticide	the set of steps a programmer follows to remove (kill) errors (bugs).
Instructions	cause a computer to take some action (e.g., print a value, perform a calculation, etc.)
Integer division	one integer operand divided by another; any decimal part will be **truncated**.
`INTENT ( IN )` arguments	used for data flowing into a program unit and may not be altered within the called program unit.
`INTENT ( INOUT )` arguments	supply useful values to a subroutine and may be modified by it — `INTENT ( INOUT )` arguments provide a two-way street for flow of data.
`INTENT ( OUT )` arguments	become undefined upon entering a subroutine. This makes unset `OUT` arguments easy to spot because any

	attempt to use them in the calling program will result in a program crash.
Interface	a black box's name and the order, purpose and type of required arguments.
Interface operator block	used to make a connection between a dot operator, `.dot-operator-name.`, and the function that defines how the operator works; the dot operator will invoke its corresponding function.
Intrinsic functions	built in to the compiler and available to all program units without declaring or describing them in any way.
`IOSTAT`	can be used to perceive when an attempt is made to read an End File record. `IOSTAT` is also used in any other I/O statement to perceive an error condition . . . one of `IOSTAT`'s great strengths is that it can report an error without the program crashing.
Iterate	to repeat.
Kind	a suffix that is used with some types to define *type* further.
Labels	strings of characters that can be stored in a spreadsheet cell.
Left justified and blank filled	means that if fewer characters are assigned to a variable than it can hold, the characters will be placed at the left of the variable and the rest of the variable will be filled with blanks.
Libraries	collections of reuseable program units that save time when writing programs by providing already-written, working subroutines.
Limit	the maximum value for `index-variable` in a fixed-limit `DO` loop.
Limited resource debugging	find and correct as many errors as possible between runs by using a program listing.
Line commenting	clarifies individual lines of a program.
Linking	the process whereby libraries are searched to supply any missing routines for a program and an executable (ready-to-run) version of the program is created.
List-directed I/O	input or output that is controlled by an I/O list; Fortran decides everything based on the types of variables/values in an I/O statement.

Local variables	variables that are needed within a program unit but aren't supplied or needed by the calling program.
Logic errors	a program successfully runs to completion but produces wrong or no answers.
Logical constants	`.TRUE.` and `.FALSE.`.
Logical expressions	conditions created with operands, relational operators and logical operators.
Logical expressions (complex)	use logical operators.
Logical expressions (simple)	don't use logical operators.
Logical operators	`.NOT.`, `.AND.`, `.OR.`, `.EQV.` and `.NEQV.`.
Loop conditions	logical expressions constructed the same as `IF` statement logical expressions; used to control how many times a loop will iterate.
Looping structures	repeat program instructions over and over in a controlled way.
Machine language	the only language a computer understands.
Main program	used to "drive" the program by calling subroutines.
Masking operation	a high-level language facility where only certain elements of an array are acted on based on an *array logical expression*.
Menubar	offers various options for manipulating a spreadsheet.
Mixed-mode expressions	ones where some operands are type `INTEGER` and others are of the real types (`REAL` or `DOUBLE PRECISION`).
Mnemonic names (identifiers)	names that imply the purpose. A variable named `Square_Root` that holds the square root of some number has a mnemonic variable name.
Modular program designs	use subroutines to implement each block of a structure chart.
Modules	Fortran program units that contain data definitions and can also be used to create functional objects.
Multialternative `IF`s	chain conditions together in an "either-or" fashion. Only one block will be executed at most. If more than one condition is true, only the block corresponding to the first true condition will be performed.

Multidimensional arrays	arrays that have more than one set of declared bounds.
$n!$	$1 \cdot 2 \cdot 3 \cdot \ldots \cdot (n-1) \cdot n$
Nested DO loops	loops that are contained within other loops. This kind of looping structure arrangement is very useful for efficiently accessing the contents of multidimensional arrays.
Nested implied DO loops	one-line looping structures that are used in data-transferring statements such as READ, WRITE, PRINT, DATA, etc.

`[(]...(list,ix=init,lmt,[incr])[,ix-2=init-2,lmt-2,[incr-2])]...`

Network	an interconnection of computers that makes sharing resources and electronic communication possible and convenient.
No Record condition	happens when an attempt is made to read a nonexistent record in a direct-access file.
Nonnumeric variables	allow other types of data to be stored and used in a program (LOGICAL, CHARACTER, etc.)
Numeric variables	allow numeric data to be stored and used in a program.
Object library	the compiled machine language version of a source library.
One-dimensional arrays	only require one subscript to uniquely identify one array position.
OPEN	prepares a file for use and describes various physical features of and access information for the file. In addition, the OPEN statement associates a UNIT number with the file. The UNIT number is used in READ, WRITE and other I/O statements to identify which file is to be accessed.
Operands	an item operated on by operators. An operand can be a constant, a variable or an expression contained in parentheses.
Operating system	a collection of sophisticated computer programs that manage the basic functions of a computer.
Operator modes	the different results operators will generate based on the *types* of operands involved in an expression. Refer to Table 4.4.

Operator order of precedence, logical expressions	(1) relational operators, (2) `.NOT.`, (3) `.AND.`, (4) `.OR.`, (5) `.EQV.` and `.NEQV.`.
Operators	perform some operation on operands. In the expression *a/b*, the "/" is a division operator that performs division on operands *a* and *b*.
Output	the results generated by a program.
Overloading function/ subroutine	using one generic interface to invoke one of several program units. The particular program unit invoked is determined by the type(s) of actual argument(s) in the generic interface.
Overloading intrinsic operators	gives the operator (such as "+") extended meaning when its operands are of a derived type.
Parallel arrays	a method of working with related data that is stored in different arrays.
Parameter	a named constant.
Parentheses	override arithmetic order of precedence. Parentheses may be nested as necessary and the innermost set will be evaluated first.
Permutations	$_nP_k = \dfrac{n!}{(n-k)!}$
Pointing device	a mouse or trackball and used to activate spreadsheet features.
`PRINT`	displays information on the screen.
Program listing, also called a **listing file**	source code (Fortran program statements) with any error messages from the compiler.
Program flow	the order that a computer executes (performs) a program's instructions.
Program stubs	simple, incomplete subroutines that allow the development of calling program units without worrying about the details of the called programs units.
Program style	the way a programmer presents her program.
Program unit	a subroutine, main program, function or module.
Prompting messages	tell users that they must take some action such as entering values from the keyboard.
`READ ( ... )`	enables program designers to get input from any "unit." Data obtained are stored into variables in the `READ` statement.

`READ *,`	temporarily stops the execution (performing of instructions) of a program so the user can enter values from the keyboard.
Record	a subdivision of files usually consisting of one line of a file.
Record pointer	points to the next position in an I/O record. For input, the record pointer points to the next available data; for output, the record pointer indicates where the next output will go.
Register a cell entry	means to finally enter spreadsheet cell contents; delete any old entry.
Relational operators	==, /=, >, <, >=, <=.
Relative cell replication	when copying formulas, cell references change relative to the location of the cells being copied.
Repeat factors	repeat descriptors or groups of descriptors. Repeat factors must be positive, literal `INTEGER` constants.
Reversion	where more values or variables are in an I/O list than a format's descriptors can deal with so the last group of repeated descriptors is used repeatedly until the I/O list is exhausted.
`REWIND`	positions file-pointer before the first record of a file.
Rows	horizontal line of adjacent cells in a spreadsheet.
Runtime errors	occur after all syntax errors are fixed. While the program is running, the program tries to do something impossible and the program crashes.
`SAVE`	preserves the values of local variables so that the next time the subroutine is called, they will have their previous values.
Scientific notation	where constants have a mantissa and a power of 10 (e.g., `6.02E+23` or `3.509341232209D-102`)
Select a range of cells	specify a rectangular group of cells by highlighting them. Operations can be performed on the selected cells.
`SELECT CASE`	a selection structure that checks a `test-expression` to see if its value is contained in any of the `set-expressions`.
Selection structures	allow a programmer to execute groups of program instructions selectively.

Sequential files	a sequential collection of records. A sequential file is created by writing records to it one after another and the file must be read the same way.
Series summation	based on the incremental assignment statement.
`Set-expressions`	must be the same type as `test-expression` and may be a range of constants or individual constants separated by commas or a combination of individual constants and ranges of constants.
Shape	the collection of an array's bounds. Shape is represented in a one-dimensional `INTEGER` array.
Sign bit	determines the sign of an integer or, in type `REAL` numbers, the sign of the mantissa and exponent. A sign bit of 1 indicates a negative value.
Size	the product of the declared bounds of an array; the number of array positions there are in an array.
Software life cycle	the evolutionary phases software goes through from testing and adding enhancements over time.
Sorted data	data arranged in some order that facilitates the data being searched.
Source library	contains the actual Fortran statements that comprise the library.
Spreadsheets	a convenient way to evaluate and graph tabular data. Often spreadsheet solutions are quicker than writing customized computer programs.
Stand-alone debugging	find an error, correct it. Find the next error, fix it; and so on. Focus on one problem at a time and don't worry about recompiling the program frequently.
Standard input (read) algorithm	provides the ability to read an unknown amount of data with an "anytime" escape from the input algorithm.
`STOP "some message"`	a Fortran statement that stops program execution and displays its optional message. Create breakpoints with the `STOP` statements; **DON'T use `STOP` as the "natural" end of a program**.
`Stride`	defines an even separation between array elements that form an array section. `Stride` may be specified in any array section reference.
Structure chart	a graphic outline of how subroutines are called.
Structure constructors	assign all parts of a derived data type in one assignment.

Subroutines	mini Fortran programs that lead to top-down designs and shorten programming efforts.
Subscripts	used to reference array positions.
Substring reference	`variable ( start-chr : end-chr ) ...` specifies a range of characters.
Syntax (grammar)	the correct and allowable ways to construct program statements.
Syntax/compile time errors	errors that prevent the program from successfully compiling.
Tabular data file	each line of the data file is the same length and its values are vertically aligned.
`Test-expression`	may be a variable or expression of type `INTEGER`, `CHARACTER` or `LOGICAL`.
Testing programs	the process where a program is run as many times as necessary with varied data to flush out any bugs.
Toolbar	menu shortcut buttons.
Top-tested or **pretested loops**	conditional loops that have the condition at the top of the loop. This is the only type of conditional loop that Fortran offers.
Traceback	a runtime error message that tells where (in which program unit) an error happened. It also lists the sequence of `CALL`s that led to the crash from the main program down to the potentially errant program unit.
Tracers/debuggers	available in some environments allowing programmers to step through their programs, create breakpoints and track the values of variables.
Two's complement	a common low-level method for representing negative values for integers.
Type	defines whether a variable or named constant is `INTEGER`, `REAL`, `DOUBLE PRECISION`, `LOGICAL`, `COMPLEX`, `CHARACTER` or a derived data type.
Unary operator	an operator with only one operand.
`USE module-name`	gives a program unit access to the definitions in `module-name`.
User testing	done after software is released and reveals bugs not uncovered by alpha and beta testing. Users are victims not volunteers of this testing.

Variable	a named container that can hold a value of a specific type.
Vector subscript	an `INTEGER` array used as a subscript. Its elements specify positions in the subscripted array. The collection of the specified elements form an array section. Vector subscripts can be used for any subscript of a multidimensional array.

`WHERE`
`[ELSEWHERE]`
`END WHERE`
	Fortran's array-masking statement.

Word	a grouping of four bytes.
Word processing	a convenient way to create, format, save and modify text.
`WRITE`	sends data to an output device.

Index

: :, 84, 88
% operator, 102, 110
&, line continuation, 44
ACCESS =, 486–487
accumulate, 166–167
accumulating a series, 80
ACTION =, 485
actual arguments, 55, 58, 234, 281
algorithm, 25–26, 29, 50, 160, 165, 234
ALLOCATE, 313
allocatable arrays, 313
alpha testing, 466, 468
ANSI, 117
antibugging, 451, 453, 460
applications, 16, 20
   database managers, 16
   spreadsheets, 16
   word processors, 16
arguments, 29, 54–58, 67, 123, 234, 241
   actual, 58
   argument number mismatch, 57
   flow of data, 54
   formal, 59
   type mismatch, 58
arithmetic expressions, 88, 127
arithmetic operators, 89. *See also* order of precedence, spreadsheet arithmetic operators
array sections, 272, 280, 490
   stride, 276
arrays, 264. *See also* parallel arrays
   allocatable, 313
   array constructors, 267, 269, 300
   array section reference, 275
   array sections, 272, 280, 490
   array variables, 267
   bounds, 266
   conformable, 269, 353
   dimension attribute, 265

   dimensionality, 266, 352
   DO loop, fixed-limit, 298, 303, 494
   dynamic allocation, 313
   formal arguments, 307, 322, 490
   implied DO loops, 299
   INTERFACE blocks, 307
   local array variables, 321
   masking operations, 300, 303, 498
   noncontiguous positions, 276
   one-dimensional, 264–266, 269
   operands, 269
   operators, 269
   shape, 266, 269, 352
   size, 352
   subscripts, 264, 266
   variables, 267
ASCII, 117
assignment statement, 79–80, 87, 130
   incremental, 80, 87, 165
   mixed numeric types, 93
assumed-shape arrays, 306, 307
attributes, 83, 88
   dimension, 265
   INTENT(OUT), 59, 62–63
automatic arrays, 318, 321

BACKSPACE, 402–404
Backus, John, 15
beta testing, 466, 468
binary, 86, 116
binary numbers, 470
binary operators, 418
binary search, 297, 298
bit, 116, 117, 120
black box, 66–67, 69, 236, 239, 257, 309
   interface, 236
BLANK =, 486
block, 157–158
block statements, 125–126, 130

blocks, 125–126, 130, 149
BOZ constants, 86, 88
breakpoint, 463, 468
bug, 32, 450–451, 455, 460
bytes, 116–117, 118, 120, 388

CALL, 23, 25, 50, 58
calling program, 50, 58
CASE, 148
CASE statement, 146, 148–149
cells, 16
central processing unit, 15
CHARACTER, 75
    concatenation, 110, 114–115
    form, 78
    left justified and blank filled, 111
    LEN, 83, 101
    strings, 78, 110–111
    strings of characters, 78
    substring references, 110, 112–114
    truncation, 111
children, 50
client/server network, 467
CLOSE, 401, 404, 491
code, 3, 31
collating sequence, 117, 120
columns, 16
combinations, 241–243
comment blocks, 42, 47, 246
comments, 46, 246
compile, 181
compile-time errors, 452, 454, 460
compiler, 15, 21
COMPLEX, 74–75
    form, 77
    precision, 77
    range, 77
complex logical expressions, 128–129, 131
computer language, 2
computer memory, 115–116. *See also* RAM
computer program, 2–3, 5
concatenation, 110, 114–115
conditional loop, 158, 162, 164
conditions, 124–125, 127, 130, 158
conformable arrays, 269, 353
constants, 72, 74, 75–78, 79, 148
    form, 75, 79
    precision, 75, 79
    range of values, 75, 79

scientific notation, 76, 79
    signed decimal, 76
controlled I/O, 390, 396, 469
controls-list, 396, 399, 492
CPU, 15
crash, 5, 37–38, 457. *See also* debugging

data, 73–74, 78
data file, 396–398
data initializations, 85–86, 88
DATA statement, 85–86
database managers (DBM), 16, 19, 21
    structured query language (SQL), 19
DEALLOCATE, 313, 316
debugger, 468
debugging, 32, 33, 41, 159, 451, 460
    alpha testing, 466
    antibugging, 451, 453
    beta testing, 466
    breakpoint, 463, 468
    bug, 32, 450–451, 455, 460
    compile-time errors, 452, 454, 640
    enhancements, 466
    error checking, 38
    error messages, 452
    generated errors, 455, 467
    insecticides, 455, 459, 462
    limited-resource, 468
    listing file, 453, 455, 467
    logic errors, 452, 461–463
    PRINT statements, 461, 462, 463
    program bugs, 452
    program crash, 457
    program listing, 467
    program stubs, 453, 460
    refinements, 466
    runtime errors, 452, 457–460
    stand-alone, 468
    STOP, 463
    syntax errors, 452, 454–457
    traceback, 459, 460
    tracer, 468
    user testing, 466
debugging PRINT statements, 460, 463
decimals, 74
declarations, 52, 55, 82–85, 87
    attribute, 83, 88
    type, 83
deconstruction, 12

Index

default, 212
default input device, 391
default output device, 391
delimiters, 267, 393
derived data types, 73, 75, 99–103, 109, 422
    percent operator, 102
    subobject, 78
design, 23–25, 32
    Five-Paragraph Essay Approach to Program Design, 24–26
    modular programs, 29–31, 32
    structure chart, 27
    The Golden Rule, 50
    top-down, 12–13
dimension attribute, 265
dimensionality, 266, 313, 322, 352, 493
direct access files, 389, 401, 405, 408
    record number, 405, 413, 493
division by zero, 124
DO loop, fixed-limit, 298, 303, 494
DO loops, implied, 268, 299
DO WHILE, 7–9, 14, 157–159
    infinite loop, 158–159
documentation, 246
documents, 18
dot operators, 128, 418
DOUBLE PRECISION, 72, 75, 93–94
    form, 77
    magnitude, 75
    precision, 77
    range, 77
drop-down menu, 188
dynamic allocation, 313

EBCDIC, 117
electronic communication, 19
electronic mail (e-mail), 20, 21
elementary data types, 75
ellipsis, 148–149
ELSE, 14
END DO, 8, 9, 14
End File record, 405, 408
ENDIF, 14
environments, 390
EOF record, 485
error checking, 38, 42, 149, 160–162, 164, 257
error messages, 33, 124, 452
    insecticides, 455, 459, 462

error trapping, 408
errors, 33. *See also* logic errors, runtime errors, syntax errors
    generated, 455, 461, 467
execution. *See* program, executable
explicit typing, 84
exponent, 119–120
expressions, 234, 269

fields, 396
FILE =, 486
file-pointer, 402–404, 474
files, 388–389, 396, 453
    ACCESS =, 486
    ACTION =, 486
    BACKSPACE, 402–404
    BLANK =, 486
    CLOSE, 401, 404, 491
    controlled I/O, 396
    controls-list, 396, 399, 485
    direct-access, 389, 405, 408
    EndFile record, 405, 408
    EOF record, 485
    fields, 396
    FILE =, 486
    file-pointer, 402–404, 474
    fixed-format, 397
    flag, 162, 165, 406
    FMT =, 487
    FORM =, 486
    free-format, 393, 397
    I/O, 389
        list, 390
        mode, 398, 400
    input, 389
    IOSTAT =, 486
    listing, 453, 455, 461, 467
    list-directed I/O, 390–395
    OPEN, 396
    output, 389
    PAD =, 486
    POSITION =, 486
    positioning, 402–404
    READ, 401
    REC =, 486
    RECL =, 486
    record number, 405, 413, 493
    records, 389, 396
    REWIND, 403–404

files, *continued*
  sequential, 389, 405, 413, 502
  standard input algorithm, 406
  STATUS =, 486
  transfer programs, 21
  UNIT =, 486
  WRITE, 401
Five-Paragraph Essay Approach to Program Design, 24, 25–26, 32
fixed-format files, 397
flag, 162, 165, 406
floating-point numbers, 418. *See also* REAL
flow control structures, 125
flow of control, 3, 6–12, 50, 55. *See also* looping structures, selection structures
flow of data, 54, 59, 63, 123
FMT =, 487
FORM =, 486
formal arguments, 55, 59, 307, 322
format descriptors, 35, 471
formats, 33–37, 41, 469, 470–471
  binary numbers, 470
  controlled I/O, 469, 470
  descriptors, 471–473
  file-pointer, 474
  formatting, 470
  I/O list, 470
  input record, 473
  label, 470, 471
  list-directed I/O, 470
  nonnumeric, 472–473
  numeric descriptors, 471–472
  output record, 473
  positional, 473
  record pointer, 473
  repeat factor, 476
  reversion, 477
  special, 473
formatting, 470
formulas, 16
Fortran, 1–3, 15, 29, 31
  compiler, 15
  program, 3
free-format data, 393, 397. *See also* delimiters
functions, 233, 241
  actual arguments, 234
  algorithm, 233
  arguments, 233, 241
  interface, 236
  intrinsic, 236
  order of precedence, 234, 235
  overloading, 428–429
  program unit, 236–237
  type, 234

gigabyte, 388
Gold Rule, The, of Program Design, 50
GUI, 185, 188

hardcopy, 467
hardware, 15
hexadecimal, 86, 117, 120, 470
high-level language, 14–15, 21

I/O, 389
  list, 390, 470
  mode, 398, 400
identifiers, 42, 45–46, 47, 52, 84
IF, 6, 14
imaginary numbers, 75
IMPLICIT NONE, 85, 88
implicit typing, 84, 88
implied DO loops, 268, 269, 299
IN, 63
incremental assignment statement, 80, 87, 165
indentations, 44
infinite loop, 158, 164
initial value, 8
INOUT, 63
input, 32, 162, 389
  algorithm, standard, 162
  record, 473
insecticides, 455, 456, 459, 462
insertion sort, 288
instructions, 3, 5, 6
INTEGER, 74
  form, 75
  precision, 75
  range, 76
  two's complement arithmetic, 118–119
integer division, 91–92, 94
INTENT, 59–63
interface, 67, 236, 257
INTERFACE blocks, 307, 420
interface overloading, 429
intrinsic functions, 236, 264
intrinsic library, 236
IOSTAT =, 406, 486, 487
isolation, of program units, 52–54, 58
iteration, 157–158, 164

kilobytes, 388
kind, 83, 88

label numbers, 35, 470
labels, spreadsheet, 16
left justified and blank filled, 111, 115
LEN, 83
libraries, 11, 14, 49, 66–68, 69, 177–180, 181
    object library, 68, 69, 178, 181
    source library, 68, 69, 177, 181
    subroutines, 142
line comments, 42, 47
linking, 68, 69, 178, 181, 236
list-directed I/O, 390–395
listing file, 453, 455, 461, 467
literal constant, 480
local data and definitions, 54
local variables, 64–65, 69
    array, 321
logic errors, 452, 461–463
LOGICAL, 75, 130
    assignment statement, 130
    form, 77
logical expressions, 127–129, 130, 158
    complex, 128, 131
    relational operators, 127, 131
    simple, 127, 131
logical operators, 128, 131
    order of precedence, 129, 131
looping structures, 8–10, 14
    infinite loop, 158–159
    looping, 8–9

machine language, 15, 21, 68, 236
magnetic media, 116
magnitude. *See* variables.
main program, 11, 27–31, 32, 42, 50, 54
mainframe computers, 467
mantissa, 75, 119
masking operations, 300–301, 303, 498
memory, 115–116
    binary, 116
    bit, 116, 120
    byte, 116–118, 119, 120
    magnetic media, 116
    RAM (random access), 115–116
    ROM (read only), 115–116
    storage, 116
    word, 116, 120
MIPS, 3

mixed-mode expressions, 91, 93, 94
mixed-type assignment, 93–94
mnemonic identifiers, 46, 84, 88, 101
modems, 389
modular programs, 29–31, 32
MODULEs, 104–105, 110, 417, 418
    binary, 418
    USE, 104
mouse, 185–186
multialternative IF, 142–143
multidimensional arrays, 266, 349

named constants. *See* PARAMETER
nested loops, 359, 361
nested parentheses, 91
networked resources, 467
networks, 19–20, 21
No Record condition, 412
nonnumeric variables, 75, 79
noncontiguous positions (in arrays), 276
numeric descriptors, 471–472
numeric variables, 75, 79

object library, 67–68, 69, 178, 181
octal, 86
one-dimensional arrays, 264–266, 269
OPEN, 400
operands, 88–89, 91, 93, 94, 127, 131, 269, 418
operating systems, 2
operators, 88–89, 94, 269
    %, 102
    binary, 418
    derived data type +, 423
    dot, 128, 418
    logical, 128, 131
    modes of, 91–92, 94
    order of precedence of, 131
    overloading, 423
    relational, 127, 131
    unary, 418
order of precedence, 269
    arithmetic, 89–90, 94
    operator, 129, 234–235
OUT, 63
output, 32, 388, 389
output record, 473. *See also* record
overloading
    functions/interface, 428–429
    operators, 422–423

PAD =, 488
parallel arrays, 322–323, 340
PARAMETER, 86–87, 88
parent, 50
percent operator, 102
permutations, 241–243
portability, 15
POSITION =, 488
pretested loop, 158, 164
PRINT, 2, 391
printout, 467
program
    bugs, 32–33
    crash, 5, 37–38, 457
    design, 24–26
    flow, 3, 6–12, 123
    listing, 461, 467
    stubs, 65, 69, 453, 460
    style, 42, 44, 46, 47
program units, 29, 42, 58, 236–237
    arguments, 54–58
    child, 50
    isolation, 54
    module, 104–105
    parent, 50
programming style, 142, 453, 460, 461
programs, 2, 24, 49
    debugging, 32–33
    errors, 32–33
    executable, 236
    friendly, 33
    generalized, 41
    modular, 29–31
    portability, 15
    refinement, 33–37
    running, 32–33, 65
    source code, 454, 455
prompting message, 7, 14
pseudocode, 33

RAM (random access memory), 115
READ, 14, 391–392, 401
REAL, 4, 75
    form, 76
    magnitude, 75
    mantissa, 75
    precision, 76
    range, 76
    scientific notation, 75, 79

REC =, 487, 488
RECL =, 486, 488
record
    count, 406
    input, 473
    number, 405, 413, 493
    output, 473
    pointer, 473
records, 389, 396
relational operators, 127, 131
repeat factor, 476
reversion, 477
REWIND, 402–404
ROM (read only memory), 115
runtime errors, 452, 457–460

SAVE, 64, 69
scanners, 389
scientific notation, 75, 79
    constants, 76, 79
search, binary, 297, 298
selection structures, 6–7, 14, 124–125
    block IF, 124, 131–132
    CASE, 148–149
    IF, 7
    multialternative IF, 142–143
    SELECT CASE statement, 146
sequential files, 389, 405, 413, 493
sequential statement structures, 6, 123
series summation, 165–168
    accumulation, 166
    incremental assignment statement, 165
server, 467
shape, arrays, 266, 269, 352
sign bit, 118–119, 120
signed decimal constants, 76
simple logical expressions, 127, 131
single precision variable, 72, 94
software, 2, 15
    life cycle, 42, 46, 467, 468
software applications. See applications
sorting, 288
    insertion sort, 288–297, 298
source code, 31, 68, 176, 178, 455
source library, 68, 69, 177, 181
spreadsheet arithmetic operators, 191, 197,
    209, 216
spreadsheets, 16–18, 21, 185
    absolute cell reference, 212, 215, 216

absolute replication, 215, 216
active cell, 188, 192, 197
anchor point, 188, 189
arithmetic order of precedence, 210–211, 216
arrow keys, 191, 197
cells, 186, 187, 191, 196
    attributes, 195, 197
    copying, 200–201
    inactive, 188, 189, 197
    moving, 199
    selecting a range, 198–199
columns, 16, 188, 189, 196
constants, 191
delimiter, 226, 229
Esc key, 192
formula area, 188, 196
formula replication, 217
formulas, 16, 186, 188, 191, 197
functions, 191, 210–211, 216
Go To command, 191, 197
GUI, 185, 188
Home key, 191, 197
I-beam, 194
importing data, 225
    free-format data, 226
    tabular data, 227–228
informational region, 187, 196
labels, 16
menu bar, 188, 196
mouse, 185–186
operands, 191, 197, 209
operators, 191, 197, 209
PgUp and PgDn keys, 191, 197
pointing device, 188, 189, 197
reltive cell reference, 212, 216
relative replication, 212–214
rows, 16, 188, 189, 197
scroll bars, 191, 197
series, 202–203
SUM, 214
time-related series, 203
    toolbar, 188, 189, 196
SQL, 19
standard input algorithm, 164, 169, 331
standard read algorithm, 164, 169
statement structures, 125–126
    CALL, 11–12, 50, 58
    looping, 8–10, 14
    selection, 6–7
    sequential, 6, 123
statements, 6–7
STATUS =, 486, 487
STOP, 463
storage, 116
stride, 276, 356
strings of characters, 78, 110–111
structure chart, 13, 14, 27–28, 32, 50, 244, 453
    blocks, 29
structure constructors, 102, 110
structured programming, 52
structured query language, 19
stubs, program, 65, 69, 453
substring reference, 110, 112–114, 115
subobjects, 78, 102
subroutines, 11–12, 14, 29, 49
    actual arguments, 55, 281
    argument number mismatch, 57
    arguments, 54, 63, 67
    flow of data, 54, 63
    formal arguments, 55
    IN, 63
    INOUT, 63
    isolation, 52
    local data and definitions, 54
    OUT, 63
    overloading, 428–429
    SAVE, 64–65
    scope, 52
    type mismatch, 57
subscripts, 264, 266
syntax errors, 452, 454, 460

terabyte, 388
text, 18
THEN, 14
toolbar, 189, 196
top-down design, 12–13, 49, 453
top-tested loop, 158, 164
traceback, 459, 460, 461
tracer, 468
truncation, 92, 111, 115
two's complement arithmetic, 118–119, 120
types
    CHARACTER, 75
    COMPLEX, 74–75
    derived data types. *See* derived data types

types, *continued*
  DOUBLE PRECISION. *See* DOUBLE PRECISION
  INTEGER. *See* INTEGER
  LOGICAL. *See* LOGICAL
  REAL. *See* REAL

unary operators, 418
undefined variables, 74, 79, 85
UNIT =, 486, 487
USE, 105, 110
user, 7, 18–19, 20
user testing, 465, 466

variables, 72, 74–75, 78–79, 262
  assignment, 79–80
  CHARACTER, 75
  COMPLEX, 74–75
  decimal parts, 74
  declarations, 87
  derived data types, 75
  DOUBLE PRECISION, 75
  INTEGER, 74
  local, 64–65, 69
  LOGICAL, 75, 130
  magnitude, 74
  nonnumeric, 75, 79
  numeric, 75, 79
  REAL, 4, 75
  undefined, 74, 79, 85
  variable list, 84, 88
vector subscripts, 359

WHILE. *See* DO WHILE
windowing environment, 185
word, 120
word processors, 16, 18–19, 21
workstations, 467
WRITE, 391, 401